T0218326

Spatial Data Mining

Deren Li · Shuliang Wang · Deyi Li

Spatial Data Mining

Theory and Application

 Springer

Deren Li
Wuhan University
Wuhan
China

Deyi Li
Tsinghua University
Beijing
China

Shuliang Wang
Beijing Institute of Technology
Beijing
China

ISBN 978-3-662-56936-8   ISBN 978-3-662-48538-5   (eBook)
DOI 10.1007/978-3-662-48538-5

Foreword I

Advances in technology have extended computers from meeting general computing needs to becoming sophisticated systems for simulating human vision, hearing, thinking and other high-level intelligence needs. Developers who envisioned building computerized expert systems in the past, however, quickly encountered the problem of establishing a suitable expert knowledge base. Subsequently, they turned to machine learning in an effort to develop a computer with self-learning functions similar to humans. But they too faced a question: from what source does the computer learn? Meanwhile, large amounts of data were becoming available more and more rapidly, and the developers came to the realization that a lot of knowledge is hidden inside the data. The information can be retrieved from databases via database management systems, and the knowledge also can be extracted from databases via data mining (or knowledge discovery), which then might further constitute a knowledge base for a computerized expert system automatically.

In 1993, Prof. Deren Li and his brother, Prof. Deyi Li, discussed the trends of data mining in the computer science community. At that time, Deren Li believed that the volume, size, and complexity of spatial databases were increasing rapidly. Spatial data for a specific problem or environment appeared to be an accumulation of primitive, chaotic, shapeless states, and using it as a source for creating rules and order was questionable. They decided to explore the possibilities of spatial data mining. At the Fifth Annual Conference on Geographical Information Systems (GIS) in Canada in 1994, Deren Li proposed the idea of knowledge discovery from a GIS database (KDG) for uncovering the rules, patterns, and outliers from spatial datasets for intelligent GIS. Subsequently, the brothers planned to co-supervise a doctoral candidate to study spatial data mining, which was achieved when Kaichang Di began to study under them in 1995. After Kaichang received his doctorate in 1999, both brothers further supervised Shuliang Wang to continue the research because he also had a strong interest and foundation in spatial data mining research. Shuliang's doctoral thesis was awarded as the best throughout China in 2005.

In the scientific community of China, an academician who is a member of the Academy of Science or the Academy of Engineering is considered the most reputable scientist. The number of academicians who are brothers is very small, and even fewer brother academicians from different subjects have co-supervised one doctoral student. Furthermore, it is rarer still for both academician brothers and their doctoral student to co-author a monograph. Deren Li and Deyi Li are an admirable example of all these achievements! They are leading and inspiring the entire spatial data mining community to move forward by their quick thinking, profound knowledge, farsighted vision, pragmatic attitude, persistent belief, and liberal treatment.

The achievements from their enduring relentless study of spatial data mining have attracted more and more researchers (i.e., academicians, professors, associate professors, post-doctoral assistants, and doctoral and master's students from geospatial information science, computer science, and other disciplines). Their research team—comprised of experienced, middle-career, and young scientists—are studying and working together. Their problem-oriented theoretical research is being practically applied to serve society and benefit the people using their own innovative research results rather than simply using existing data mining strategies. Most of their research results have engendered invitations to make presentations at many international conferences to the acclaim of many international scholars. Also, their research is published in international journals, and respected institutions are utilizing their work. Thanks to their great efforts, spatial data mining has become a great crossover, fusion, and sublimation of geo-spatial information science and computer science. This monograph is the first summary of the collective wisdom from their research after more than ten years in the field of spatial data mining, as well as a combination of their major national innovative research and the application of horizontal outcomes.

In this monograph, there are interesting theoretical contributions, such as the Deren Li Method for cleaning spatial observed data, the data field method for modelling the mutual interactions between data, the cloud model method for transforming between a qualitative concept and quantitative data, and the mining view method for depicting different user demands under different mining tasks. The methodology is presented as a spatial data mining pyramid, and the mining mechanism of rules and outliers are well explained. GIS and remote sensing data are highlighted in their applications along with a spatial data mining prototype system. Having engaged in theoretical studies and applications development for many years, the authors did their best to focus closely on the characteristics of spatial data, both in theory and practice, and solving practical problems. The interdisciplinary characteristics of this monograph meet the requirements of geo-spatial information science academics looking to expand their knowledge of computer science and for computer science academics seeking to more fully understand geo-spatial information science. To produce this excellent book, they invited many people from different fields to proofread the draft many times. All of the helpful comments provided to them were acknowledged as they revised and refined the manuscript. Their meticulous approach is commendable and produced a book worth learning.

In this era of data and knowledge economy, spatial data mining is attracting increasingly more interest from scholars and is becoming a hot topic. I wish success to the authors with this book as well their continuing research. May they attain even greater achievements! May more and more brightly colored flowers open upon the theory and application of spatial data mining.

January 2005   Shupeng Chen

A member of the Chinese Academy of Science
A Scientist in Geography, Cartography, and Remote Sensing

Foreword II

Rapid advances in the acquisition, transmission, and storage of data are producing massive quantities of big data at unprecedented rates. Moreover, the sources of these data are disparate, ranging from remote sensing to social media, and thus possess all three of the qualities most often associated with big data: volume, velocity, and variety. Most of the data are geo-referenced—that is, geospatial. Furthermore, internet-based geospatial communities, developments in volunteered geographic information, and location-based services continue to accelerate the rate of growth and the range of these new sources. Geospatial data have become essential and fundamental resources in our modern information society.

In 1993, Prof. Deyi Li, a scientist in artificial intelligence, talked with his older brother, Prof. Deren Li, a scientist in geographic information science and remote sensing, about data mining in computer science. At that time, Deren believed that the volume, variety, and velocity of geospatial data were increasing rapidly and that knowledge might be discovered from raw data through spatial data mining. At the Fifth Annual Conference on Geographic Information Systems (GIS) in Canada in 1994, Deren proposed the idea of knowledge discovery from GIS databases (KDG) for uncovering the rules, patterns, or outliers from spatial datasets in intelligent GIS. Subsequently, both brothers were farsighted in co-supervising two doctoral research students, Mr. Kaichang Di (from 1995 to 1999) and Mr. Shuliang Wang (from 1999 to 2002), to study spatial data mining by combining the principles of data mining and geographic information science. Shuliang's doctoral thesis was awarded one of China's National Excellent Doctoral Thesis prizes in 2005.

It is rare that Deren and Deyi, brothers but from different disciplines, collaborated and co-supervised a graduate student, Shuliang. It is even rarer for them to have co-authored a monograph. Together, they have fostered a research team to undertake their pioneering work on the theories and applications of spatial data mining.

Their knowledge and wisdom about spatial data mining is manifested in this book, which offers a systematic and practical approach to the relevant topics and is designed to be readable by specialists in computer science, spatial statistics, geographic information science, data mining, and remote sensing. There are various new concepts in this book, such as data fields, cloud models, mining views, mining pyramids, clustering algorithms, and the Deren Li methods. Application examples of spatial data mining in the context of GIS and remote sensing also are provided. Among other innovations, they have explored spatiotemporal video data mining for protecting public security and have analyzed the brightness of night-time light images for assessing the severity of the Syrian Crisis.

This monograph is an updated version of the authors' books published earlier by Science Press of China in Chinese: the first edition in 2006 and the second in 2013. The authors tried their best to write this book in English, taking nearly 10 years to complete it. I understand that Springer, the publisher, was very eager to publish the monograph, and one of its Vice Presidents came to China personally to sign the publication contract for this English version when the draft was ready.

After reading this book, I believe that readers with a background in computer science will have gained good knowledge about geographic information science; readers with a background in geographic information science will have a greater appreciation of what computer science can do for them; and readers interested in data mining will have discovered the unique and exciting potential of spatial data mining. I am pleased to recommend this monograph.

August 2015   Michael Frank Goodchild

Emeritus Professor of Geography at the University of California, Santa Barbara
Former Director of National Center for Geographic Information and Analysis
Member of the American National Academy of Sciences
Member of the American Academy of Arts and Sciences
Foreign Member of the Royal Society of Canada
Foreign Member of the Royal Society of the British
Corresponding Fellow of the British Academy

Foreword III

The technical progress in computerized data acquisition and storage has resulted in the growth of vast databases, about 80 % of which are geo-referenced (i.e., spatial data). Although a computer user now has less difficulty understanding the increasingly large amounts of data available in these spatial databases, there is an impending bottleneck where the data are excessive while the knowledge to manage it is scarce. In order to overcome this bottleneck, spatial data mining, proposed under the umbrella of data mining, is receiving growing attention.

In addition to the common shared properties of data mining, spatial data mining has its own unique characteristics. Spatial data include not only positional and attribute data, but also the spatial relationships between spatial entities. Moreover, the structure of spatial data is more complex than the tables in ordinary relational databases. In addition to tabular data, there are vector and raster graphic data in spatial databases. However, insufficient attention has been paid to the spatial characteristics of the data in the context of data mining.

Professor Deren Li is an academician of the Chinese Academy of Science and an academician of the Chinese Academy of Engineering in Geo-informatics. He proposed knowledge discovery from geographical databases (KDG) at the International Conference on GIS in 1994. Professor Deyi Li is an academician of the Chinese Academy of Engineering in computer science, specifically for his contributions to data mining. The brothers co-supervised Dr. Shuliang Wang, whose thesis was honored as one of the best national Ph.D. theses in China in 2005. Moreover, their research group has successfully applied and finished many sponsored projects to explore spatial data mining, such as the National Natural Science Fund of China (NSFC), the National Key Fundamental Research Plan of China (973), the National High Technique Research and Development Plan of China (863), and the China Postdoctoral Science Foundation. Their research work focuses on the fundamental theories of data mining in computer science in combination with the spatial characteristics of the data. At the same time, their theoretical and technical results are being applied concurrently to support and improve spatial data-referenced decision-making in the real world. Their work is widely accepted by scholars worldwide, including me.

In this monograph, there are several contributions. Some new methods are proposed, such as cloud model, data field, mining view, and pyramid of spatial data mining. The discovery mechanism is believed to be a process of uncovering a form of rules plus exceptions at hierarchal mining-views with various thresholds. Their spatial data cleaning algorithms are also presented in this book: the weighted iteration method, i.e., Deren Li Method. Three clustering techniques are also demonstrated: clustering discovery with cloud models and data fields, fuzzy clustering under data fields, and mathematical morphology-based clustering. Mining image databases with spatial statistics, inductive learning, and concept lattice are discussed and application examples are explored, such as monitoring landslides near the Yangtze River, deformation recognition on train wheels, land classification based on remote-sensed image data, land resources evaluation, uncertain reasoning, and bank location selection. Finally, a prototype spatial data mining system is introduced and developed.

As can be seen in the information above, *Spatial Data Mining: Theory and Application* is the fruit of their collective work. The authors not only approached spatial data mining as an interdisciplinary subject but pursued its real-world applicability as well.

I thank the authors for inviting me to share their monograph in its draft stage and am very pleased to recommend it to you.

July 2005   Lotfi A. Zadeh

Founder of fuzzy set
Member of the American Academy of Engineering
Professor in Department of EECS
University of California, Berkeley

Foreword IV

The rapid growth in the size of data sets brought with it much difficulty in using the data. Data mining emerged as a solution to this problem, and spatial data emerged as its main application. For a long time, the data mining method was used for spatial data mining without recognition of the unique characteristics of spatial data, such as location-based attributes, topological relationships, and spatiotemporal complexity.

Professor Deren Li is an expert in geospatial information science, and he conceived the "Knowledge Discovery from GIS databases (KDG)." Professor Deyi Li is an expert in computer science who also studied data mining early. In essence, they combined the principles of data mining and geospatial information science to bring the spatial data mining concept to fruition. They had the foresight to co-supervise two doctoral students: Kaichang Di and Shuliang Wang, and thereafter fostered a spatial data mining research team to continue carrying out their pioneering research. They have completed many successful projects, such as the National Nature Science Foundation of China, the results of which are successfully being applied in a wide variety of areas, such as landslide disaster monitoring. Many of their achievements have received the acclaim of scholars in peer-peer exchanges and have attracted international academic attention.

This book is the collaborative collective wisdom of Deren Li, Deyi Li, and Shuliang Wang regarding spatial data mining presented in a methodical fashion that is easy to read and maneuver. The many highlights of this book center on the authors' innovative technology, such as the cloud model, data field, mining view, mining pyramid, and mining mechanisms, as well as data cleaning methods and clustering algorithms. The book also examines spatial data mining applications in various areas, such as remote sensing image classification, Baota landslide monitoring, and checking the safety of train wheels.

After reading this book, readers with a computer science background will know more about geospatial information science, readers from the geospatial information science area will know more about computer science, readers with data mining experience will discover the uniqueness of spatial data mining, and readers who are doing spatial data mining jobs will find this book a valuable reference.

It was a great honor to have been asked to read this book in its earlier draft stages and to express my appreciation of its rich academic achievements. I wholeheartedly recommend it!

August 2006   Jiawei Han

Abel Bliss Professor, University of Illinois at Urbana-Champaign
Author of "Data Mining: Concepts and Techniques",
3rd ed., 2011; 2nd ed., 2006; 1st ed., 2000

Preface

Data fundamentally serve in the development of a variety of knowledge bases, from understanding the real world to creating and enjoying the information world. Initially, humans used stones and shells to count according to the principle of one-to-one and kept records by tying knots. Later, picture notes were conceived to historically preserve more accurate records by using simple graphics along with a perceptual cue. Once pictures became relatively fixed by shape symbols and associated with words in the human language, text was created. The advent of text abstracted and generalized the world, promoted cultural understanding, and prepared the necessary foundation for the development of science. To break through the restrictions that the written symbols depended on, artificial transcription, carving or engraving, and human-operated machines were used after the industrial revolution to batch mechanization production and subsequently improve the efficiency of cultural transmission.

Thereafter, the computer was given the ability to listen to high-speed computing and spin off software within the hardware, which enabled the dissemination of information "electronically" and "automatically." The internet then focused on networks by interrelating computers, essentially breaking through the local information limitation. Mobile communication released humans from their computers by enabling the computer to follow the user's movements. The Internet of Things cares for the application of human-machine interoperation by automatically identifying objects, and allowing information sharing between humans and things. Cloud computing centralized these information services by consolidating the expertise and optimizing the allocation of resources. All the while, data have been continuously created and accumulated in big data centers that are closely associated with various applications for which spatial data mining is a necessity.

Spatial data account for the vast majority of data mining because most objects are now associated with their geo-spatial positions. Data mining attempts to discover the previously unknown, potentially useful, and ultimately understandable patterns of big data. The complex types, intrinsic relationships, and implicit auto-correlations in spatial big data make it more difficult to extract the useful patterns from spatial datasets than from conventional numeric and categorical datasets. To penetrate the volume, variety, velocity, and veracity of the values in big data, at times it may be necessary for spatial data mining to take advantage of population data instead of sample data.

In this monograph, we present our novel theories and methods of spatial data mining as well as our successful applications of them in the realm of big data. A data field depicts object interactions by diffusing the data contribution from the universe of samples to the universe of population. The cloud model bridges the mutual transformation between qualitative concepts and quantitative data; and the mining view of spatial data mining hierarchically distinguishes the mining requirements with different scales or granularities. The weighted iteration method is used to clean spatial data of errors using the principles of post-variance estimation. A pyramid of spatial data mining visually illustrates the mining mechanism. All of these applications concentrate on the bottlenecks that occur in spatial data mining in areas such as GIS and remote sensing.

We were urged by scholars throughout the world to share our innovative interdisciplinary approach to spatial data mining; this monograph in English is the result of our 10-year foray into publishing our collective wisdom. The first Chinese edition was published in 2006 by Science Press and was funded by the National Foundation for Academy Publication in Science and Technology (NFAPST) in China. Simultaneously, we began writing an English edition. The first Chinese edition, meanwhile, was well received by readers and sold out in a short time, and an unplanned second printing was necessary. In 2013, writing a second Chinese edition was encouraged for publication in Science Press on the basis of our new contributions to spatial data mining. As a result, the English edition, although unfinished, was reorganized and updated. In 2014, Mr. Alfred Hofmann, the Vice President of Publishing for Springer, came to Beijing Institute of Technology personally to sign the contract for the English edition for worldwide publication. In 2015, the second Chinese edition won the Fifth China Outstanding Publication Award, which is a unique award by Science Press; Ms. Zhu Haiyai, the President of Publishing on Geomatics for Science Press, also personally wrote a long article on the book's publication process, which appeared in the Chinese Publication Newspaper. Following the criteria of contribution to the field, originality of the research, practicality of research/results, quality of writing, rigor of the research, substantive research and methodology, the data field method was awarded the Fifth Annual InfoSci®-Journals Excellence in Research Awards of IGI. The contributions were further reported by VerticalNews journalists, collected in *Issues in Artificial Intelligence, Robotics and Machine Learning* by ScholarlyEditions, *Developments in Data Extraction, Management, and Analysis* by IGI Global.

To finish writing the monograph in English as soon as possible, we made it a priority. We tried many translation methods, which taught us that only we, the authors, could most effectively represent the monograph. Although it took nearly 10 years to finish it, the final product was worth the wait!

Deren Li
Shuliang Wang
Deyi Li

Acknowledgments

The authors are most grateful to the following institutions for their support: the National Natural Science Foundation of China (61472039, 61173061, 60743001, 40023004, 49631050, 49574201, 49574002, 71201120, 70771083, 70231010, and 61310306046), the National Basic Research Program of China (2006CB701305, 2007CB310804), the National High Technology Research and Development Plan of China (2001AA135081), China's National Excellent Doctoral Thesis prizes (2005047), the New Century Excellent Talents Foundation (NCET-06-0618), Development of Infrastructure for Cyber Hong Kong (1.34.37.9709), Advanced Research Centre for Spatial Information Technology (3.34.37.ZB40), the Doctoral Fund of Higher Education (No. 20121101110036), Yunan Nengtou project on New techniques of big energy data and so on.

Knowing this book would not have become a reality without the support and assistance of others, we acknowledge and thank our parents, families, friends, and colleagues for their assistance throughout this process. In particular, we are very grateful to Professors Shupeng Chen, Michael Goodchild, Lotfi A. Zadeh, and Jiawei Han for taking the time to write the forewords for this book; the following colleagues for providing their assistance in various ways: Professors Xinzhou Wang, Kaichang Di, Wenzhong Shi, Guoqing Chen, Kevin P. Chen, Zongjian Lin, Ying Chen, Yijiang Zou, Renxiang Wang, Chenghu Zhou, Jianya Gong, Xi Li, Kun Qin, Hongchao Ma, Yan Zou, Zhaocong Wu, Benjamin Zhan, Jie Shan, Xiaofang Zhou, Liangpei Zhang, Wenyan Gan, Xuping Zeng, Yangsheng You, Liefei Cai, etc.; and the following students for their good work: Yasen Chen, Jingru Tian, Dakui Wang, Yan Li, Caoyuan Li, Likun Liu, Hehua Chi, Xiao Feng, Ying Li, Linglin Zeng, Wei Sun, Hong Jin, Jing Geng, Jiehao Chen, Jinzhao Liu, and so on.

The State Key Laboratory Engineering in Surveying Mapping and Remote Sensing at Wuhan University, the International School of Software at Wuhan University, the School of Software at Beijing Institute of Technology, and the School of Economics and Management at Tsinghua University provided computing resources and a supportive environment for this project.

We appreciate Karen Hatke for her proofreading and copyediting.

It has been a pleasure working with the helpful staff at Springer, especially Mr. Alfred Hofmann, Dr. Celine Chang, Ms. Jane Li, and Mr. Narayanasamy Prasanna Kumar. We also appreciate the assistance of Haiyai Zhu, Lili Miao, and Peng Han at Science Press.

We extend great appreciation to our family for their support.

Finally, the authors wish to remember these individuals who left us during the process of writing the English edition: Professors Shupeng Chen and Xinzhou Wang and Mingzhen Peng, the mother-in-law of Shuliang Wang.

Deren Li
Shuliang Wang
Deyi Li

Contents

About the Authors

Deren Li, Ph.D. a scientist in photogrammetry and remote sensing, is a member of the Chinese Academy of Sciences, the Chinese Academy of Engineering, and the Euro-Asia International Academy of Science; a Professor and Ph.D. Supervisor at Wuhan University; Vice President of the Chinese Society of Geodesy, Photogrammetry and Cartography; and Chairman of the Academic Commission of Wuhan University and the National Laboratory for Information Engineering in Surveying, Mapping and Remote Sensing (LIESMARS). His entire career has focused on research and education in spatial information science and technology represented by remote sensing, global positioning systems, and geographic information systems. His major areas of expertise are analytic and digital photogrammetry, remote sensing, mathematical morphology and its application in spatial databases, theories of object-oriented GIS, and spatial data mining in GIS, as well as mobile mapping systems. He served as Comm. III and Comm. VI president of ISPRS in 1988–1992 and 1992–1996; was elected an ISPRS fellow in 2010, and an ISPRS Honorary Member in 2012 (There may not be more than ten living Honorary Members at any given time); worked for CEOS in 2002–2004; and was president of the Asia GIS Association in 2003–2006. He received a Dr.h.c. from ETH in 2008.

Shuliang Wang, Ph.D. a scientist in data science and software engineering, is a Professor at Beijing Institute of Technology in China. His research interests include spatial data mining and software engineering. For his innovative study of spatial data mining, he was awarded the Fifth Annual InfoSci®-Journals Excellence in Research Awards of IGI Global in 2012, the IEEE Outstanding Contribution Award for Granular Computing in 2013, one of China's National Excellent Doctoral Thesis prizes in 2005, and, Wu Wenjun Artificial Intelligence Science and Technology Innovation Award in 2015.

Deyi Li, Ph.D. a scientist in computer science and artificial intelligence, conceived the cloud model method. He is now a professor in Tsinghua University in China, a member of the Chinese Academy of Engineering, and a member of the Euro-Asia International Academy of Science. His research interests include networked data mining, artificial intelligence with uncertainty, cloud computing, and cognitive physics. For his contributions, he has received many international and national prizes or awards, the Premium Award by IEE Headquarters, the IFAC World Congress Outstanding Paper Award, and the National Science and Technology Progress Award.

Abstract

In the context of numerous and changeable big data, the spatial discovery mechanism is believed to be based on a form of rules plus exceptions at hierarchal mining views with various thresholds. This monograph explores the spatiotemporal specialties of big data and introduces the reader to the data field, cloud model, mining view, and Deren Li methods. The data field method captures the interactions between spatial objects by diffusing the data contribution from a universe of samples to a universe of population, thereby bridging the gap between the data model and the recognition model. The cloud model is a qualitative method that utilizes quantitative numerical characters to bridge the gap between pure data and linguistic concepts. The mining view method discriminates the different requirements by using scale, hierarchy, and granularity in order to uncover the anisotropy of spatial data mining. Finally, the Deren Li method performs data preprocessing to prepare it for further knowledge discovery by selecting a weight for each iteration in order to clean the observed spatial data as much as possible. In these methods, the spatial association, distribution, generalization, and clustering rules are extracted from geographical information system (GIS) datasets while simultaneously discovering knowledge from the images to conduct image classification, feature extraction, and expression recognition.

The authors' spatial data mining research explorations are extensive and include projects such as evaluating the use of spatiotemporal video data mining for protecting public security, analyzing the brightness of nighttime light images for assessing the severity of the Syrian Crisis, and indicating the nighttime light dynamics in the Belt and Road for helping promote the beneficial cooperation of the countries in the global sustainability. All of their concepts and methods are currently being applied in practice and are providing valuable results for a wide variety of applications, such as landslide protection, public safety, humanitarian aid, recognition of train wheel deformation, selection of new bank locations, specification of land uses, and human facial recognition. *GISDBMiner*, *RSImageMiner*, and *EveryData* are computerized applications of these concepts and methods.

This book presents a rich blend of novel methods and practical applications and also exposes the reader to the interdisciplinary nature of spatial data mining.

A prerequisite for grasping the concepts in this book is a thorough knowledge of database systems, probability and statistics, GIS, and remote sensing. It is suitable for students, researchers, and practitioners interested in spatial data mining, both as a learning text and as a reference book; and professors and instructors will find it suitable for curricula that include data mining and spatial data mining.

Chapter 1
Introduction

Spatial data mining (SDM) extracts implicit knowledge from explicit spatial datasets. In this chapter, the motivation for SDM and an overview of its processes are presented. The contents and contributions of this monograph are also summarized.

1.1 Motivation for SDM

With the rapid development of advanced techniques and instruments to acquire, store, and transmit massive data, spatial data are constantly expanding beyond what ever could have been imagined. These data play an important role in the sustainable development of natural resources and human society, not only to meet human demands to study Earth's resources and environments but to broaden the usable sources of information as well. However, spatial data are far beyond common transactional data. The content is richer in depth and breadth, the types are more varied, and the structure is more complex; the rapid growth of spatial data features has also rapidly increased their volume. Traditional methods to process spatial data have comparatively lagged behind the growth of big data. A large number of spatial data cannot be interpreted and applied in a timely manner and subsequently often are put aside. On the other hand, people are eager to recognize the unknown world and promote sustainable development by using spatial data.

Spatial datasets contain accumulated knowledge and experience and the precepts of geomatics; they also have previously unknown, potentially useful, and ultimately understandable rules. These rules are deeply implicit in a database where a dataset of objects has spatial attributes and non-spatial attributes, which are often irretrievable with traditional query techniques. This conundrum motivated the inception of SDM by Li Deren (Li and Cheng 1994) when he proposed the Knowledge Discovery from Geographical Information Systems databases (KDG) approach, which continues to assist SDM in the further development of data mining and geomatics.

© Springer-Verlag Berlin Heidelberg 2015
D. Li et al., *Spatial Data Mining*, DOI 10.1007/978-3-662-48538-5_1

1.1.1 Superfluous Spatial Data

The applied cycle of new technologies is shortening. Humans are different from animals in their ability to walk upright and make tools; human social properties and natural properties may be separated based on physical performance, skills, and intelligence. Human intelligence broadens the scope of activities in society, economics, and culture and is constantly shortening the applied cycle of new technical innovations or inventions. Under this umbrella of evolving intelligence, the instruments used in scientific and socio-economic fields to generate, collect, store, process, and transmit spatial data are qualitatively changing constantly, which has greatly shortened the cycle of their use (Fig. 1.1). For example, the integration degree of a computer chip, the power of a central processing unit (CPU), and the transfer rate of communications channels are doubling every 18 months. Reaching 50 million users in the United States took 38 years for radio broadcasting, 13 years for television, and only four years for the internet. As a resource gateway for information interdependence and interaction, the network space of the internet and the World Wide Web (WWW) has greatly accelerated the computerized accumulation of interactive data.

In the real world, many of these vast datasets are spatially referenced (Li and Guan 2000) and provide geo-support for decision-making in areas such as transportation services, urban planning, urban construction, municipal infrastructure, resource allocation, hazard emergency response, capital optimization, product marketing, and medical treatment. This spatial data infrastructure has accumulated

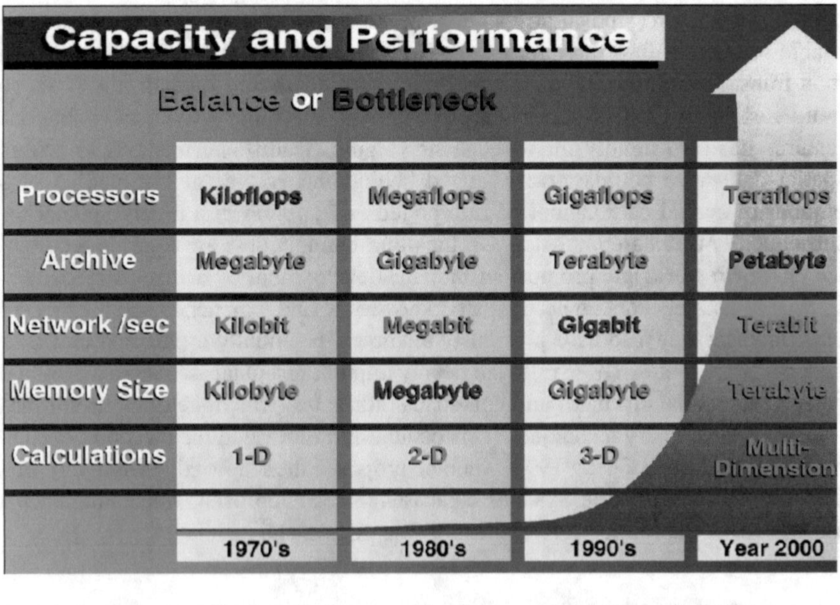

Capacity and Performance
Balance or Bottleneck

Processors	Kiloflops	Megaflops	Gigaflops	Teraflops
Archive	Megabyte	Gigabyte	Terabyte	Petabyte
Network /sec	Kilobit	Megabit	Gigabit	Terabit
Memory Size	Kilobyte	Megabyte	Gigabyte	Terabyte
Calculations	1-D	2-D	3-D	Multi-Dimension
	1970's	1980's	1990's	Year 2000

Fig. 1.1 Instruments and equipment changes with data accumulation (Li et al. 2006)

a wide variety of databases containing electronic maps and planning networks, as well as engineering, geological, and land use data for regulatory planning, municipal construction, cadastral land, and land-use protection on farmland. In addition to the data collected from conventional surveying and mapping operations, increasingly more data are being gathered from census records, land investigations, scanning maps, and statistical charts. This rapid accumulation is due to new measurement equipment and techniques, such as radar, infrared, photoelectric, satellites, multispectral scanners, digital cameras, imaging spectrometers, total stations, telescopes, television cameras, electronic imaging, computerized tomography (CT) imaging, global positioning systems (GPS), remote sensing (RS), geographic information systems (GIS), as well as other macro-sensors or micro-sensors. The internet-based geospatial community of volunteered geographic information (Goodchild 2007) and location-based sensing services are contributing to increasing the complexity of spatial relationships. The accumulation rate of spatial data now greatly exceeds the rate of common transactional data.

Most spatial data are remotely sensed from earth observations (Fig. 1.2). RS technology is playing an important role in capturing a wide range of spatio-temporal information about the dynamic resources and environments that surround humans (Wynne et al. 2007). Remotely sensed data are now a fundamental part of social, political, and economic decision-making for human development processes. The RS techniques are now multi-purpose, multi-sensor, multi-band, multi-resolution,

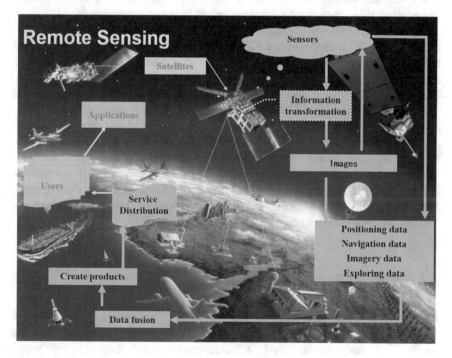

Fig. 1.2   Remotely sensed spatial data from earth observations

and multi-frequency and are being marketed with high space, high dynamics, high spectral efficiency, and high data capacity. Based on these characteristics, a three-dimensional global network of earth observations is becoming a reality (Wang and Yuan 2014); this is comprised of large, medium, and small satellites that orbit the earth at high, medium, and low levels. The sensors on the satellites enable the observation of spatial entities on earth by using multi-level, multi-angle, all-weather, or all-around methods in a pan-cubic network. The observed resolution of satellite imagery may be coarse or fine. The ground resolutions can range from kilometer-band to centimeter-band, the sensed spectrum can range from ultraviolet rays to infrared rays, the sensed time interval can range from once every 10 days to three times a day, and the detected depth can range from a few meters to over 10,000 m.

Furthermore, the capacity of sensed data is growing larger and larger while the cycle gets shorter and shorter. The National Aeronautics and Space Administration (NASA) Earth Observing System (EOS) is projected to generate some 50 Gbits of remotely sensed data per hour. Only the sensors of EOS-AM1 and PM1 satellites can collect remotely sensed data up to terabyte level daily. Landsat satellites obtain global coverage of satellite imaging data every two weeks, which has accumulated more than 20 years of global data. Some satellites may give more than one type of imagery data; for example, EOS may provide MODIS imaging spectral data, ASTER thermal infrared data, and measurable four-dimensional simulation of the CERES data, MOPIT data, and MISR data. Nowadays, remotely sensed data from new types of satellites, such as Quick Bird, IRS, and IKONOS, are being marketed and applied; these computerized data are stored in magnetic mediums.

In order to define, manipulate, and analyze these data, spatial information systems such as GIS were developed for a personal computer. Because the local limitation of a personal computer is not an issue within the network, data for the same objects also can be globally networked, analyzed, or shared with authorization. The Internet of Things (Höller et al. 2014) enables the internet to enter the real world of physical objects. Ubiquitous devices are mutually connected and extensively used (e.g., laptops, palmtops, cell phones, and wearable computers), with an enormous quantity of data being created, transmitted, maintained, and stored. Digital Earth is becoming a smart planet covered by a data skin.

The above mentioned spatial data highly match the potential demand of humans to recognize and use natural resources for sustainable development. Moreover, these data are continuously increasing and are being amassed instantaneously with both attribute depth and the scope of objects.

1.1.2 Hazards from Spatial Data

Humans are being inundated by massive data. On average, it takes about 30–45 min for an adult to browse 24 pages of a newspaper. However, the same article by the same author may be simultaneously published on both the "social rule of law" page and the "youth topic" page of the same issue of the same newspaper when not detected by any editor. What's more, when the advantages and disadvantages of VCDs, CVDs, and DVDs are compared on the home appliances page, the living page, and the technology page of the same issue of a daily newspaper on the same day, three different conclusions may be drawn and introduced to the readers.

The situation is much worse for spatial data because they are more complex, more changeable, and larger than common transactional datasets. Spatial data include not only positional data and attribute data but also the spatial relationships among spatial entities in the universe, on the earth's surface, and in other space. Moreover, the database structure for spatial data is more complex than the tables in ordinary relational databases. In addition to tabular data, there are raster images and vector graphics in spatial databases, the attributes for which are not explicitly stored in the database. Furthermore, spatial data are heterogeneous, uncertain, time-sensitive, and multi-dimensional; spatial data for the same entity may be distributed in different locations and departments with different structures and standards. Handling these data is complicated. For example, faced with the endless images coming from satellite reconnaissance, the U.S. Department of Defense has been unable to fully deal with these vast remotely sensed data.

Conventionally, RS focuses on the fundamental theories and methods of mathematical model analysis, which continues to be the primary stage of spatial information processing. In this stage, data become information instead of knowledge, and the amount of processed data is also very limited. Current commercial software on image processing (e.g., ENVI, PCI, and ERDAS) cannot meet the need to take advantage of new types of imagery data from RS satellites. Because these software programs lack new techniques to deal with the new types of imagery data (partially or completely), it is impossible for them to achieve intelligent mixed-pixel segmentation, automatic spectral match, and automatic feature extraction of the target objects.

We now face the challenge of processing excessive quantities of spatial data while humans are clamoring for maximum use of the data. With the continuous expansion of spatial data in terms of its scale, scope, and depth, more magnetic mediums for storage, appropriate instruments for processing, and trained technicians for maintaining them are continually needed. Although they exist in this sea of spatial data, humans cannot make full use of it to aid spatially referenced applications. Thus, the huge amounts of computerized datasets have far exceeded human ability to completely interpret and use them appropriately, as shown in Fig. 1.3 (Li et al. 2006).

Eventually, new bottlenecks will appear in geomatics, such as how to distinguish the data of interest from the meaningless data, how to understand spatial data, how to extract useful information from the data, how to uncover the implicit patterns in spatial data, and how to summarize uncovered patterns into useful knowledge. It is also very difficult to answer time-sensitive questions such as the following using these massive spatial data: Will it rain in the coming year? Where will an earthquake happen tomorrow? The excessive data seem to be an unsolvable problem. In the words of Naisbitt (1991), "We are drowning in information but starved for knowledge."

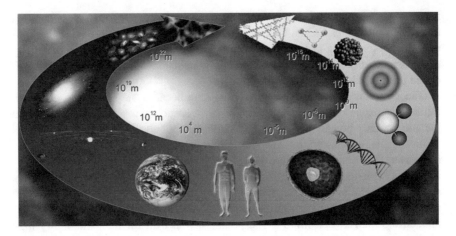

Fig. 1.3 The capacity of spatial data production is excessively larger than the capacity available for data analysis (Li et al. 2006)

1.1.3 Attempts to Utilize Data

Since the beginning of humankind, people have been individually accumulating and processing data in the interest of finding useful rules to improve their living and working conditions. For example, a hunter caught prey animals by following their trails, a farmer set his budget based on estimating the yield of his crops, a policeman solved a case from the evidence and witnesses, and a boy behaved according to the reaction of his girlfriend when they were passionately in love. Obviously, probability and mathematical statistics are applied, along with error estimation, to learn how often an event may happen.

Then, database techniques were developed. An integrated and shared collection of logically related data constitutes a database, which is saved and stored on disks. A database is designed to meet the information needs of an organization, which also can be used simultaneously by many departments and users. To control the database, a database management system (DBMS) has been developed to accelerate the speed of data query. As the database engine, the DBMS system software is a collection of programs that enables users with an interface to define, create, and manipulate multiple databases as well as maintain and provide controlled access to them. For example, structured query language (SQL) summarizes data in a simple reordering of data to ad hoc queries while its query and reporting tools describe the contents of a database. The query retrieves the information and the reporting tool presents it. Data retrieval explores the data and the information of interest from archives and databases and provides a preliminary statistical analysis of the data.

While the DBMS is general in purpose, a database system is developed to support the operations of a specific organization or a specific set of applications. That is, a database system consists of an application-specific database, the DBMS that maintains that database, and the application software that manipulates the database, along with some simple functions for data analysis and reporting (Connolly 2004). GIS is one such specific software that has a spatial database. In support of the management decision-making process, a subject-oriented, integrated, time-variant, and non-volatile collection of data constitutes a data warehouse (Inmon 2005). Efficient organization of the data in a data warehouse, coupled with efficient and scalable tools, allows the data to be used correctly and efficiently to support decision-making. During the process of data utilization, the industry-standard data warehouses and online analysis and process (OLAP) platforms are directly integrated. OLAP explains why certain relationships exist using a graphical multi-dimensional hypercube and can provide useful information for databases with a small number of variables, but problems arise when there are tens or hundreds of variables (Giudici 2003).

Artificial intelligence (AI) and machine learning (ML) were proposed to replace the manual calculation of the data by using computers. There are three main schools of thought on the fundamental theories and methodologies for AI: symbolism, connectionism, and behaviorism. Symbolism maintains that the basic element of human cognition is a symbol and the cognitive process is equal to the operating process of the symbol. Symbolism is based on the physical symbol system hypothesis—that is, a symbol operating system and limited rationality theory. Connectionism maintains that the basic element of the human brain is the neuron rather than a symbol, and the working mode of the brain is to use the neural network, which is connected by ample multilayered parallel neurons. Connectionism concerns itself with artificial neural networks and evolutional computation. Behaviorism believes that intelligence depends on perception and behavior and emphasizes the interactions between the real world and its surroundings. The three strategies are all simulations of human deterministic intelligence. To automatically acquire the knowledge from expert systems, ML was introduced to simulate human learning. Using centralized datasets, computerized devices learn the ability to improve their performance from their past performance. However, it is necessary to establish an auxiliary computer expert system and to be prepared to work hard to train self-learning to the computer. Regarding the differences between AI and ML, the issues that they are dealing with is only a small category of human intelligence or human learning (Li and Du 2007). Thus, datasets cannot be used sufficiently by independently utilizing statistics, a database system, AI, or ML. Moreover, besides the conventional querying and reporting of explicit data, new demands on spatial data are increasing steadily.

First, most techniques focus on a database or data warehouse, which is physically located in one place. However, the rapid advance of web techniques broadens the data scope of the same entity—that is, from local data to global data. As increasingly more dynamic sources of data and information become available online, the web is becoming an important part of the computing community and many data may be distributed on heterogeneous sites. This growth has an accompanying need for providing improved computational engines in a cost-effective way with parallel multi-processor computer technology.

Second, before data are used for a given task, the existing data first should be filtered by abandoning the meaningless data, leaving only the data of interest for ultimate use and thereby improving the quality and quantity of the spatial data used in depth and width.

Third, the increasing heterogeneity and complexity of new forms of data, such as those arriving from the web and earth observation systems, require new forms of patterns and models, together with new algorithms to discover such patterns and models efficiently. Obviously, the vast amounts of accumulated spatial data have extremely exceeded human capability to make full use of them. For example, contemporary GIS analysis functionalities are not intelligent enough to offer a further implicit description and future trends beyond the explicit spatial datasets automatically.

Therefore, it is necessary to search for new techniques that will automatically take advantage of the growing wealth of spatial databases and supervise the use of data intelligently.

1.1.4 Proposal of SDM

In 1989, after attending a session discussing the technical convergence of databases, AI, mathematical statistics, and visualization at the International Joint Conference on Artificial Intelligence in Detroit, Michigan, it became apparent that a lot of the knowledge attendees were yearning for was already implicitly hidden in the accumulated data stored throughout a large number of repositories. The Knowledge Discovery in Databases (KDD) was first proposed to meet the challenge of dealing with large amounts of datasets (Sridharan 1989). In 1994 at the International Conference on GIS held in Ottawa, Canada, Li Deren introduced the Knowledge Discovery in GIS Database (KDG) (Li and Cheng 1994). Using the same datasets with spatial characteristics, the implicit knowledge could be uncovered with KDG while the explicit information could be retrieved or reported by the database system. Using KDG, the limited GIS data is changed into the unlimited knowledge, which could make GIS more intelligent. In 1995, Data Mining (DM) was formally introduced by Usama Fayaad at the First International Conference on Knowledge Discovery and Data Mining in Montreal, Canada (Fayyad and Uthurusamy 1995). In 1997, *Data Mining and Knowledge Discovery*, an international journal, was founded on the basis of KDG (Di et al. 1997; Di 2001). Nowadays, DM is generally thought of as a synonym of KDD instead of a component of the KDD process (Han et al. 2012); the same presumption was true with SDM (Li et al. 2006). By utilizing network-based cloud computing, SDM also broke through the restrictions on local spatial data and now can be implemented on a global multi-dimensional data hypercube under a complex network, such as spatial online analytical mining (SOLAM) and spatial networked data mining (Wang 2002; Li et al. 2006).

In this monograph, SDM refers to the process of discovering knowledge from large spatial databases. The process first identifies the initial aims and ultimately applies the decision rules to the pre-processing of data, selection of data, modeling of objects, extrapolating the knowledge, interpreting the knowledge, and applying the knowledge. The knowledge is previously known, potentially useful, and ultimately understandable. The datasets hide the spatial distribution, topological relationships, geographical features, moving trajectory, images, and graphics (Li et al. 2006; Leung 2009; Miller and Han 2009; Giannotti and Pedreschi 2010).

1.2 The State of the Art of SDM

Since its introduction, SDM has continued to attract increasing interest from individuals and organizations due to its various contributions to the location-referenced field of academic activities, theoretical techniques, and real-world applications.

1.2.1 Academic Activities

The academic activities surrounding SDM are rapidly growing. SDM is an important theme and a hot topic that draws large audiences for papers, monographs, demos, competitions, tools, and products. Besides its specific fields of application, SDM has penetrated related fields, such as DM, knowledge discovery, and geo-spatial information science. It is a key topic in the Science Citation Index (SCI). Many well-known international publishing companies are also becoming increasingly interested in SDM.

- *Academic organizations: Knowledge Discovery Nuggets, the American Association for Artificial Intelligence (AAAI), the International Society for Photogrammetry and Remote Sensing (ISPRS).*
- *Conferences: IEEE International Conference on Spatial Data Mining and Geographical Knowledge Services (ICSDM), IEEE International Conference on Data Mining (ICDM), International Conference on Knowledge discovery in databases (KDD), Symposium on Spatio-Temporal Analysis and Data Mining (STDM), Advanced Spatial Databases, among others.*
- *Journals: Data Ming and Knowledge Discovery, WIRS Data Ming and Knowledge Discovery, International Journal of Data Warehousing and Mining, IEEE Transactions on Knowledge and Data Engineering, International Journal of Very Large Databases, International Journal of Geographical Information Science, International Journal of Remote Sensing, Artificial Intelligence, Machine Learning, among others.*
- *Organizations: Spatial Data Mining Lab at Wuhan University, and Beijing Institute of Technology in China, Spatial Data Mining and Visual Analytics Lab at University of South Carolina in USA, Spatial Database and Spatial Data Mining Research Group at University of Minnesota in USA.*

Since the inception of SDM, Li Deren has devoted himself and his team to exploring SDM theories and applications funded by organizations such as the National Science Foundation of China, the National High Techniques Research and Development Plan, and the National Basic Research Plan. The authors have proposed novel techniques such as the cloud model, data fields, and mining views, which are successfully being implemented to forecast landslide hazards, monitor train wheel deformation, protect public safety, assess the severity of the Syrian crisis, promote the beneficial cooperation of Asian countries, and understand Mars images, to name a few. Li Deren and his team have been invited to chair international conferences, give keynote speeches, edit special issues, and write overview papers. They also founded the International Conference of Advanced Data Mining and Applications (ADMA). During the continuing process of research and application of SDM, students have received doctorates, and one thesis was awarded "China's National Excellent Doctoral Thesis Prizes" by the Chinese government. Their contributions to SDM are inspiring more and more institutions and industries.

1.2.2 Theoretical Techniques

SDM is a process to support decision-making. Its basic knowledge is in the form of rules plus exceptions. Theoretical techniques directly decide whether the discovered knowledge is good for the application at hand. In the face of mining objects, the SDM methods currently used are often based on crisp data or uncertain data, such as probability theory, evidence theory, spatial statistics, rules induction, clustering analysis, spatial analysis, fuzzy sets, cloud models, data fields, rough sets, geo-rough space, neural networks, genetic algorithms, visualization, decision trees, and online SDM (Ester et al. 2000).

Set theory was introduced by George Contor, the founder of modern mathematics. Under the umbrella of crisp set theory, probability theory (Arthurs 1965), and spatial statistics, Cressie (1991) focused on SDM with randomness. Evidence theory is an extension of probability theory (Shafer 1976). Rule induction, clustering analysis, and spatial analysis originated from spatial statistics (Pitt and Reinke 1988). Clustering algorithms may be based on partition, hierarchy, or location (Kaufman and Rousseew 1990) and implemented with supervised or unsupervised information (Tung et al. 2001).

Set theory was further developed for when an indeterminate entity is depicted—that is, from a crisp set to an uncertain set. For SDM, fuzzy sets (Zadeh 1965), rough sets (Pawlak 1991), and cloud models (Li and Du 2007) were introduced. In a crisp set on certainty, there are only two resulting values along with its characteristic function—that is, $\{0, 1\}$ for $\{NO, YES\}$. In a fuzzy set on fuzziness, the resulting values range in a universe along with a fuzzy membership function—that is, $[0, 1]$ instead of $\{0, 1\}$. The calculation is implemented on a continuous interval of fuzzy membership in fuzzy sets but not two discrete elements in crisp sets, which obviously solved the "specious" ambiguity. In a rough set on incompleteness, the resulting value range in a set combined the crisp set and the fuzzy interval—that is, {0, (0, 1), 1}, along with a precise pair of upper and lower approximations. The unknown decision-making is approached using the known background attributes. In a cloud model, the randomness and fuzziness are compatible with each other—that is, a random case with fuzziness or a fuzzy set with randomness, which bridges qualitative knowledge and its quantitative data.

In addition to the set methods for SDM, neural networks (Miller 1990) and genetic algorithms (Buckless and Petry 1994) are connected to bionics. Visualization is a visual tool in which many abstract datasets for complex entities are depicted in specific graphics and images with human perception (Maceachren 1999). The decision tree characterizes spatial entities via a tree structure. The rules are generated by using up-down expansion or down-up amalgamation (Quinlan 1993). The spatial data warehouse refines spatial datasets for effective management and public distribution on SDM (Inmon 2005). Based on the data warehouse, online SDM may be implemented on multi-dimensional views and efficient and timely responses to user commands. The network broadens the scope of SDM (e.g., web mining). The space of the discovery state provides a methodological framework for implementing SDM (Li and Du 2007).

The state of the art of SDM has appeared many times in the literature. Grabmeier and Rudolph (2002) reviewed clustering techniques under GIS. Koperski et al. (1999) summarized the association rules discovered in RS, GIS, computerized mapping, environmental assessment, and resource planning. After the framework of SDM based on concepts and techniques was proposed (Li 1998), Li Deren provided an overview of the theories, systems, and products (Li et al. 2001). Moreover, he wrote a monograph entitled "Spatial Data Mining Theory and Application," which systematically presented the origin, concepts, data sources, knowledge to discover, usable techniques, system development, practical cases, and future perspectives (Li et al. 2006, 2013). This monograph was praised as a landmark in SDM (Chen 2007). Han et al. (2012) extended the concepts and techniques of DM into SDM at an international conference focusing on geographic knowledge discovery (Han 2009). Geostatistics-based spatial knowledge discovery also was summarized (Leung 2009). The above-mentioned techniques are not isolated from the practical application of them. The techniques are used comprehensively and fully draw from the mature techniques of related fields, such as ML, AI, and pattern recognition.

1.2.3 Applicable Fields

SDM is a technique that supports decision-making with geo-referenced datasets. It can provide decision-makers with valuable knowledge by guided data interpretation, database reconstruction, information retrieval, relationship discovery, new entity discovery, and spatiotemporal optimization. According to reports from *Knowledge Discovery Nuggets* (http://www.kdnuggets.com), DM has begun to penetrate human socio-economic activities and even the U.S. Federal Bureau of Investigation (FBI). During the process of realizing results, SDM not only receives the characteristics of the data mined but also develops its specific properties. SDM is active in fields with large-scale datasets such as space exploration, energy discovery, environment protection, resource planning, weather forecasting, emergency response, business location, communication prediction, financial fraud, real estate, urban planning, weather forecasting, medical imaging, road transport, target navigation, and crop yield estimation (Eklund et al. 1998; Han et al. 2012; Wang 2002; Li et al. 2006, 2013).

The discovered knowledge is yielding many immeasurable benefits (Han et al. 2012, Wang 2002), as shown by the following examples. The beer-diaper association rule discovered from a barcode dataset accelerated the product sales of Walmart stores. The Sky Image Cataloging and Analysis Tool (SKICAT) found 16 new quasars that are extremely far away. The accuracy of image classification of space stars was improved from 75 % to 94 % by using the Palomar Observatory Sky Survey (POSS). The Magellan study identified the volcanoes on phosphorous stars by mining about 30,000 radar images with high resolution. CONQUEST discovered the sampled knowledge of how an ozone hole was generated in the atmosphere via spatiotemporal mining. Mars image mining helps humans understand the images of the Mars surface collected by the rovers, which further directed the Mars rovers (e.g., Spirit and Opportunity) to travel under the Exploration missions. CASSIOPEE—a system to control the quality of aircraft engines produced by American manufacturer General Electric (GE) and French manufacturer SNECMA—can diagnose and predict the malfunctions of Boeing aircraft with the support of DM. British Broadcasting Corporation (BBC) rationally arranged their television program schedule in the context of the knowledge it gained from audience rating prediction, which had significant economic benefits for BBC. By virtue of DM, American Express increased the interest rate of credit cards by 10 %–15 %. AT&T telecommunication solved the worrisome phenomena of international telephone fraud quickly. DM can even uncover the rulings of criminal cases for directing public security. DM in bioinformatics may improve human health.

SDM plays an important role in everyday life as well, and its scope of application is ever-expanding under the remotely sensed aids of our Smart Planet. For example, RS cannot be substituted for human monitoring of the sustainable development of the earth; however, it is one of the current areas that RS is capable of doing automatically and its sensed images would be valuable. SDM provides new access to automatically sensed knowledge-based object identification and classification from huge RS high-resolution images. In hyperspectral images of the earth's surface, there is a triple data gain of spatial, radial, and spectral information. By mining hyperspectral images, the objects with fine spectral features may be intelligently analyzed, extracted, identified, matched, and classified with high accuracy and thereby accelerate the application time of RS images. SDM may greatly improve the speed of knowledge-based information processing. From nighttime light images, the severity of the Syrian crisis was assessed in time for humanitarian aid, and the countries in Asia will be helped to promote beneficial cooperation in global sustainability.

1.3 Bottleneck of SDM

The spatial real world is an uncertain system of multi-parameter, nonlinear, and time-varying data that can be observed and depicted by a variety of spatial data. Spatial data that are spatial, temporal, multi-dimensional, massive, complex, and uncertain are capable of identifying the geographic location of the features, boundaries, and relationships of the spatial entities on the earth. Although SDM has achieved some of its research and application goals, there are technical difficulties that need to be addressed (Koperski 1999), such as excessive data, high-dimensional data, polluted data, uncertain data, mining differences, and problems to represent the discovered knowledge.

1.3.1 Excessive Spatial Data

As mentioned in Sect. 1.1.1, the amount, size, complexity, and transmission of spatial data are growing rapidly. Spatial data's rate of growth is far beyond that of conventional transactional data. With the rapid development of RS, digital techniques, networks, multimedia, and images have become important data sources. The new types of satellite sensors available are high-resolution, highly dynamic, and can collect data at a very rapid speed and in a short cycle. Under normal circumstances, the number of sensed images is so large that the capacity of the data is on the order of magnitude of more than a gigabit. For example, only the sensed data coming from EOS-AM1 (Terra) and EOS-PM1 (Aqua) every day are on the order of magnitude of more than a terabit (Li and Guan 2000). It is universally easy now to obtain an image via digital instruments—the cycle of which is becoming significantly shorter.

The rapid growth of spatial datasets is one reason why DM was developed, but it has brought about some challenges. Some traditional methods appear to be powerless in the face of several megabits, a few gigabits, or even a greater number of bytes of spatial data, such as the enumeration method of data analysis. Solving this problem will require that DM is capable of summarizing and understanding the dataset population via a dataset sample under a user-defined discovery mission.

1.3.2 High-Dimensional Spatial Data

The volume of spatial data is also rapidly increasing—not only by its horizontal instances but vertical attributes as well. To depict a spatial entity accurately and completely in a computerized world, increasingly more types of attributes need to be observed in order to truly represent the objects in the real world. Only in the spatial data infrastructures of Digital Earth are there attributes such as images, graphics, digital terrain models (DTMs), transportation networks, river systems, and entity boundaries. At the same time, spatial data have a wide range of sources for developing techniques to acquire attributes. For instance, the spatial data in Digital Earth may be from satellites, aircraft, observation, investigation, maps, etc. Moreover, various spatial or non-spatial relationships exist among spatial entities or between two attributes. For example, all the entities in Digital Earth are networked in the context of the Internet of Things, different attributes are integrated in an entity, and the same attributes are topologically interacted between two entities. These attributes increase the number of dimensions, which contributes to the growth in the size of spatial data.

However, the current spatial information system, GIS, has been inadequate to describe the geographical elements within a multi-dimensional structure. As a result, it is difficult to generally analyze and master multi-dimensional datasets, which has seriously restricted the development of geo-spatial information science. In the realm of discovery knowledge from high-dimensional databases, some issues are challenging, such as how to efficiently organize and store multi-dimensional elements, how to quickly search and retrieve the target data, how to sharply decrease the number of dimensions, and how to clearly summarize multi-dimensional knowledge.

1.3.3 Polluted Spatial Data

Spatial data are the root sources for SDM. Poor data quality may directly result in unreliable knowledge, inferior service, and wrong decision-making (Shi and Wang 2002). However, spatial data collected from the real world are contaminated, which means that SDM frequently faces data problems such as incompleteness, dynamic changes, noise, redundancy, and sparsity (Hernàndez and Stolfo 1998). Therefore, the United States (U.S.) National Center for Geographic Information and Analysis has named the accuracy of GIS as its priority theme, and error analysis of spatial data was set as the foremost problem in the 12th working group. The National Data Standards Committee for Digital Mapping in the U.S. determined the following quality standards for spatial data: position accuracy, attribute accuracy, consistency, lineage, and integrity. However, it is not enough to study and apply spatial data cleaning and not understand the errors in multi-source SDM.

(1) Incompleteness. Observational data are sampled from their data population. When compared to its population, the limited samples are incomplete; the observational data are also inadequate to understand a true entity. For example, when monitoring the displacement of a landslide, it is impossible to monitor every point in the landslide. Therefore, representative points under the geological properties are chosen as the samples. Here, it is incomplete to monitor the displacement of the landslide on the basis of the observed data on the representative points. Each observation is not independent, and it may affect every point with different weights in the universe of

discourse. Different data contribute differently to the tasks of SDM. To better understand the data population by their data samples, it is necessary to mathematically model the diffusion of the contribution on a mining task from the sampled subspace to expand their population space. In addition, some attributes may be null in individual records in the database. Moreover, some specific records that are prerequisite to SDM may be completely non-existent. Some natural or artificial hazards may be damaged in some of the spatial data when they are observed, saved, processed, and transformed, which has resulted in incomplete spatial data.

(2) Dynamic changes. The world space is always changing. Accordingly, its described spatial data also are changing every moment. Data-dynamic changes are a key feature of most databases, such as RS images from observational satellites, stream data from community charge coupled devices (CCD) cameras, transactional data from supermarkets, and web logs from networks. An online system should be able to guarantee that the data provided will not lead to changes in the error occurrence.

(3) Noises. Spatial data with noise will affect the resulting accuracy and reliability of SDM. Noise may be from humans, the natural environment, and computerized infrastructure. When spatial data are manually or subjectively observed, input, processed, and output, man-made errors sometimes occur, which cause noise in the data. Natural phenomena may contain ambiguous objects or useless information when spatial objects are remotely observed, such as fog, rain, snow, clouds, wind, and brume. If computerized infrastructures are polluted, there is noise during data manipulation; for example, when a CCD camera is polluted, all the images will be fogged.

(4) Data redundancy. The same data on the same object may be replicated if they are collected from more than one source or they are stored in different places simultaneously. In order to keep the data safe, the same data may be further saved as a hierarchical backup in different places. A view of a database is a logical mirror image, which becomes a physical repetition once it has been saved. When data are normalized, the redundant attributes are maintained for fast retrieval. Redundant data may result in disaccord or error during the process of DM; and even the discovered knowledge may be redundant and not of interest to users.

(5) Sparse data. A large number of spatial entities are not uniformly distributed and are not easily classified distinctly to make use of them. The user must abstract the important characteristics of a real spatial entity for representing its computerized entity, such as point, line, and polygon. If the computerized entity is depicted by using a determined model (Burrough and Frank 1996), such as an object model, the acquired data cannot cover the entire world. Therefore, the data stored in databases are sampled from the data population. Familiar data that are easy to acquire may be sampled frequently, but it is difficult to observe data such as landslide hazards and hurricanes. Although the amount of familiar spatial data is excessive, the density of the actual data records is very sparse.

1.3.4 Uncertain Spatial Data

It is an unavoidable fact that there is data uncertainty in SDM. First, approximate sample data and abstract mathematical models bring about spatial data uncertainty. In classical data-processing methods, it is often hypothesized that the entity is ideally certain in the physical space, and a series of discrete points, lines, and polygons are able to describe the spatial distribution of all entities. Usually, spatial data are assumed to have been carefully checked and verified during the process of data collection, entry, editing, and analysis to assure there are no gross errors.

It is basically feasible to attain a distinctly defined spatial entity. However, spatial entities are often mixed with one another in the complicated and ever-changing real world and experience randomness, fuzziness, default, chaos, and other uncertain factors. Sometimes, the boundary between two entities is indeterminate, which is difficult to define, and no pure points, lines, or polygons exist. Data sampling is one of the methods to achieve a true entity (Wang and Yuan 2014).

In addition, it is difficult to collect a large number of true values of spatial data for an entity; some truths or absolute truths of spatial data may not exist at all. There also are systematic errors, stochastic errors, and man-made gross errors. One, two, or all three types of errors are cumulated when the entity is observed (Li et al. 2006). Furthermore, when the observed data enter the computerized system for SDM, part of the data are discarded or deleted (e.g., generalized cartography). As a result, the observed data obtained and used cannot completely extract all of the real-world attributes. There inevitably exist differences between the real entity and the entity described by the spatial observed data. However, the research, development, and application of SDM are all based on the computerized data in spatial databases or spatial data warehouses with data uncertainty (Zhang and Goodchild 2002).

The conversion between the spatial concept and the spatial data is the cornerstone of the conversion between quality and quantity, which includes uncertainty. At present, some methods are commonly used in the conversion between quality and quantity, such as analytical hierarchical analysis (AHP) quantified with weights, experts scoring, and qualitative analysis mixed with mathematical models and quantitative calculation. Cybernetics controls the uncertainties by eliminating one error on the other error between a given target and its actual behavior. However, none of them can take into account the randomness and fuzziness of spatial data simultaneously. An alternative is a cloud model to integrate the randomness and fuzziness in the conversion between the qualitative concept and the quantitative data (Li and Du 2007).

Finally, although the uncertainty in spatial data quality is attracting increased attention (Shi and Wang 2002), it is inadequate to study and use spatial data uncertainty profoundly and completely. At present, the theoretical methods for SDM (e.g., probability theory, GIS models, and sensitivity analysis) are generally applied to certain data with crisp set. Paying less attention to spatial data uncertainty, the user may think that every single attribute is relevant to a spatial entity and that there is a determined boundary between two attributes. This approach is obviously inconsistent with reality in the complicated and ever-changing real world. Some studies addressing spatial data uncertainty have focused more on positional uncertainty than attribute uncertainty.

1.3.5 Mining Differences

SDM is utilized to understand and grasp spatial datasets in different cognitive hierarchies. Using the same datasets, users who are the same with different perspectives or different users with the same perspective are likely to obtain results that are different from the results of the discovered knowledge; but each of these different results has its distinct uses under a given requirement. It is said that the benevolent see benevolence and the wise see wisdom. In different fields, different requirements may make it necessary to restrict the knowledge discovered from the same spatial datasets. In the same fields, the requirements may be hierarchical. Using landslide monitoring data as an example, a high-level decision-maker may require the most generalized knowledge for grasping the macro-decisions on the whole landslide (e.g., a sentence, a figure); a middle-level decision-maker may be interested in more knowledge in order to carefully check the deformation of each landslide section; and a lower-level decision-maker may ask for the most specific rules for technically mastering the deformation of each monitoring point that is preset on the landslide. For that example using SDM, it is essential to reflect the hierarchical decision-maker's awareness of these differences, to determine a hierarchy of concluding rules from the landslide monitoring data, and to convert the knowledge between the different levels for the same dataset. The data field and cloud model may produce a comprehensive solution for uncovering and bridging different levels of cognitive knowledge.

1.3.6 Problems to Represent the Discovered Knowledge

Knowledge representation is a key issue in SDM. The acquired knowledge is summarized or deduced from spatial datasets—most of which are qualitative knowledge or a combination of qualitative and quantitative knowledge. One of the best methods to depict such knowledge is natural language, or the linguistic value is used to express the conceptual knowledge. By using natural language, the knowledge is more understandable, adequate information is transformed with less delivery cost, and more flexible decisions are reasoned depending on the complexity of the entities. However, there are problems, such as how to mathematically model the conversion between a qualitative description of the language and its quantitative values, how to achieve a mutual conversion between a numerical value and its symbolic value at any time, how to reflect the uncertainty in mapping between a qualitative concept and quantitative data, and how to measure to what degree the discovered knowledge is supportable, reliable, and of interest.

In summary, the above-mentioned difficulties may have a direct impact on the accuracy and reliability of the resulting knowledge in SDM. They further make it difficult to discover, assess, and interpret a large amount of important knowledge. Sometimes, they disturb some development of SDM; however, if the difficulties are reasonably resolved, decision-making errors using SDM may be avoided. Additionally, spatial uncertainty in measuring the supporting information may reflect the degree of confidence in the discovered knowledge. When the uncertainties in the process of SDM are ignored, the resulting knowledge may be incomplete or prone to errors even if the techniques used are adequate.

1.3.7 Monograph Contents and Structures

It is possible to overcome SDM's current technical difficulties by developing novel theories and methods that focus on its core purpose as well as developing software to meet the requirements of geomatics. This monograph proposes just such new approaches, which include the following. The *data field* diffuses the data contribution from the universe of sample to the universe of population. The *cloud model* bridges the mutual transformation between qualitative concepts and quantitative data. The *mining view* hierarchically distinguishes the mining requirements with different scale or granularity. A *pyramid of SDM* visually reveals the discovery mechanism. A *weighted iteration* is used to clean the errors in spatial data according to the principles of post-variance estimation. *GIS data mining* and *remote sensing image mining* are focused on applications.

This monograph is organized according to the various aspects of SDM (see Fig. 1.4). The first part, this chapter, is an overview. The second part, which includes Chaps. 2–4, presents the fundamental concepts. The third part (Chaps. 5–7) presents the methods and techniques. The fourth part, including Chaps. 8 and 9, demonstrates the applications. The fifth and final part, Chap. 10, introduces the developed software. A brief introduction to each chapter follows.

Chapter 1: Introduction to SDM and the state of the art of SDM.

Chapter 2: SDM principles. SDM is an interdisciplinary subject that addresses the discovery of a variety of knowledge. While closely related to many other subjects, SDM is different from them. It is a process of discovery instead of proofing. The SDM pyramid visualizes the process. The SDM view depicts the various purposes of the special DM tasks. Generally, spatial knowledge is represented as "spatial rule plus exception."

Chapter 3: SDM data sources. Spatial data are presented from the aspects of contents, characteristics, acquirements, structure, and model. Spatial databases and data warehouses are discussed. NSDI, GGDI, Digital Earth, Smart Earth, and big data contribute to the capabilities of SDM.

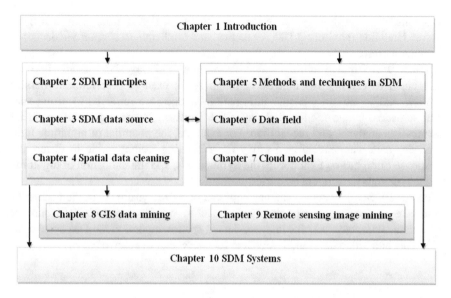

Fig. 1.4   Content architecture of the monograph

Chapter 4: Spatial data cleaning. Whether spatial data are qualified or not, they may influence the level of confidence in discovered knowledge directly. The errors that may occur in spatial data are summarized along with cleaning techniques to remove them.

Chapter 5: Usable methods and techniques in SDM. Crisp set theory includes probability theory, evidence theory, spatial statistics, spatial analysis, and data fields. The extended set theory includes fuzzy sets, rough sets, and cloud models. By using artificial neural networks and genetic algorithms, SDM simulates human thinking and evolution.

Chapter 6: Data fields. Data fields bridge the gap between the mining model and the data model. Informed by the physical field, a data field depicts the interaction between data objects. Its field function models how the data are diffused in order to contribute to the mining task. All the equipotential lines depict the topological relationships among the interacted objects.

Chapter 7: Cloud model. The cloud model bridges the gap between the qualitative concept and quantitative data. Based on the characters $\{Ex, En,$ and $He\}$, there are cloud model generators to create all types of cloud models for reasoning and controlling in SDM.

Chapter 8: GIS data mining. SDM promotes the automation and intelligent application of discovered knowledge, thereby enabling more effective use of the current and potential value of spatial data in GIS. Spatial association rules are discovered with an Apriori algorithm, concept lattice, and cloud model. Spatial distribution rules are determined with inductive learning. Decision-making knowledge is discovered with a rough set. Clustering focuses on fuzzy comprehensive clustering, mathematical morphology clustering, and hierarchical clustering with data field. Baota landslide-monitoring DM is practical and creditable to support the decision-making hierarchically.

Chapter 9: RS image mining. Inductive learning and Bayesian classification are combined to classify RS images. A rough set is used to describe and classify images and to extract thematic information. Images are retrieved by using spatial statistics. Cloud models and data fields perform image segmentation, facial expression analysis, and recognition. For example, the potential of nighttime light imagery data was studied as a proxy for evaluating freight traffic in China, the response of nighttime light during the Syrian crisis, and the indication of nighttime light dynamics in different countries.

Chapter 10: SDM systems. The development of SDM systems is in the context of software engineering and professional demands. We developed *GISDBMiner* for GIS data, *RSImageMiner* for imagery data, a spatiotemporal video DM system for GIS data and video data, and EveryData for every data in every day.

1.4 Benefits to a Reader

This monograph is a rich blend of novel methods and practical applications of SDM studied from an interdisciplinary approach. For readers to benefit from this monograph, knowledge of database systems, probability and statistics, GIS, and RS would be helpful.

The information in this monograph will be of interest to scientists and engineers engaged in SDM, computer science, geomatics, GIS, RS, GPS, data analysis, AI, cognitive science, spatial resources planning, land science, hazard prevention and remediation, management science and engineering, and decision-making. It is also suitable for undergraduate students, graduate students, researchers, and practitioners interested in SDM both as a learning text and as a reference book. Professors can readily use it for course sections on DM and SDM.

References

Arthurs AM (1965) Probability theory. Dover Publications, London

Buckless BP, Petry FE (1994) Genetic algorithms. IEEE Computer Press, Los Alamitos

Burrough PA, Frank AU (eds) (1996) Geographic objects with indeterminate boundaries. Taylor & Francis, Basingstoke

Chen SP (2007) A landmark masterpiece of spatial data mining: review of "Theory and Application of Spatial Data Mining". Science Bulletin 52(21): 2577

Connolly TM (2004) Database systems: a practical approach to design, implementation, and management, 4th edn. Addison-Wesley

Cressie N (1991) Statistics for spatial data. Wiley, New York

Di KC (2001) Spatial data mining and knowledge discovery. Wuhan University Press, Wuhan

Di KC, Li DR, Li DY (1997) Framework of spatial data mining and knowledge discovery. Geomatics Inf Sci Wuhan Univ 4:328–332

Eklund PW, Kirkby SD, Salim A (1998) Data mining and soil salinity analysis. Int J Geogr Inf Sci 12(3):247–268

Ester M et al (2000) Spatial data mining: databases primitives, algorithms and efficient DBMS support. Data Min Knowl Disc 4:193–216

Fayyad UM, Uthurusamy R (eds) (1995) Proceedings of the first international conference on knowledge discovery and data mining (KDD-95), Montreal, Canada, Aug 20–21, AAAI Press

Giannotti F, Pedreschi D (eds) (2010) Mobility, data mining and privacy: Geographic knowledge discovery. Springer, Berlin

Giudici P (2003) Applied data mining: statistical methods for business and industry. Wiley, Chichester

Goodchild MF (2007) Citizens as voluntary sensors: spatial data infrastructure in the world of web 2.0. Int J Spat Data Infrastruct Res 2:24–32

Grabmeier J, Rudolph A (2002) Techniques of clustering algorithms in data mining. Data Min Knowl Disc 6:303–360

Han JW, Kamber M, Pei J (2012) Data mining: concepts and techniques, 3rd edn. The Morgan Kaufmann Publishers Inc, Burlington

Hernàndez MA, Stolfo SJ (1998) Real-world data is dirty: data cleansing and the merge/purge problem. Data Min Knowl Disc 2:1–31

Höller J, Tsiatsis V, Mulligan C, Karnouskos S, Avesand S, Boyle D (2014) From machine-to-machine to the internet of things: introduction to a new age of intelligence. Elsevier

Inmon WH (2005) Building the data warehouse, 4th edn. Wiley, New York

Leung Y (2009) Knowledge discovery in spatial data. Springer, Berlin

Kaufman L, Rousseew PJ (1990) Finding groups in data: an introduction to cluster analysis. Wiley, New York

Koperski K (1999) A progressive refinement approach to spatial data mining. PhD thesis, Simon Fraser University, British Columbia

Li DR (1998) On the interpretation of "GEOMATICS". J Surveying Mapp 27(2):95–98

Li DY, Du Y (2007) Artificial intelligence with uncertainty. Chapman and Hall/CRC, London

Li DR, Cheng T (1994) KDG-Knowledge discovery from GIS. In: Proceedings of the Canadian Conference on GIS, Ottawa, Canada, June 6-10, pp 1001–1012

Li DR, Guan ZQ (2000) Integration and implementation of spatial information system. Wuhan University Press, Wuhan

Li DR, Wang SL, Shi WZ, Wang XZ (2001) On spatial data mining and knowledge discovery (SDMKD). Geomatics Inf Sci Wuhan Univ 26(6):491–499

Li DR, Wang SL, Li DY, Wang XZ (2002) Theories and technologies of spatial data mining and knowledge discovery. Geomatics Inf Sci Wuhan Univ 27(3):221–233

Li DR, Wang SL, Li DY (2006) Theories and applications of spatial data mining. Science Press, Beijing

Li DR, Wang SL, Li DY (2013) Theories and applications of spatial data mining, 2nd edn. Science Press, Beijing

Maceachren AM et al (1999) Constructing knowledge from multivariate spatiotemporal data: integrating geographical visualization with knowledge discovery in database methods. Int J Geogr Inf Sci 13(4):311–334

Miller WT et al (1990) Neural network for control. MIT Press, Cambridge

Miller HJ, Han JW (eds) (2009) Geographic data mining and knowledge discovery, 2nd edn. The Chapman and Hall/CRC, London

Naisbitt J (1991) Megatrends 2000. Avon Books, New York

Pawlak Z (1991) Rough sets: theoretical aspects of reasoning about data. Kluwer Academic Publishers, London

Pitt L, Reinke RE (1988) Criteria for polynomial time (conceptual) clustering. Mach Learn 2(4):371–396

Quinlan JR (1993) C4.5: programs for machine learning. Morgan Kaufmann, San Mateo

Shafer G (1976) A mathematical theory of evidence. Princeton University Press, Princeton

Shi WZ, Wang SL (2002) GIS attribute uncertainty and its development. J Remote Sens 6(5):393–400

Sridharan NS (ed) (1989) Proceedings of the 11th international joint conference on artificial intelligence, Detroit, MI, USA, August 20–25, Morgan Kaufmann

Wang SL (2002) Data field and cloud model based spatial data mining and knowledge discovery. Ph.D. Thesis, Wuhan University

Wang SL, Yuan HN (2014) Spatial data mining: A perspective of big data. Int J Data Warehouse Min 10(4):50–70

Wynne HW, Lee ML, Wang JM (2007) Temporal and spatio-temporal data mining. IGI Publishing, New York

Tung A et al. (2001) Spatial clustering in the presence of obstacles. IEEE Transactions on Knowledge and Data Engineering, 359–369

Zadeh LA (1965) Fuzzy sets. Inf Control 8(3):338–353

Zhang JX, Goodchild MF (2002) Uncertainty in geographical information. Taylor & Francis, London

Chapter 2
SDM Principles

The spatial data mining (SDM) method is a discovery process of extracting generalized knowledge from massive spatial data, which builds a pyramid from attribute space and feature space to concept space. SDM is an interdisciplinary subject and is therefore related to, but different from, other subjects. Its basic concepts are presented in this chapter, which include manipulating space, SDM view, discovered knowledge, and knowledge representation.

2.1 SDM Concepts

SDM aims to improve human ability to extract knowledge and insights from large and complex collections of digital data. It efficiently extracts previously unknown, potentially useful, and ultimately understandable knowledge from these huge datasets for a given task with constraints (Li et al. 2001, 2006, 2013; Wang 2002; Han et al. 2012; Wang and Yuan 2014). To implement a credible, innovative, and interesting extraction, the SDM method not only relies on the traditional theories of mathematical statistics, machine learning, pattern recognition, neural networks, and artificial intelligence, but it also engages new methods, such as data fields, cloud models, and decision trees.

2.1.1 SDM Characteristics

SDM extracts abstract knowledge from concrete data. The data are explicit while, in most circumstances, the knowledge is implicit. The extraction may be a repeated process of human–computer interactions between the user and the

© Springer-Verlag Berlin Heidelberg 2015
D. Li et al., *Spatial Data Mining*, DOI 10.1007/978-3-662-48538-5_2

dataset. The original data are raw observations of spatial objects. Because the data may be dirty, they need to be cleaned, sampled, and converted in accordance with the SDM measurements and the user-designated thresholds. The discovered knowledge consists of generic patterns acting as a set of rules and exceptions. The patterns are further interpreted by professionals when they are utilized for data-referenced decision-making with various requirements.

SDM's source is spatial data, which are real, concrete, and massive in volume. They already exist in the form of digital data stored in spatial datasets, such as databases, data markets, and data warehouses. Spatial data may be structured (e.g., instances in relational data), semi-structured (e.g., text, graphics, images), or non-structured (e.g., objects distributed in a network). Advances in acquisition hardware, storage capacity, and Central Processing Unit (CPU) speeds have facilitated the ready acquisition and processing of enormous datasets, and spatiotemporal changes in networks have accelerated the velocity of spatial data accumulation. Noises and uncertainties exist in spatial data (e.g., errors, incompleteness, redundancy, and sparseness), which can cause problems for SDM; therefore, polluted spatial data are often preprocessed for error adjustment and data cleaning.

SDM aims to capture knowledge. The knowledge may be spatial or non-spatial; it is previously unknown, potentially useful, and ultimately understandable under the umbrella of spatial datasets. This knowledge can uncover the description and prediction of the patterns of spatial objects, such as spatial rules, general relationships, summarized features, conceptual classification, and detected exception. These patterns are hidden in the data along with their internal relationships and developing trends. However, SDM is a much more complex process of selection, exploration, and modeling of large databases in order to discover hidden models and patterns, where data analysis is only one of its capabilities. SDM implements high-performance distributed computing, seamless integration of data, rational knowledge expression, knowledge updates, and visualization of results. SDM also supports hyperlink and media quotas among hierarchical document structures when mining data.

The SDM process is one of discovery instead of proofing. Aided by SDM human–computer interaction, the process is automatic or at least semi-automatic. The methods may be mathematic or non-mathematic, and the reasoning may be deductive or inductive. The SDM process has the following requirements. First, it is composed of multiple mutually influenced steps, which require repeated adjustment to spiral up in order to extract the patterns from the dataset. Second, multiple methods are encouraged, including natural languages, to present the process of discovery and its results. SDM brings together all of the available variables and combines them in different ways to create useful models for the business world beyond the visual representation of the summaries in online analysis and processing (OLAP) applications. Third, SDM looks for the relationships and associations between phenomena that are not known beforehand. Because the discovered knowledge usually only needs to answer a particular spatial question, it is not necessary to determine the universal knowledge, the pure mathematical formula, or a new scientific theorem.

2.1.2 Understanding SDM from Different Views

SDM is a repeated process of spatial data-referenced decision-making. It can be understood best from the following five perspectives:

(1)   **Discipline**. SDM is an interdisciplinary subject that matches the multidisciplinary philosophy of human thinking and suitably deals with the complexity, uncertainty, and variety present when informing data and representing rules. SDM is the outcome in the stage at which some technologies develop, such as spatial data access technology, spatial database technology, spatial statistics, and spatial information systems; therefore, it brings together the fruits of various fields. Its theories and techniques are linked with data mining, knowledge discovery, database systems, data analysis, machine learning, pattern recognition, cognitive science, artificial intelligence, mathematical statistics, network technology, software engineering, etc.

(2)   **Analysis**. SDM discovers unknown and useful rules from huge amounts of data via a set of interactive, repetitive, associative, and data-oriented manipulations. It mainly utilizes certain methods and techniques to extract various patterns from spatial datasets. The discovered patterns describe the existent rules or predict a developing trend, along with the certainty or credibility to measure the confidence, support, and interest of the conclusions derived from analysis. They can help users to make full use of spatial repositories under the umbrella of various applications, stressing efficient implementation and timely response to user commands.

(3)   **Logic**. SDM is an advanced technique of deductive spatial reasoning. It is discovery, but not proofing, in the context of the mined data. As a part of deductive inference, it is a special tool for spatial reasoning that allows a user to supervise or focus on discovering the rules of interest. The reasoning can be automatic or semi-automatic. Induction is used to discover knowledge, while deduction is used to evaluate the discovered knowledge. The mining algorithms are a combination of induction and deduction.

(4)   **Actual object operation**. The data model can be hierarchical, network, relational, object-oriented, object-related, semi-structured, or non-structured. The data format may be vector, raster, or vector-raster spatial data. The data repositories are file systems, databases, data markets, data warehouses, etc. The data content may involve locations, graphics, images, texts, video streams, or any other data collections organized together, such as multimedia data and network data.

(5)   **Sources**. SDM is implemented on original data in databases, cleaned data in data warehouses, detailed commands, information from users, and background knowledge from applicable fields, which include the raw data provided by spatial databases and the corresponding attribute databases or the manipulated data stored in the spatial data warehouse, the advanced instructions sent by the user to the controller, and the various expert knowledge of different fields stored in the knowledge base. With the help of the network, SDM can

break down the local restrictions of spatial data, using not only the spatial data in its own sector, but larger scopes or even all of the data in the field of space and related fields. It also can be available to discover more universal spatial knowledge and to implement spatial online data mining (SOLAM). To meet the needs of decision-making, SDM makes use of decentralized heterogeneous data sources, with timely and accurately extracted information and knowledge through data analysis using the query and analysis tools of the reporting module.

2.1.3 Distinguishing SDM from Related Subjects

SDM's interdisciplinary nature integrates machine learning, computer visualization, pattern recognition, statistical analysis, database systems, and artificial intelligence. SDM and its related disciplines are related but different.

SDM is a branch of data mining with spatial characteristics. The general mining object of SDM is a conventional structured relational database. SDM objects are spatial datasets in which there are not only attribute and location data (e.g., maps, remote sensing images, and urban spatial planning), but also spatial relations and distance data as well. Unstructured spatial graphics and images may be vector or raster in a number of layers. Moreover, its spatial data storage structures, query methods, data analysis, and database operations are different from a conventional database. As for granularity, data mining is transactional data; and SDM data may be a point, line, polygon, pixel, or tuple. Taking a vector object as granularity, SDM can use the location, form, and spatial correlation of spatial objects to discover knowledge. Taking a raster pixel as granularity, SDM can use the pixel location, multi-spectral value, elevation, slope, and other information to extract image features for fine to coarse granularity. As to scale, SDM adds scale dimension to represent the geometric transformation of spatial data from large scale to small scale. The larger the scale is, the finer are the spatial patterns of the presented object.

Machine learning obtains or reproduces information via a training computer, focusing on the design and development of new algorithms as empirical input. It is a training technique via analysis so the specific data prepared for machine learning need not have significance in the real world (Witten and Frank 2000). In machine learning, pattern recognition is the assignment of a label to a given input value. According to the output, pattern recognition may be supervised or unsupervised. Based on the known properties learned from the training data, machine learning and pattern recognition pay attention to prediction. SDM is an extracted process via interacted discovery, such as task understanding, data conversion, data cleaning, dimension reduction, and knowledge interpretation. SDM also highlights the discovery of the unknown properties of the data.

Artificial intelligence is the academic basis of data mining generation and is based on the study of how human beings acquire and use knowledge. Mainly based on interpretation, artificial intelligence is a positive step toward understanding the world. While SDM uses machines to simulate human intelligence for the discovery of knowledge from data, machines also are used to reproduce human understanding, which is the reverse way to understanding the world. The artificial intelligence of space-based data mining systems, which have human-like thinking and cognitive ability, can discover new knowledge to complete a new task. Discovering knowledge through SDM is making progress as far as constructing expert systems and generating knowledge bases to achieve a new entity model for the cognitive science of artificial intelligence.

SDM has the highest information capacity and is the most difficult to implement. Query and reporting have the lowest information capacity and are the easiest to implement. Query and reporting tools help explore data at various levels. Data retrieval extracts interesting data and information from archives and databases with preset criteria by using preliminary statistical analysis. The query and reporting tools describe what a database contains; however, OLAP, which can create multidimensional reports, is used to explain why certain relationships exist. This suggests a trade-off between information capacity and ease of implementation. The user-made hypothesis about the possible relationships between the available variables is checked and confirmed by analyzing a graphical multi-dimensional hypercube from the observed data. Unlike SDM, the research hypotheses are suggested by the user and are not uncovered from the data. Furthermore, the extrapolation is a purely computerized procedure, and no use is made of modeling tools or summaries provided by the statistical methodology. OLAP can provide useful information for databases with a small number of variables; however, problems arise when there are tens or hundreds of variables. Then, it becomes increasingly difficult and time-consuming to find a good hypothesis and analyze the database with OLAP tools to confirm or deny it. OLAP and SDM are complementary; when used together, they can create useful synergies. OLAP can be used in the preprocessing stages of data mining, which makes understanding the data easier because it becomes possible to focus on the most important data, identifying special cases, or looking for principal interrelations. The final data mining results, expressed using specific summary variables, can be easily represented in an OLAP hypercube. As a web-based authentication of SDM, SOLAM supports multi-dimensional data analysis, verifying the set assumptions under the guidance of users (Han et al. 2012).

2.1.4 SDM Pyramid

In accordance with its basic concept, SDM's process includes data preparation (understanding the prior knowledge in the field of application, generating target datasets, cleaning data, and simplifying data), data mining (selecting the data mining functions and algorithms; searching for the knowledge of interest in the form of certain rules and exceptions: spatial associations, characteristics, classification, regression, clustering, sequence, prediction, and function dependencies), and post-processing of data mining (interpretation, evaluation, and application of the knowledge). During the SDM process, every move to the next stage deepens the awareness and understanding of the spatial entities, transforming the spatial data into information and then into knowledge. The more abstract, coherent, and general the description is, the more advanced the technologies need to be.

As a result, when explaining the concept of data mining, Piatetsky-Shapiro (1994) proposed the concept of the Data, Information, and Knowledge Pyramid (DIKP). However, the pyramid merely distinguished the specific concept elements of the different levels in data mining and failed to make an association between the concepts of data mining with the process. Han et al. (2012) depicted the process of data mining visually by different graphics, but only the process of data mining was emphasized; therefore, they failed to clearly illustrate the role of all the other elements and their distinctions. Because spatial data are far more complex than common data, it was necessary to combine DIKP and the process of SDM in order to clearly explain the concept and roles of SDM. After they were combined, clearly the SDM pyramid was more specific and complete (Fig. 2.1).

It can be seen in Fig. 2.1 that under the effects of external forces, such as access to the spatial data, storage of the spatial data, network sharing, calibration of target data, data cleaning, data mining, interpretation, and evaluation, the entities of the

Fig. 2.1 SDM pyramid

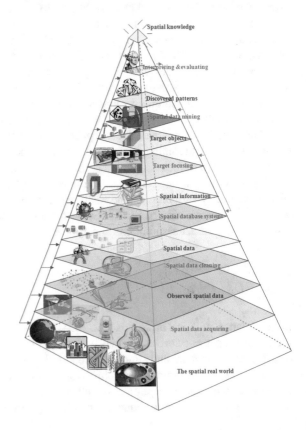

SDM pyramid more closely matched the reality of the physical world experience as far as the spatial concept, spatial data, spatial information, changes in spatial size, and increases in spatial scale, which finally become spatial knowledge. In different disciplines, the definitions of some basic concepts in Fig. 2.1, such as spatial data, spatial information, and spatial knowledge, are probably different.

2.1.5 Web SDM

The Web is an enormous distributed parallel information space and valuable information source. First, it offers network platform resources, such as network equipment, interface resources, computing and bandwidth resources, storage resources, and network topology. Second, a variety of data resources use the platform as a carrier, such as text, sound, video data, network software, and application software. The rapid development of network technology provides a new opportunity for a wide range of spatial information sharing, synthesis, and knowledge discovery.

While the high level of network resources makes the information of the network suitable, user-friendly, general, and reusable, how can these distributional, autonomous, heterogeneous data resources be used to obtain, form, and use the required knowledge in a timely manner? With the help of a network, the local restrictions are brought to the spatial dataset and can make use of not only the internal data of a department, but also on a greater scale or even all of the spatial data in the field of space or space-related fields. The discovered knowledge thus becomes more meaningful. Because of the inherent open, distributed, dynamic, and heterogeneous features of a network, it is difficult for the user to accurately and quickly obtain the required information.

Web mining is the extraction of useful patterns and implicit information from artifacts or activities related to the World Wide Web under the internet (Liu 2007). As the internet broadens the available data resources, web mining may include web-content mining, web-structure mining, and web-usage mining. Web-content mining discovers knowledge from the content of web-based data, documents, and pages or their descriptions. Web-structure mining uncovers the knowledge from the structure of websites and the topological relationships among different websites (Barabási and Albert 1999). Web-usage mining extracts Web-user behavior or modeling and predicting how a user will use and interact with the Web (Watts and Strogatz 1998). One of the most promising mining areas being explored is the extraction of new, never-before encountered knowledge from a body of textual sources (e.g., reports, correspondence, memos, and other paperwork) that now reside on the Web. SDM is one of those promising areas. Srivastava and Cheng (1999) provided a taxonomy of web-mining applications, such as personalization, system improvement, site modification, business intelligence, and usage characterization. Internet-based text mining algorithms and information-based Web server log data mining research are attracting increasingly more attention. For example, some of the most popular Web server log data are growing at a rate of tens of megabits every day; discovering useful model, rules, or the visual structures from them is another area of research and application of data mining. To adapt to the distributed computing environment of networks, SDM systems should also make changes in the architecture and data storage mode, providing different levels of spatial information services such as data analysis, information sharing, and knowledge discovery on the basis of display, service, acquisition, and storage of spatial information on the distributed computing platform. SOLAM and OLAP in multidimensional view with SDM on a variety of data sources (Han et al. 2012), stress the efficient implementation and timely response to user commands for more universal knowledge.

In the course of the development of a network, users come to realize that the internet is a complex, non-linear network system. Next-generation internet will gradually replace the traditional internet and become the information infrastructure of the future by integrating existing networks with new networks that may appear in the future. As a result, the internet and its huge information resources will be the most important basic strategic resource of a country. By efficiently and reasonably using those resources, the massive consumption of energy and material resources can be reduced and sustainable development can be achieved. The authors believe that networked SDM must take into account the possible space, time, and semantic inconsistencies in the databases distributed in networks, as well as the difference in spatiotemporal benchmarks and standards for the semantics. We must make use of spatial information network technology to build a public data assimilation platform on these heterogeneous federal spatial databases. Studies are underway to address solutions to these issues.

2.2 From Spatial Data to Spatial Knowledge

SDM is a gradual process of sublimation from data to knowledge, experiencing spatial numerical values, spatial data, spatial information, and spatial knowledge (see Fig. 2.1). These basic concepts are distinguishable and related. Based on geo-spatial information science, SDM meanings are assigned to these concepts, referencing the existing interdisciplinary definitions from many literature sources (Li et al. 2001; Frasconi et al. 1999; Han et al. 2012).

2.2.1 Spatial Numerical

A spatial numerical value is a numerical value with a measurement unit, which assigns spatial meanings to numbers. The number may be a natural number (1, 2, 3…), a rational number (-1, 0, 6.2, …), or an irrational number (e, π, …). The role of numbers is public; for example, the number 6,000 may represent a land area of 6,000 m^2, a monthly salary of \$6,000, or a distance of 6,000 km. When a number is restricted to a specific meaning, a numerical value appears. Furthermore, when it is restricted in the field of spatial information, a numerical value becomes the symbol to represent a spatial numerical value privately. For example, the number 6,000 is restricted to representing the distance of 6,000 km. Once a numerical value becomes a spatial numerical value, it is distinguished from the common number. The privatized numerical value is able to characterize spatial objects. In the process of verification of various theories and algorithms on computers, a variety of spatial numerical values are actually computerized via numbers. The unit of measurement is used to assign different spatial meanings. A spatial numerical value is a carrier when the objects are collected, transformed, stored, and applied in a computer system.

2.2.2 Spatial Data

Spatial data are important references that help humans to understand the nature of objects and utilize that nature by numerically describing spatial objects with the symbol of attributes, amounts, and positions and their mutual relationships. Spatial data can be numerical values, such as position elevation, road length, polygon coverage, building volume, and pixel grayscale; character strings, such as place name and notation; or multimedia information, such as graphics, images, videos, and voices. Spatial data are rich in content from the microcosmic world to the macroscopic world, such as molecular data, surface data, and universal data. Compared to common data, spatial data contain more specifics, such as spatiotemporal change, location-based distribution, large volume, and complex relationships. In spatial datasets, there are both spatial data and non-spatial data. Spatial data describe a geographic location and distribution in the real world, while non-spatial data consist of all the other kinds of data. Sometimes, a spatial database is regarded as a generic database—one special case of which is a common database. Spatial data can be divided into raw data and processed data or digital data and non-digital data. Raw data may be numbers, words, symbols, graphics, images, videos, language, etc. Generally, SDM utilizes digital datasets after they are cleaned.

2.2.3 Spatial Concept

A spatial concept defines and describes a spatial object, along with its connotation and extension. Connotation refers to the essence reflected by the concept, while extension refers to the scope of the concept. Generally, the definition of spatial concept and its interpretation of connotation and extension are applied to explain the phenomena and the states of spatial objects, as well as to resolve problems in the real world. Take the concept of remote-sensing image interpretation as an example. The connotation of "green" refers to the type of land covered by green plants, whose gray value of low-level eigenvalue is in a particular range. Its extension includes all the image pixels whose gray values are in that particular range. The scope of concept extension will decrease if its connotation scope increases. For example, if the attribute of "texture characteristics" is added to the connotations of "green," its extension will be greatly reduced. As such, it only refers to the greens that have certain texture features, such as grassland, woodland, and forest land. The concept is hierarchical. Although "green" is a concept and "grassland," "forest land," or "woodland" are also concepts, these concepts are at different hierarchies. The concept layer of "forest land" is lower than that of "green" but higher than that of "woodland."

Spatial objects represented by spatial concepts sometimes are uncertain (e.g., random and ambiguous). Conceptual space refers to the universe of discourse on the same kind of concept. When discussing the different meanings of a conceptual variable, it is necessary to specify their connotation and extension, as well as their mutual similarity or affiliation in the universe of discourse. For example, a variable of "distance" may be "3000 km or so," "around a 12-h train journey," or "very far." In this way, the technology supported by a spatial concept can be meaningful and therefore leap from the unknown world to the known world. Spatial concepts should be defined along with granularity and scale. The same concept can have different spatial meanings of granularity under a different scale, and scale reflects the scaling of the concept granularity. For example, the concept of "the distance from Wuhan to Beijing is 1,100 km" is a close distance on global scale but a long distance on the scale of Hubei Province, China.

2.2.4 Spatial Information

Spatial information characterizes the features of spatial objects in the form of a summarizing dataset. It is the product of data processing that provides a new result and fact in the storage, transmission, and conversion of object data. Wiener, the founder of cybernetics, stated that "information is the content of exchange between human beings and the outer world in the process of their interaction." Shannon, a founder of information theory, argued that "information is what is constant in the process of reversible re-encoding on communication." Philosophically, information indicates the ways of existence and the moving laws of substances, spreading from person to person for communication and interaction. Sometimes, information is treated as a resource to directly help decision-makers solve problems. When it supports the decision-makers only after reprocessing, data are still data or a piece of a message but not information. Spatial information provides useful interpretation to eliminate spatial uncertainty in a meaningful form and order after data processing in specific environments. Incomplete information results in uncertainty, while complete information results in determination. For example, "South Lake" may remind us of various information, such as "the South Lake in Wuhan City," "the South Lake in Nanning," "the South Lake Airport" and "the area of South Lake," etc. because of incomplete information. After more specific information is added, it is clearly known the information is referring to "the closed South Lake Airport in Wuhan."

A variety of geometric features and spectrum features are extracted from remotely sensed images. Thematic information from investigation and observation can create a multisource and multi-dimensional thematic map of a substantial amount of thematic information for one location, such as location, shape, size, quality, height, slope, noise, pollution, transportation, land cover, moisture, distribution, and relationships of surface objects such as rivers, hills, forests, and soil. The same data under a different background may represent different information; for example, "600 m^2" can be either the area of a land parcel or the construction area of a building. Different data under the same background can represent the same information; for example, "around 1,100 km," "around a 12-h train journey," "far away," etc. can all be information about "distances from Beijing to Wuhan." The quality of spatial information can be represented by a percentage or a rating description of "excellent, good, common, or poor," which makes it possible to sublimate SDM concepts. Spatial information should be collected and updated in a timely manner because its change is dynamically spatiotemporal, such as the rise and fall of sea water, the movement of the desert, the melting of a glacier, and a change of land use. The position identity of spatial information is linked with spatial data. Because satellite positioning and remote sensing systems can uninterruptedly access spatial data for a long period of time, they are and will continue to be an effective and convenient way to collect spatial information.

2.2.5 Spatial Knowledge

Spatial knowledge is a useful structure for associating one or more pieces of information. The spatial knowledge obtained by SDM mainly includes patterns such as correlation, association, classification, clustering, sequence, function, exceptions, etc. As a set of concepts, regulations, laws, rules, models, functions, or constraints, they describe the attributes, models, frequency, and cluster and predict the trend discovered from spatial datasets, such as an association of "IF the road and river intersect, THEN the intersection is a bridge on the river with 80 % possibility." Spatial knowledge is different from isolated information, such as "Beijing, the capital of China." In many practical applications, it is not necessary to strictly distinguish information from knowledge. It should be noted that data processing to obtain professional information is a basis of SDM but is not equal to SDM, such as image processing, image classification, spatial query, and spatial analysis. The conversion from spatial data to spatial information—a process of data processing—can be achieved by a database system. The conversion from spatial information to spatial knowledge is another cognition process, along with human–machine interaction and specific SDM techniques. The value of information and knowledge is that it can be converted into productive power or used to extract new information and to uncover new knowledge. Spatial information that is not being used, despite being meaningful, is of no value. Only if it is accumulated systematically with a purpose can it become knowledge, just like commodities that are not exchanged have no value.

2.2.6 Unified Action

Spatial numerical values act as digital carriers during the process of object collection, object transmission, and object application. Objects in the spatial world are first transferred to forms of data by macro- and micro-sensors and other equipment based on a certain model of the concept models, approximately described according to some theories and methods, and finally stored as physical models in the physical medium (e.g., hard drives, disks, tapes, videos) of a database in a spatial information system (e.g., GIS) or as a separate spatial database. In fact, during the process of data mining, the numerical data participate in the actual calculations, and the unit of measurement in a certain space is only used to give those numerals different spatial meanings.

A spatial numerical value is a kind of spatial data, while spatial data are the spatial information carriers, which refers to the properties, quantities, locations, and relationships of the spatial entities represented by spatial symbols, such as the numerical values, strings, graphics, images, etc. Spatial data represent the objects, and spatial information looks for the content and interpretation of spatial data. Spatial information is the explanation of the application values of spatial data in a specific environment, and spatial data are the carrier of spatial information. The conversion from spatial objects to spatial data, and then to spatial information, is a big leap in human recognition. The same data may represent different information on different occasions, while different data may represent the same information on the same occasion when closely related to the spatial data. For example, "the landslide displaces 20 mm southward" is spatial data, while "the landslide displaces about 20 mm southward" is a spatial concept.

Spatial data are also the key elements to generate spatial concepts. A spatial concept is closely related to spatial data; for example, "around 3,000 km" is a spatial concept, but "3,000 km" is spatial data. The conversion between spatial concept and spatial data is the cornerstone of the uncertain conversion between the qualitative and the quantitative. A spatial concept is the description and definition of a spatial entity. It is an important method used to represent spatial knowledge. Although spatial data and spatial information are limited, spatial knowledge is infinite. The applicable information structure that is formed by one or more pieces of associated information is spatial knowledge, which is the relationship among spatial entities on different cognitive levels with a high degree of cohesion and distillation of the spatial data and spatial information. It is more general and abstract and can be used directly by users. Spatial knowledge has different spatial meanings of the granularity on different scales. Different spatial concepts of granularity should be defined under a specific scale.

2.3 SDM Space

With the generalization of a large spatial dataset, SDM runs in different spaces, such as attribute space to recognize attribute data, feature space to extract features, and concept space to represent knowledge. SDM organizes emergent spatial patterns according to data interactions by using various techniques and methods. Fine patterns are discovered in the microcosmic concrete space, and coarse patterns are uncovered in the macrocosmic abstract space.

2.3.1 Attribute Space

Attribute space is a raw space that is composed of the attributes to depict spatial objects. An object in human thinking is described in a nervous system with an entity (e.g., case, instance, record, tuple, event, and phenomenon), and an object in SDM is represented as an entity with attributes in a computerized system. The dimension of attribute space comes from the attributes of spatial objects. Spatial objects with multiple attributes create a multi-dimensional attribute space. Every attribute in the attribute space links to a dimensional attribute of the object. When it is put into attribute space, an object becomes a specific point with a tuple of its attribute data in each dimension (e.g., position, cost, benefit). Thousands of spatial objects are projected to the attribute space as tens of thousands of points. In the context of specific issues, attribute data are primitive, chaotic, shapeless accumulations of natural states, but they are also the sources to generate order and rules. Going through the disorganized and countless appearance of attribute data, SDM uncovers the implied rules, orders, outliers, and relevance.

2.3.2 Feature Space

Feature space is a generalized space that highlights the object essence on the basis of attribute space. A number of different features of a spatial entity create a multi-dimensional feature space. The feature may be single attribute, composite attribute, or derived attribute. It gradually approximates the nature of a great deal of objects with many attributes. The more generalized the dataset, the more abstract the feature is. For example, feature-based data reduction is an abstraction of spatial objects by sharply declining dimensions and greatly reducing data processing. Based on the features, attribute-based object points are freely spaced, the whole of which generates several object groups. The objects in the same group are often characterized with the same feature. With gradual generalization, datasets are summarized feature by feature, along with the distribution change of objects in feature space. Diverse distributions result in various object compositions and even re-compositions. The combination varies with different features in the context of the discovery task. Jumping from attribute space to feature space, attribute-based object points are summarized as feature-based object points or clusters. When further jumping from one microscopic feature space to another macroscopic feature space, object points are partitioned into more summarized clusters. Furthermore, the feature becomes more and more generic until the knowledge is discovered (i.e., clustering knowledge).

2.3.3 Conceptual Space

Conceptual space may be generated from attribute space or feature space when concepts are used in SDM. The objective world involves physical objects, and the subjective world reflects the characteristics of the internal and external links between physical objects and human recognition. From the existence of the subjective object to the existence of self-awareness, each thinking activity targets a certain object. The concept is a process of evolution and flow on objects, linking with the external background. When there are massive data distributed in the conceptual space, various concepts will come into being. Obviously, concepts are more direct and better understood than data. Reflecting the intention and extension, all kinds of concepts create a conceptual space. All the data in the conceptual space contributes to a concept. The contribution is related to the distance between the data and the concept and the value of the data. The greater the value of the data is, the larger the contributions of the data are. Given a set of quantitative data with the same scope of attributes or features, how to generalize and represent the qualitative concepts is the basis of knowledge discovery. By combining and recombining the basic concepts in various ways, the cognitive events further uncover the knowledge. In conceptual space, various concepts to summarize the object dataset show different discovery states.

2.3.4 Discovery State Space

Discovery state space is a three-dimensional (3D) operation space for the cognition and discovery activity (Li and Du 2007). Initially, it is composed of the attribute-oriented dimension, entity-oriented dimension, and template-oriented dimension, in terms of the breadth and depth of data mining. The attribute-oriented operation is for the relationships and rules among various attributes, the entity-oriented operation is for the consistency and difference in the multi-attribute model of entities, and the template-oriented operation is for knowledge discovery that generalizes or details the knowledge, changing from a microscopic knowledge template to a macroscopic template. Both the attribute-oriented operation and the entity-oriented operation create a two-dimensional knowledgebase, which are the operations for a specific knowledge template. A template-oriented operation addresses the attributes and the entities as a whole and promotes the degree of knowledge abstraction under induction. When the degree of abstraction increases, the generality in the dimension of both the attribute and the entity increases and the physical size of the knowledge template becomes increasingly smaller. When introduced into geo-spatial information sciences, scale dimension is appended in discovery state space, and as a result, forms a four-dimensional discovery state space. A scale-oriented operation addresses the measurement transformation of knowledge discovery when there are various scales, such as 1:500, 1:1,000, and 1:10,000.

In addition to the scale of measurement, the discovery state space is also related to granularity in resolution and hierarchy in cognition. In fact, an increase in the degree of abstraction is an increase of granularity size, a decrease in the measuring scale, and/or the increase of the degree of induction of the cognition hierarchy. Specifically, if each dimension of attributes and entities in the knowledge-template tends toward generalization and the physical size of the knowledge template decreases, then the data of the spatial entity are making combinations or enrichment differently (Fig. 2.2) according to their different tasks in attribute space, conceptual space, or feature space.

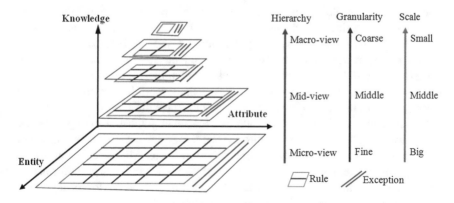

Fig. 2.2   Discovery state space = {attribute space → conceptual space → feature space | cognitive level (granularity and/or scale)}. Reprinted from Wang and Shi (2012), with kind permission from Springer Science+Business Media

2.4 SDM View

SDM view is a mining perspective which assumes that different knowledge may be discovered from the same spatial data repositories for various mining purposes. That is, different users with different backgrounds may discover different knowledge from the same spatial dataset in different applications by using different methods when changing measurement scales at different cognitive hierarchies and under different resolution granularity.

View-angle enables the dataset to be illuminated by the purpose lights from different angles in order to focus on the difference when SDM creates patterns innovatively or uncovers unknown rules. The difference is accomplished via the elements of SDM view, which include the internal essential elements that drive SDM (i.e., the user, the application, and the method) and the external factors that have an impact on SDM (i.e., hierarchy, granularity, and scale). The composition of the elements and their changes result in various SDM views. As a computerized simulation of human cognition by which one may observe and analyze the same entity from very different cognitive levels, as well as actively moving between the different levels, SDM can discover the knowledge not only in worlds with the same view-angle but also in worlds with different view-angles from the same datasets for various needs. The choice of SDM views must also consider the specific needs and the characteristics of the information system.

2.4.1 SDM User

A user is interested in SDM with personality. There are all kinds of human users, such as a citizen, public servant, businessman, student, researcher, or SDM expert. The user also may be an organization, such as a government, community, enterprise, society, institute, or university. In an SDM organization, the user may be an analyst, system architect, system programmer, test controller, market salesman, project manager, innovative researcher, or president. The background knowledge context of the SDM user may be completely unfamiliar, somewhat knowledgeable, familiar, or proficient. The realm indicates the level of human cognition of the world; different users with different backgrounds have different interests. When a SDM-referenced decision is made, the users are hierarchical. Top decision-makers macroscopically master the entire dataset for a global development direction and therefore ask for the most generalized knowledge. Middle decision-makers take over from the above hierarchy level and introduce and manage information. Bottom decision-makers microscopically look at a partial dataset for local problem resolution and therefore ask for the most detailed knowledge.

2.4.2 SDM Method

When users input external data into a system, summarize datasets, and build a knowledge base, the conventional methods encounter problems because of the complexity and ambiguity of the knowledge and the difficulties in representing it. Fortunately, this is not the case for and is the major advantage of SDM. The SDM model mainly includes dependency relationship analysis, classification, concept description, and error detection. The SDM methods are closely related to the type of discovered knowledge, and the quality of the SDM algorithms directly affects the quality of the discovered knowledge. The operation of the SDM algorithms is supported by techniques that include rule induction, concept cluster, and association discovery. In practice, a variety of algorithms are often used in combination. The SDM system can be an automatic human–computer interaction using its own spatial database or external databases from GIS databases. Development can be stand-alone, embedded, attached, etc. Various factors need to be taken into account in SDM; therefore, SDM's theories, methods, and tools should be selected in accordance with the specific needs of the user. SDM can handle many technical difficulties, such as massive data, high dimension, contaminated data, data uncertainty, a variety of view-angles, and difficulties in knowledge representation (Li and Guan 2000).

2.4.3 SDM Application

SDM is applied in a specific field with constraints and relativity. SDM uncovers the specific knowledge in a specific field for spatial data-referenced decision-making. The knowledge involves spatial relationships and other interesting knowledge that is not stored in external storage but is easily accepted, understood, and utilized. SDM specifically supports information retrieval, query optimization, machine learning, pattern recognition, and system integration. For example, SDM will provide knowledge guidance and protection to understand remote sensing images, discover spatial patterns, create knowledge bases, reorganize image databases, and optimize spatial queries for accelerating the automation, intelligence, and integration of image processing. SDM also allows the rational evaluation of the effectiveness of a decision based on the objective data available.

SDM is therefore a kind of decision-making support technology. The knowledge in the decision-making system serves data applications by helping the user to maximize the efficient use of data and to improve the accuracy and reliability of their production, management, operation, analysis, and marketing processes. SDM supports all spatial data-referenced fields and decision-making processes, such as GIS, remote sensing, GPS, transportation, police, medicine, transportation, navigation, and robotics.

2.4.4 SDM Hierarchy

Hierarchy depicts the cognitive level of human beings when dealing with a dataset in SDM, reflects the level of cognitive discovery, and describes the summarized transformation from the microscopic world to the macroscopic world (e.g., knowledge with different demands). Human thinking has different levels based on the person's cognitive stature. Decision-makers at different levels and under different knowledge backgrounds may need different spatial knowledge. Simultaneously, if the same set of data is mined from dissimilar view-angles, there may be some knowledge of different levels. SDM thoroughly analyzes data, information, and concepts at various layers by using roll-up and drill-down. Roll-up is for generalizing coarser patterns globally, whereas drill-down is for detecting finer details locally. Hierarchy conversion in SDM sets up a necessary communication bridge between hardware platforms and software platforms. Their refresh and copy technology include communication and reproduction systems, copying tools defined within the database gateway, and data warehouse designated products. Its data transmission and transmission networks include network protocol, network management framework, network operating system, type of network, etc. Middleware includes the database gateway, message-oriented middleware, object request broker, etc.

Human thinking is hierarchal. The cognitive activity of human beings may arouse some physical, chemical, or electrical changes in their bodies. Reductionism in life science supposes that thinking activities can be divided into such brain hierarchies as biochemistry and neural structure. However, the certain relationships among thinking activities and sub-cellular chemistry and electrical activities cannot be built up, nor can the kind of neural structure be determined that will produce a certain kind of cognitive model. A good analogy here is the difficulty of monitoring e-mail activities on computer networks by merely detecting the functions of the most basal silicon CMOS (Complementary metal-oxide-semiconductor) chip in a computer. As a result, reductionism is questioned by system theory, which indicates that the characteristics of the system as a whole are not the superposition of low-level elements. An appropriate unit needs to be found to simulate human cognition activities. The composition levels of the matter can be seen as the hierarchy. For example, if the visible objects are macroscopic, celestial bodies are cosmologic. Objects that are smaller than atoms and molecules are called microscopic objects. The atomic hierarchy is very important because the physical model of atoms is one of the five milestones for human beings to recognize the world. Cognitive level is also related to resolution granularity and measurement scale.

Depending on the objects to be mined, SDM has the hierarchies of objects distributed throughout the world, from the analysis of protein interactions in live cells to global atmospheric fluctuations. Spatial objects may be stars or satellites distributed in the universe; various natural or manmade features on the Earth's surface are also projected reflected in computerized information. They also can be the structure of proteins or chromosomes in molecular biology or the trajectory of electrons that move around the nucleus in atoms.

Various concepts may create a hierarchical structure on a linguistic variable. In the conceptual space, the concepts may be large or small, coarse or fine, as well as the equivalent relationship or subordinate relationships between them. A concept is divided into a number of sub-concepts. All of the concepts in the same linguistic variable naturally form a hierarchal structure—that is, the conceptual tree. The traditional conceptual tree has a clear concept boundary, and each leaf sub-concept only belongs to one father concept. However, it neglects the uncertainty between a qualitative concept and its quantitative data. Sometimes, one concept may interact with another concept and there may be an indeterminate boundary between mutual concepts. A pan-conceptual tree is presented in Fig. 2.3 to match the actual characteristic of concepts (Wang 2002). Comparatively, the concepts interact with each other in the same hierarchy, showing the randomness and ambiguity of qualitative concepts. When the hierarchy jumps up or down, each sub-concept may belong to more than one father concept. In a pan-conceptual tree considering uncertainty, the structure is also pan-hierarchy due to the concepts with different granularity sizes. Figure 2.3 is such a pan-conceptual tree of the linguistic variable of "landslide displacement" at different cognitive hierarchies in the conceptual space.

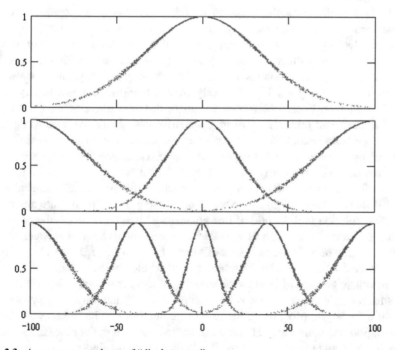

Fig. 2.3 A pan-conceptual tree of "displacement"

2.4.5 SDM Granularity

Granularity depicts the resolution of a dataset on spatial objects in SDM. It reflects the precision of SDM's interior detailing and describes the combined transformation from the fine world to the coarse world (e.g., an image with different pixels). Granularity, which is the average metric of particle or size in physics, can be descriptions of the pixel size, database tuples, the division of program modules, cognitive level for spatial data, etc. Granularity may be macro-size, meso-size, and micro-size under the abstract degree of spatial information and spatial knowledge. Based on granularity, knowledge can be grouped as generalized knowledge with coarse granularity, multiple knowledge with various granularities, and raw knowledge with fine granularity.

The granularity of general data mining is directly obtained from the field or derived by simple mathematical or logical operators. Spatial relationships, graphic features, and other granular properties, in general, are not directly stored in the database but rather are implied in images with multiple layers. Such properties can be obtained by spatial operation and spatial analysis using vector or raster. For example, the altitude of spatial objects is determined from the overlay analysis, the adjacent objects from topology analysis, the distance between the objects from the buffer analysis and distance analysis, and the slope and direction of the DEM terrain analysis.

Humans think in granular terms. In artificial intelligence and cognitive science, researchers recognize that human intelligence can freely observe and analyze the same problem in worlds of different granularities. The granularities may be very different from each other, but each granularity matches a specific purpose. Not only can people solve problems a world of the same granularity, but they also can change their thinking quickly and freely from a world of one granularity to the world of another. Even problems in worlds of different granularities can be analyzed at the same time. The size of granularity that should be used to discover knowledge from spatial data is not permanent, as there is no permanent optimal distance of observation and it depends on the research problem being addressed. For example, the granularity size of a spatial data warehouse is multiple. How much information will be reserved in the world of fine granularity after abstraction? If there is no solution in the world of coarse granularity, is there any corresponding solution in the world of fine granularity? It is an issue worthy of further research. Generally, the size of granularity reflects the difference between spatial objects in thickness, level, and point of view. That is to say, for the observation of an abstracted target, high-level problems from whole angles of view are a reflection of coarse granularity, and the observation of a concrete target, low-level problems from single angle of view is a reflection of fine granularity. In the process of spatial information processing, the size of granularity can be linked with spatial information processing units. If robust function is needed in the processing, then it is in the large granularity; otherwise, it is in the small granularity.

2.4.6 SDM Scale

Scale depicts the measurement of a dataset on spatial objects in SDM. It reflects the measurement of the exterior geometry of DMKD and describes the zoomed transformation from the big world to the small world, such as observing a map with the manipulation of zoom-in or zoom-out. Scale helps SDM extract the generalities and personalities from datasets. Generality masters objects at a high cognitive hierarchy and is more profound than personality. Personality understands objects at a low cognitive hierarchy and provides more information than generality. When SDM runs from big scale to small scale, little of the main personality is kept in generality, and much more is generalized or abandoned. However, neither generality nor personality can replace each other as each one meets a specific demand. For example, when land use is surveyed from a big scale to a small scale, the results change from personality to generality, each of which has a distinct application (i.e., land parcel, farmland, village land, town land, county land, city land, province land, or national land). A small scale helps data miners increase the distance to observe objects via datasets. Ignoring the nuances, datasets are generalized to determine the generality; a large scale helps miners shorten the distance to observe objects via datasets. Datasets are distinguished in order to determine the personality; the nuances hidden in complex phenomena are uncovered carefully and accurately.

Scale is closely related to the geometric transformation of multiple scales. The larger the scale is, the more detailed the SDM runs. With a large scale, spatial data are summarized to obtain the detailed knowledge of some objects, which may decrease the mining efficiency because the space required for data storage and computerization becomes enormous. On the other hand, it may keep the object details, resulting in the problem interpreting and resolving large amounts of data under spatial knowledge. The smaller the scale is, the more generalized SDM runs, and tiny attributes or features become invisible or abandoned. With small scale, spatial data are summarized to discover the common knowledge of all objects, which may increase the mining efficiency because less space is required for data storage and computerization. However, details may be lost, which decreases the problem of large amounts of data to interpret and resolve under spatial knowledge. In discovery state space, the scale-oriented operation changes from fine data to coarse knowledge (e.g., map generalization in cartography). Houses and rivers, for example, in a large-scale map are turned into points and lines, respectively, but these features become too tiny to keep in a small-scale map.

2.4.7 Discovery Mechanism

Humans think by searching the outer appearance to perceive the essence of an object. SDM operates in the same way through a computer by discovering patterns from datasets. To simulate human thinking activities, SDM must find approaches to establish relationships between the human brain and the computer. Logically, SDM is an inductive process that goes from concrete data to abstract patterns, from special phenomena to general rules. The process of discovering the implicit knowledge from the explicit dataset is similar to the human cognitive process of uncovering the internal nature from the external phenomena. As a discovery process, SDM changes from concrete data to abstract patterns and also from particular phenomena to general laws. Therefore, the feature interactions among objects may help knowledge discovery.

The essence of SDM is to observe and analyze the same dataset by reorganizing datasets at various distances with mining views. The change from the elements of view results in various SDM views. For example, the change of one or all of the hierarchy, granularity, and scale will make SDM different in the angle aspects of high cognition or low cognition, coarse resolution or fine resolution, small measurement or big measurement. In the SDM process, the spatial concept first is extracted from its corresponding dataset, and then the preliminary features are extracted from the concept; finally, the characteristic knowledge is induced from feature space. At a close distance, the knowledge template is at a low cognitive hierarchy with fine granularity and a large scale; however, the discovered knowledge is a detailed personality for microscopically distinguishing object differences carefully. The tiny features may be uncovered for looking at the typical examples, such as sharpening an image for local features. At a far distance, the knowledge template is at a high cognitive hierarchy with coarse granularity and a small scale; the discovered knowledge is summarized generally for macroscopically mastering all of the objects as a whole. The subtle details are neglected for grasping the key problems, such as smoothing an image for local features.

Regular rules and exceptional outliers are discovered simultaneously. A spatial rule is a pattern showing the intersection of two or more spatial objects or space-dependent attributes according to a particular spacing or set of arrangements (Ester et al. 2000). In addition to the rules, during the discovering process of description or prediction, there may be some exceptions (also called outliers) that deviate very much from other data observations (Shekhar et al. 2003). The following approaches identify and explain exceptions (surprises). For example, spatial trend predictive modelling first discovers the centers that are local maximal of a certain non-spatial attribute and then determines its theoretical trend when moving away from the centers. Finally, a few deviations are found in that some data are far from the theoretical trend, which may arouse the suspicion that they are noise or are generated by a different mechanism. How are these outliers explained? Traditionally, the detection of outliers has been studied using statistics. A number of discordancy tests have been developed, most of which treat outliers as noise and then try to eliminate their effects by removing them or by developing some outlier-resistant method (Hawkins 1980). These outliers actually prove the rules; in the context of data mining, they are meaningful input signals rather than noise. In some cases, outliers represent unique characteristics of the objects that are important to an organization. Therefore, a piece of generic knowledge is virtually in the form of a rule plus exception.

An exception (outlier) is also a useful pattern in SDM. In the implementation process of SDM, one or more specific elements in the target dataset may be selected (e.g., object attributes, personality circumstances). There also will be some special circumstances when generalizing attributes, promoting concepts, or introducing features from lower reorganization levels to higher reorganization levels. That is, there are always some spatial patterns that are described in the world of a fine knowledge template, which cannot be involved in the world of a coarse knowledge template and therefore exists as an exception.

Thus, SDM is the process of discovering and extracting patterns from spatial data and converting them to spatial knowledge with view-angles. The knowledge is the pattern of "rule plus exception" or "class plus outlier." SDM view utilizes its integrated human cognition to change individual objects to the general knowledge-base and from concrete data to abstract patterns. It enables the user to thoroughly manipulate a spatial dataset as a whole for grasping data essence.

2.5 Spatial Knowledge to Discover

A variety of knowledge can be discovered from spatial datasets (Table 2.1). The discovered knowledge follows the pattern of "rule plus exception" with different SDM views (Fig. 2.2). Rules mainly include both generality and individuality rules, such as association, characteristics, discrimination, clustering, classifications, serials, predictions, and functional dependence. Exceptions refer to the bias of the rules. At the same time, the type of knowledge also depends on the type of SDM task (i.e., the problem that SDM can solve). The knowledge is not isolated from either. A variety of rules are required at the same time when solving practical problems.

2.5.1 General Geometric Rule and Spatial Association Rule

Generally, some types of geometric rules can be discovered from GIS databases, such as geometric shape, distribution, evolution, and bias (Di 2001). This general knowledge of geometry refers to the common geometric features of a certain group of target objects (e.g., volume, size, shape). The objects can be divided into three categories: points (e.g., independent tree, settlements on a small-scale map), lines (e.g., rivers, roads), and polygons (e.g., residents, lakes, squares). The numbers or sizes of these objects are calculated by using the methods of mathematical probability and statistics. The size of linear objects is expressed by length and width, and the size of a polygon object is represented by its area and perimeter. The morphological features of objects are expressed by a quantitative eigenvalue

Table 2.1 Spatial knowledge to be discovered

Knowledge	Interpretations	Examples
Association rule	A logic association among different sets of entities that associate one or more objects with other objects for studying the frequency of items occurring together in databases	Rain (location, amount of rain) $\Rightarrow$ Landslide (location, occurrence), support is 76 %, confidence is 98 %, interest is 51 %
Characteristics rule	A common feature of a kind of entity, or several kinds of entities, for summarizing similar features of objects in a target class	Characterize similar ground objects in a large set of remotely sensed images
Discriminate rule	A different feature that distinguishes one entity from another entity and for comparing the core features of objects between a target class and a contrasting class	Compare land prices in a suburban area with land prices in urban center
Clustering rule	A segmentation rule that groups a set of objects by virtue of their similarity without any knowledge of what causes the grouping and how many groups exist. A cluster is a collection of data objects that are similar to one another within the same cluster and are dissimilar to the objects in other clusters	Group crime locations to find distribution patterns
Classification rule	A rule that defines whether an entity belongs to a particular class with a defined kind and variables, or a set of classes	Classify remotely sensed images based on spectrum and GIS data
Serial rule	A temporal constrained rule that relates entities or the functional dependency among the parameters in a time sequence, for analyzing the sequential pattern, regression, and similar sequences	During summer, landslide disasters often occur
Predictive rule	An inner trend that forecasts future values of some variables when the temporal or spatial center is moved to another one, or predicts some unknown or missing attribute values based on other seasonal or periodical information	Forecast the movement trend of landslide based on available monitoring data. Identify the segments of a population likely to respond similarly to given events
Exception	An outlier that is isolated from the common rules or is derived from other data observations substantially, used for identifying anomalous attributes and objects	A monitoring point detecting exceptional movement that predicts landslides or detecting fraudulent credit card transactions

that is easily achieved by a computer through an intuitive and visual graph. The morphological features of the linear objects are characterized by twists and turns degree (complexity) and the direction; polygon objects are characterized by the intensity, twists and turns degree of the boundaries, and the major axis direction; and point objects have no morphological features. However, the morphological features of point objects gathered together as a cluster can be determined by methods similar to those for polygon objects. Generally, GIS databases only store some geometric features (e.g., length, area, perimeter, geometric center), while the calculation of morphological features requires special algorithms. Some statistical values of geometric features (e.g., minimum, maximum, mean, variance, plural) can be calculated; if there are enough samples, the feature histogram data can be used as priori probability. Therefore, general geometric knowledge at a higher level can be determined according to the background knowledge.

Spatial association refers to the internal rules among spatial entities that are present at the same time and describes the conditional rules that frequently appear in the feature data of spatial entities in a given spatial database. First, an association rule may be adjacent, connective, symbiotic, and contained. A one-dimensional association includes a single predicate, and a multi-dimensional association includes two or more spatial entities or predicates. Second, the association may be general and strong. General association is a common correlation among spatial entities, and a strong association appears frequently (Koperski 1999). The meanings of strong association rules are more profound and their application range is broader. They also are known as generalized association rules. Third, association rules are descriptive rules that provide a measurement of the support, confidence, and interest. For example, "(x, road) $\rightarrow$ close to (x, river) (82 %)" is a description of Chengde associated with roads and rivers. Describing association rules in a form that is similar to structured query language (SQL) brings SDM into standard language and engineering. If the attributes of objects in SDM are limited to Boolean type, the association rules can be extracted through the conversion type in the database that contains the categorical attributes by combining some information of the same objects. Fourth, association rules are temporal and transferable and may ask for additional information, such as a valid time and transferable condition. For example, the association rules on Yellow River water—"IF it rains, THEN the water level rises (spring and summer)" and "IF it rains, THEN the water level recedes (autumn and winter)"—cannot be exchanged when they are used to prevent and reduce the flooding of the Yellow River. For example, the association rules obtained in the first round of election may be different from or even contrary to the rules determined in the second round of election as the voters' willingness may be transferred under the given conditions. Rationally predicating this transfer can help candidates adjust the strategy of election, thereby increasing the possibility of winning. Finally, an association rule is a simple and practical rule in SDM that attracts many researchers for normalization, query optimization, minimizing a decision tree, etc. in spatial databases.

2.5.2 Spatial Characteristics Rule and Discriminate Rule

Spatial characteristic rules summarize the generality of distribution and attribute characteristics of a single kind or multiple kinds of spatial entities and are the generalized description of the concept and spatial class. If there are enough samples, the map, histogram, pie chart, bar graph, curve, data cube, and data table of spatial characteristics can be converted to the priori probability knowledge. For example, "the direction of most roads in Beijing is east–west or north–south" and "most roads in Beijing are relatively straight" are two spatial characteristics rules that describe the common features of Beijing's roads, which are also general geometry knowledge.

Spatial discrimination distinguishes the feature difference between two types of spatial objects or among multiple types of spatial objects. The feature difference determines the target class and the contrastive class under the given feature from spatial datasets. Discriminate rules are discovered by comparing the characteristics between the target class and the contrastive class. For example, "the direction of most roads in Beijing are east–west or north–south, while the direction of most roads in Tianjin are parallel or perpendicular to the rivers" and "most roads in Beijing are relatively straight, while most roads in Tianjin are flexural" are two discriminate rules that describe the general differences in the direction and shape of roads in Beijing and Tianjin. Discriminately, spatial objects are distributed in vertical, horizontal, or vertical-horizontal directions, such as the vertical distribution of alpine vegetation, the difference between the infrastructures of a city and a village, or exotic features distributed along a slope and their exposure. For example, "the majority of Chinese rice-growing areas are located south of the Huaihe River and the Qinling Mountains" or "most of the Chinese wheat-growing areas are located southeast of the mountain, north of the Huaihe River and the Qinling Mountains" are two spatial distribution rules of crops.

2.5.3 Spatial Clustering Rule and Classification Rule

Spatial clustering rules group a set of data in a way that maximizes the feature similarity within clusters and minimizes the feature similarity between two different clusters. Sequentially, spatial objects are partitioned into different groups via the feature similarity by making the difference in data objects between different groups as large as possible and the difference between data objects in the same group as small as possible (Grabmeier and Rudolph 2002). According to the different criteria of similarity measurement and clustering evaluation, the commonly used clustering algorithms may be based on partition, hierarchy, density, and grid (Wang et al. 2011). Clustering rules further help SDM to discretize data, refine patterns, and amalgamate information. For example, continuous datasets are partitioned into discrete hierarchical clusters and multisource information are amalgamated for the same object.

Spatial classification classifies an object by mapping it to a specific class according to its discriminate rules. It attempts to assign each input value to one of a given set of classes. Classification reflects the generality to characterize the difference among different kinds of objects and the similarity inside the same kind of objects. Classification rules may be described in the form of a decision tree, concept lattice, and predicate logic. For example, the value searching of a decision tree from the root to the branches and leaf nodes will be able to determine the category or predict the unknown value of an object. In GIS, object-oriented classification rules refer to the subclass structure of the objects and the general knowledge.

Clustering rules for spatial datasets without cluster labels are different from classification rules with class labels. Clustering rules do not preset cluster labels. Before clustering datasets, the clusters are unknown as far as amount, content, name, etc. At the same time, partitioned clusters may run the preparation of classification generation, and classified classes may supervise clustering.

2.5.4 Spatial Predictable Rule and Serial Rule

Predictable spatial rules can forecast an unknown value, label, attribute, or trend by assigning a spatial-valued output to each input. These rules can determine the internal dependence between spatial objects and their impact variables in the future, such as a regressive model or a decision tree. Before forecasting, correlation analysis can be used to identify and exclude attributes or entities that are useless or irrelevant. For example, the future occurrence of a volcano is extracted from the structure, the plate movement, and the gravity field of Earth in the database. When predicting future values of data based on the trends changing with time, the specificity of the time factor should be fully considered.

A spatial serial rule summarizes the spatiotemporal pattern of spatial objects changing during a period of time. It links the relationships among spatial data and time over a long period of time. For example, in a city over the years, banks and their branches store their operating income and expenditure accounting records and the police department records security cases, from which financial and social trends can be discovered under their geographical distribution. The time constraints can be depicted with a time window or adjacent sequences. Time-constrained serial rules also are called evolution rules. Although they are related to other spatial rules, serial rules mining concentrates more on historical datasets of the same object in different times, such as series analysis, sequence match, time reasoning, etc. Spatial serial rule mining is utilized when the user wants to obtain more refined information excavated for only a period of implicit models. Only by using a series of values of the existing data changing with time can SDM better predict future trends based on the mined results. When little change has occurred in a database, gradual sequence rule mining may speed up the SDM process by taking advantage of previous results.

2.5.5 Spatial Exception or Outlier

Outlier detection, in addition to the commonly used rules, is used to extract interesting exceptions from datasets in SDM via statistics, clustering, classification, and regression (Wang 2002; Shekhar et al. 2003). Outlier detection can also identify system faults and fraud before they escalate with potentially catastrophic consequences. Although outlier detection has been used for centuries to detect and remove anomalous observations from data, there is no rigid mathematical definition of what constitutes an outlier. Ultimately, it is a subjective exercise to determine whether or not an observation is an outlier. There are three fundamental approaches to outlier detection (Hodge and Austin 2004):

(1)   Determine the outliers without prior knowledge of the data, which processes the data as a static distribution, pinpoints the most remote points, and flags them as potential outliers. Essentially, it is a learning approach analogous to unsupervised clustering.

(2)   Model both normality and abnormality, which is analogous to supervised classification and requires pre-labeled data tagged as normal or abnormal.

(3)   Model only normality (or in a few cases, abnormality), which may be considered semi-supervised as the normal class is taught but from which the algorithm learns to recognize abnormality. It is analogous to a semi-supervised recognition or detection task.

Normal distribution of the data is assumed in order to identify observations that are deemed unlikely on the basis of the mean and standard deviations. Distance-based methods frequently use the distance to the nearest neighbors to label observations as outliers or non-outliers (Ramaswamy et al. 2000). In the sequence rule, outlier detection is a heuristic method, which recognizes data that cause a sudden severe fluctuation in the sequential data as an exception by using linear deviation detection. Lee (2000) used a fuzzy neural network to estimate the rules for dealing with distribution abnormality in spatial statistics.

Spatial exceptions or outliers are the deviations or independent points beyond the common features of the most spatial entities. An exception is an abnormality. If manmade factors have been ruled out, an exception is often the presence of sudden changes (Barnett 1978). Deviation detection—a heuristic approach to data mining—can identify the points that have sudden fluctuations in the sequence data as exceptions (Shekhar et al. 2003). Spatial exceptions are the object features that are inconsistent with the general actions or universal models of the data in spatial datasets. They are the descriptions of analogical differences, such as the special case in a standard class, the isolated points out of various classifications, the difference between a single attribute value and a set of attribute values in time serials, and a significant difference between the actual value of an observation and the system forecasting value. A lot of data mining methods ignore and discard exceptions as noise or abnormality. Although excluding such exceptions may be conducive to highlighting the generality, some rare spatial exceptions may be much more significant than normal spatial objects (Hawkins 1980). For example, near a notable feature of the displacement in a large landslide point, there may be a potential landslide hazard, which is the decisive knowledge of landslide prediction. Spatial exceptional knowledge can be discovered with data fields, statistical hypothesis testing, or identifying feature deviations.

2.6 Spatial Knowledge Representation

Knowledge representation is a key issue in SDM. At present, the common knowledge representation methods include natural language (e.g., "the Baota of the Three Gorges landslide moved towards the micro-west south during the monitoring, and was accompanied by a small amount of settlement"), predicates logic, a function model, a characteristic table (relationship table), a generalized rule, a semantic network, a framework, a script, a process, a Petri net, and visualization, among others. The introduction of natural language in knowledge representation is a general recognition of the uncertainty in thinking and perception. The language value increases the flexibility of knowledge, making the discovered knowledge more reliable and easier to understand. In actual applications, they all have their advantages and disadvantages. Different methods are suitable for different knowledge. The same knowledge generally can be represented by a number of methods, which can be mutually converted.

2.6.1 Natural Language

Natural language is one of the best methods to describe datasets based on human thinking and communicating with each other. As a carrier of human thinking, natural language helps to achieve a powerful tool for thinking to display and retain the subject for thought and to organize the process of thinking. It is the foundation of a variety of other formal systems or languages, which are derived from a special language, such as computer language, or some specific symbol languages, such as mathematical language. The formal systems constituted by these symbols further become a new formal system.

The basic language value of natural language is a qualitative concept, corresponding to a group of quantitative data. Seen from the process of the atomic model that evolved from the Kelvin model, the Thomson model, the Lenard model, the Nagaoka model, and the Nicholson model to Rutherford's atomic model with nuclei, it is a universal and effective methodology to work out the model of material composition. The concept maps the object from the objective world to subjective cognition. As far as concept generation, regardless of whether the characteristic table theory or the prototype theory is used, all conceptual samples are reflected by a set of data. The smallest unit of natural language is the language value to describe the concepts. The most basic language value represents the most basic concept—that is, the linguistic atom. As a result, the linguistic atom forms the atomic model when human thinking is modeled with the help of natural language.

The difference between spatial knowledge and non-spatial knowledge lies in spatial knowledge having spatial concepts and spatial relationships; furthermore, the abstract representation of these spatial concepts and spatial relationships is most appropriately expressed by language values. Mastering the quantitative dataset with qualitative language values conforms to human cognitive rules. Obtaining qualitative concepts from a quantitative dataset reflects the essence of objects more profoundly, and subsequently fewer resources are spent to deliver adequate information and make efficient judgments and reasoning of complex things. When representing the definitive properties of discovered knowledge, soft natural language is more universal, more real, more distinct, more direct, and easier to understand than exact mathematical language. A lot of knowledge obtained by SDM is qualitative knowledge after induction or abstraction, or a combination of qualitative and quantitative knowledge. The more abstract the knowledge is, the more suitable is natural language. However, the concept represented by natural language inevitably has uncertainty commonly, is even blind and undisciplined, and therefore is a bottleneck to freely transform between quantitative data and qualitative concept.

2.6.2 Conversion Between Quantitative Data and Qualitative Concept

A conversion model inevitably needs to be set up between qualitative language values and quantitative numerical values in order to realize the conversion between numerical values and symbol values at any time, to establish the interrelated and interdependent map relationship between quality and quantity, and to reflect the uncertainty of mapping between quality and quantity (particularly randomness and fuzziness). At present, the commonly used qualitative and quantitative conversion methods include the analysis hierarchy process, quantizing weightings, experts grading, qualitative analysis combined with some mathematical models, and quantitative calculation. The basic of cybernetics in dealing with uncertainty is to eliminate errors depending on the actual goal and practice. Therefore, none of the above methods are perfect because they fail to take both randomness and ambiguity into consideration.

The cloud model is an uncertainty conversion model between quantitative data and qualitative concepts. The numerical characteristics of the cloud fully integrate the fuzziness and randomness; together, they constitute a mapping between quality and quantity, which can be used as the basis of knowledge representation. When expanding from one-dimensional to two-dimensional and multi-dimensional, the cloud model can express spatial concepts and spatial relationships. For example, the "far," "close" language value uses a one-dimensional model of cloud to express "northeast" and "southwest" language value used in a two-dimensional model. A one-dimensional cloud model is used to express quantity language values such as "a small number of," "basically," "the majority," "almost all," etc. (Wang 2002). The language values indicated by the cloud model can be used to express the qualitative concept. The integration of the cloud model and the traditional expression method is an enhancement of the qualitative concept expression and the qualitative, quantitative conversion of these knowledge expression methods.

2.6.3 Spatial Knowledge Measurement

Spatial knowledge measurement is the certainty or credibility of knowledge, such as support, confidence, expected confidence, lift, and interest (Agrawal and Srikant 1994; Reshef et al. 2011). Because not all the discovered patterns are interesting or meaningful, measurements can be used to supervise or limit the progress of SDM. Here, association rules are taken as an example to introduce measurements.

Support that dataset D is a collection of T (Transaction), $I = (i_1, i_2,..., i_n)$ is the set of Items, e.g., object, entity, attribute, feature, and $T \subseteq I$. Each T has an identifier of TID. A, B is the item set, $A \subseteq T$, $B \subseteq T$, and $A \cap B = \emptyset$. An association rules $A \Rightarrow B$ is given with measurements in Eq. (2.1).

$$((A_1{}^\wedge A_2{}^\wedge ...{}^\wedge A_m) \Rightarrow (B_1{}^\wedge B_2{}^\wedge ...{}^\wedge B_n))|([s],[c],[ec],[l],[i]) \qquad (2.1)$$

Where, A_1, A_2,..., A_m; B_1, B_2,..., B_n is a set of spatial or non-spatial predicates, [...] is optional, $[s]$, $[c]$, $[ec]$, $[l]$, $[i]$ measure the certainty of support, confidence, expected confidence, lift, and interest, respectively.

- *Support measures the level of certainty that describes the probability $P(A \cup B)$ of the union of A and B happening in D, i.e., $s(A \Rightarrow B) = P(A \cup B)$*
- *Confidence measures the level of certainty that describes the conditional probability $P(B/A)$ of the intersection of A and B happening in D; i.e., $c(A \Rightarrow B) = P(B/A)$ or $c(A \Rightarrow B) = P(A \cap B)$*
- *Expected Confidence measures the certainty that describes the expected probability $P(B)$ of B happening in D; i.e., $ec(A \Rightarrow B) = P(B)$. It describes the probability that B happens without any conditions.*
- *Lift measures the certainty of the ratio of confidence and expected confidence; i.e., $l(B/A) = c(A \Rightarrow B)/ec(B)$. It indicates how the probability that A happens impacts the probability that B happens. The bigger Lift is, the more the happening of A will accelerate the happening of B.*
- *Interest measures the level of certainty that users are interested in SDM patterns when making decisions. According to its definition, association rules exist between any two sets of items. If users do not consider whether or not they are useful, a lot of rules may be uncovered but not all will be meaningful.*

Support and confidence are two important indicators that reflect the usable level of the rules: support is the measurement of a rule's importance, whereas confidence is the measurement of a rule's accuracy. The support illustrates how it is representative of the rule in all transactions; the larger the support is, the more important the associations rule becomes. A rule with a high confidence level but very small support indicates that it rarely happens, along with having little practical value.

In the actual SDM, it is important to define the thresholds of spatial knowledge measurements (e.g., minimum support and minimum confidence). Under normal circumstances, when the lift of the useful association rule is more than 1, it suggests that the confidence of the association rules is greater than the expected confidence. If its lift is not more than 1, the association rule is meaningless. The general admission threshold value is based on experience, but users can also be determined by statistics. Only when their measurements are larger than the defined thresholds can the rules be accepted as the interesting ones (also called strong rules) for application. Moreover, the defined thresholds should be appropriate under the given situations. If the thresholds are too small, a large number of useless rules will be discovered, which not only affect the efficiency of the implementation and waste system resources, but also may flood the main objective. If the value is too large, the results may not be rules at all, the number of rules is too small, or the expected rules may be filtered out.

2.6.4 Spatial Rules Plus Exceptions

SDM knowledge can be expressed in a number of ways. A more reasonable expression idea is "spatial rules plus exception," accompanied with measurements at different levels. The patterns of "rule plus exceptions" have extensive applications in intelligent spatial analysis and interpretation. To match different tasks, a group of spatial datasets can be understood from different mining views, enabling many kinds of knowledge on spatial objects to be summarized. The summarized knowledge may technically support decision-making to resolve and interpret natural and human problems at personalized levels. They identify the links between mutual records to generate a summary of a spatial database and to create a prediction model and a classification model for spatial expert systems or decision support systems. They also can restrict, support, and supervise the interpretation of remote sensing images to address the phenomenon in the same spectrum with different objects, and different objects with the same spectrum, in order to reduce doubt in the classification results. SDM enhances the reliability, accuracy, and speed of the interpretation; for example, the rules of a road connected to a town or village or a bridge at the intersection of a road and a river can improve the accuracy of the classification and update the spatial database when they are applied in image classification. The "grass $\Rightarrow$ forest" rule from the image database states that grass and forest often appear at the same time, but the grass on the forest boundary is assigned a larger weight to provide a defined boundary in order to improve the accuracy of classification.

References

Agrawal R, Srikant R (1994) Fast algorithms for mining association rules. In: Proceedings of international conference on very large databases (VLDB), Santiago, Chile, pp 487–499

Barnett V (1978) Outliers in statistical data. Wiley, New York

Barrabbasi AL, Albert R (1999) Emergence of scaling in random networks. Science 286:509–512

Di KC (2001) Spatial data mining and knowledge discovering. Wuhan University Press, WuHan

Ester M et al (2000) Spatial data mining: databases primitives, algorithms and efficient DBMS support. Data Min Knowl Disc 4:193–216

Frasconi P, Gori M, Soda G (1999) Data categorization using decision trellises. IEEE Trans Knowl Data Eng 11(5):697–712

Grabmeier J, Rudolph A (2002) Techniques of clustering algorithms in data mining. Data Min Knowl Disc 6:303–360

Han JW, Kamber M, Pei J (2012) Data mining: concepts and techniques, 3rd edn. Morgan Kaufmann Publishers Inc., Burlington

Hawkins D (1980) Identifications of outliers. Chapman and Hall, London

Hodge VJ, Austin J (2004) A survey of outlier detection methodologies. Artif Intell Rev 22(2):85–126

Koperski K (1999) A progressive refinement approach to spatial data mining. Ph.D. thesis, Simon Fraser University, British Columbia

Lee ES (2000) Neuro-fuzzy estimation in spatial statistics. J Math Anal Appl 249:221–231

Li DR, Guan ZQ (2000) Integration and Implementation of spatial information system. Wuhan University Press, Wuhan

Li DR, Wang SL, Shi WZ, Wang XZ (2001) On spatial data mining and knowledge discovery (SDMKD). Geomatics Inf Sci Wuhan Univ 26(6):491–499

Li DR, Wang SL, Li DY (2006) Theory and application of spatial data mining, 1st edn. Science Press, Beijing

Li DR, Wang SL, Li DY (2013) Theory and application of spatial data mining, 2nd edn. Science Press, Beijing

Li DY, Du Y (2007) Artificial intelligence with uncertainty. Chapman and Hall/CRC, London

Liu B (2007) Web data mining: exploring hyperlinks, contents, usage data, 2nd edn. Springer, Heidelberg

Piatetsky-shapiro G (1994) An overview of knowledge discovery in databases: recent progress and challenges. In: Ziarko Wojciech P (ed) Rough sets, fuzzy sets and knowledge discovery. Springer, Berlin, pp 1–10

Ramaswamy S, Rastogi R, Shim K (2000) Efficient algorithms for mining outliers from large data sets. In: Proceeding SIGMOD '00 proceedings of the 2000 ACM SIGMOD international conference on management of data, pp 427–438

Reshef DN et al (2011) Detecting novel associations in large data sets. Science 334:1518

Shekhar S, Lu CT, Zhang P (2003) A unified approach to detecting spatial outliers. GeoInformatica 7(2):139–166

Srivastava J, Cheng PY (1999) Warehouse creation-a potential roadblock to data warehousing. IEEE Trans Knowl Data Eng 11(1):118–126

Wang SL (2002) Data field and cloud model based spatial data mining and knowledge discovery, PhD thesis, Wuhan University, Wuhan

Wang SL, Shi WZ (2012) Data mining and knowledge discovery. In: Kresse Wolfgang, Danko David (eds) Handbook of geographic information. Springer, Berlin

Wang SL, Yuan HN (2014) Spatial data mining: a perspective of big data. Int J Data Warehouse Min 10(4):50–70

Wang SL, Gan WY, Li DY, Li DR (2011) Data field for hierarchical clustering. Int J Data Warehouse Min 7(4):43–63

Watts DJ, Strogatz SH (1998) Collective dynamics of 'small world' networks. Nature 393:400–442

Witten I, Frank E (2000) Data mining, practical machine learning tools and techniques with java implementation. San Francisca: Morgan Kaufman Publishers

Chapter 3
SDM Data Source

SDM would be like water without a source or a tree without roots if it was separated from its data resources. The development of new techniques promotes the service features of spatial data. This chapter will explain spatial data based on their contents and characteristics; review the techniques for acquiring spatial data; introduce the structure of spatial data based on vectors, raster structures, and their integration; discuss the process of modeling spatial data; explain spatial databases and data warehouses in the context of seamless organization and fusion; and introduce the National Spatial Data Infrastructures (NSDI) of nations and regions, highlighting China's NSDI; and based on NSDI, discriminate Digital Earth, Smart Earth, and Big Data, in which SDM plays an important role.

3.1 Contents and Characteristics of Spatial Data

Spatial data depict the geo-referenced distribution of natural or artificial objects in the real world, as well as its changes. Based on the locations of spatial data, non-spatial data indicate the connotation of objects. Compared to common data, spatial data are characterized not only as being spatial, temporal, multi-dimensional, massive, and complicated, but uncertain as well.

3.1.1 Spatial Objects

Spatial objects are the core of an object-oriented data model. They are abstracted representations of the entities (ground objects or geographical phenomena) that are natural or artificial in the real world (e.g., building, river, grassland). In a spatial dataset, a large number of spatial objects are classified as a point, line, or area or complex objects.

© Springer-Verlag Berlin Heidelberg 2015
D. Li et al., *Spatial Data Mining*, DOI 10.1007/978-3-662-48538-5_3

1. Point objects refer to points on Earth's surface—that is, single points (chimneys, control points), directed points (bridges, culverts), and group points (street lamps, scattered trees). A point object contains a spatial location without shape and size and occupies only one location data point in a computerized system.

2. Linear objects refer to the spatial curve of Earth's surface. They are either a point-to-point string or an arc string that can be reticulated, but they should be interconnected (e.g., rivers, river grids, roads). They often have a shape but no size; the shape is either a continuous straight line or a curve when projected on a flat surface. In a computerized system, a set of elements are required to fill the whole path.

3. Areal objects refer to the curved surface of Earth's surface and contain shape and size. A flat surface is comprised of the compact space surrounded by its boundaries and a set of elements that fill the path. Areal objects are composed of closed but non-crossing polygons or rings, which may contain a number of islands that represent objects such as land, housing, etc.

4. Complex objects consist of any two or more basic objects (point, line, or area) or the more complicated geo-objects. A simple ground object can be defined as the composition of a single point, a single line, or a single area. Likewise, a complex ground object can be defined as a composition of several simple ground objects (e.g., the centerline, surface, traffic lights, overpasses, etc. on a road) or elemental objects such as roads, houses, trees, and water towers. Elemental objects are depicted with geometric data and their interrelated semantics.

3.1.2 Contents of Spatial Data

For SDM, a spatial dataset is comprised of a position, attribute, graph, image, web, text, and multimedia.

Positional data describe the specific position of a spatial object in the real world and are generally the quantitative coordinate and location reference observed by certain equipment and methodologies. Survey adjustment refers to the process of confirming the estimate and accuracy of unknown parameters through observed data obtained according to the positional data. To reduce data redundancy, the essential organized data can be made up of nodes and arcs; that is, only the node and arc contain the spatial location and other spatial objects have node-arc structures as well. Modern systems to acquire spatial data are now able to provide spatial data to the micron resolution, which can result in discovering a large number of positional coordinates.

Attribute data are the quantitative and qualitative description index of substances, characteristics, variables, or certain geographic objects. They can be regarded as the facts of some point, point set, or features (Goodchild 1995). The attribute data can be divided into quantitative data and qualitative data, which can be discrete values or continuous values. A discrete value is the limited elements of a finite set, while a continuous value is any value in a certain interval (Shi and Wang 2002). The attribute data of GIS describe the attributes of a point, line, polygon, or remote sensing images.

Graphics and images are the visual representation and manipulation of spatial objects on a certain surface. Their main texture or objects constitute the body of spatial objects. Generally, graphic data are vector structures and imagery data are raster structures. In recent years, with the increase in various satellites, the amount of available remote-sensing images has grown dramatically. Finding interesting objects from complex remote-sensing images for tasks such as environmental inspection, resources investigation, and supervision of ground objects, which all rely on remote sensing, is becoming increasingly cumbersome and time-consuming. Remote-sensing images may be gray or pseudo-color and include the geometric and spectral data of the ground objects. Images at different scales are often used in different fields. For example, a small-scale image for environment inspection is mainly about texture and color, and a large-scale image for ground target supervision is chiefly about shapes and structures. The similarity of image features reflects the similarity between the entire image and its main objects. In addition to the feature extraction that common image retrieval provides, remote-sensing image retrieval also considers the comprehensive metadata, including geographical location, band, different sensor parameters, the relationship between the scale factors and the image content, as well as the storage costs and efficiency of inquiries that require expansive and numerous details. Consequently, content-based remote-sensing image retrieval has its own characteristics in terms of feature selection, similarity comparison, inquiry mechanism, system structure, and many other aspects.

Network data are in a distributed and paralleled data space. The development of network technology and its extensive use have made it possible to exchange data in different countries, different regions, different departments in the same regions, and the interior parts of a department. Web-based applications are infiltrating all aspects of human life. In the context of a network, human–computer interaction has become more electronic, informative, and massive, including e-commerce, e-government, banking networks, network industries, network cultures, and a series of new things. Web-based services have become distributed, shared, and collaborative, and the quantity and quality of network data consumption are improving. Because a network is open, dynamic, distributed, and heterogeneous, it is difficult for its users to acquire data accurately and in a timely manner. The spatial data in a network provide SDM, which originated from bar-code technology and massive storage technology, with a whole new field for research and application. In addition, there are records of network access and the log data of the use of the web server, from which useful patterns, rules, or topological relationships can be found.

Text data may be plain or rich documents to describe spatial objects. Plain text is a pure sequence of character codes that are public, standardized, and universally readable as textual material without much processing. Rich text is any text representation containing plain text completed by data, such as a language identifier, font size, color, or hypertext link. In SDM, text data are electronic for any document that is readable in digital form as opposed to such binary data as encoded integers, real numbers, images, etc. The integration of text data and sequential data can result in the extraction of more useful knowledge.

Multimedia data cover a broad range of text, sound, graphics, images, animations, and forms. The problems of multimedia databases include the effective storage of different types of data as well as targeting different types of objects into a single framework. Users need to have the ability to input different types of data and to use a direct approach to scan heterogeneous data. In SDM, multimedia databases make use of their methods of data storage and processing to address the integrated relationship between the multimedia data and the spatial data with the spatial objects as the framework and the multimedia data attached to the objects.

3.1.3 Characteristics of Spatial Data

Spatial data mainly describe the location, time, and theme characteristics of various objects on Earth. Location is the geographic identification of the data associated with a region. Time indicates the temporal factor or sequential order of the data occurrence or process (i.e., dynamic changes). Theme is the multi-dimensional structure of the same position with a number of subjects and attributes.

Location refers to the position, azimuth, shape, and size of a spatial object as well as the topological relationship of its neighboring objects. The position and topological relationship only relate to a spatial database system. Location can be described by different coordinate systems, such as latitude and longitude coordinates, standard map projection coordinates, or any set of rectangular coordinates. Through a given coordinates conversion software, it is possible to realize the conversion of different coordinate systems. There are two types of topological relationships between spatial objects: the first is the elemental relationship between point, line, and polygon (e.g., polygon–polygon, line–line, point–point, line–polygon, point–line, and point–polygon), which illustrates the topological structure among geometric elements; the second type is the object relationship between spatial objects (e.g., disjoint, inclusion, exclusion, union, and intersection), which is implicitly represented through positional relationships and queried by related methods. GIS software has the functions to generate and query spatial topology. In general, the positioning of a certain object is confirmed by determining the location relationship between it and the referenced objects, especially the topological relationships, rather than memorizing spatial coordinates. For example, location-based service (LBS) is a general class of computer program-level services that are used to include specific controls for location and time data as control features in computer programs.

Time is a specific time or a particular period of time when spatial data are collected or computed. Spatial data may be treated as the functions of time on spatial objects. Some of the data change slowly and others rapidly. It is hard to collect the changes at any time by ordinary methods. Satellite remote sensing is an alternative method for capturing the dynamic change features of spatial objects with time effectively and continuously. For example, every 30–60 min, data from a geostationary meteorological satellite can be used to forecast weather and ocean storms. The high resolution of time provided by satellite remote sensing is of great significance in the study of the dynamic changes in the natural process.

Theme is a subject-oriented aspect of attribute data and refers to the characteristics apart from location and time, such as terrain slope, terrain exposure, annual precipitation, land endurance, pH value, land cover, population density, traffic flow, annual productivity, per capita income, disease distribution, pollution level, and resource distribution. Such themes can be extracted through aviation and aerospace remote-sensing images manually, automatically, or semi-automatically and stored and handled in other database systems, such as a relational database management system (RDBMS). Further development of remote sensing technology is expected to obtain more thematic data.

3.1.4 Diversity of Spatial Data

SDM has a wide range of data sources, including observation, maps, remote sensing, and statistics. These sources can be divided into direct data and indirect data. Direct data are the first-hand raw data from original spatial objects—that is, data observed with senses, such as vision and touch. Indirect data, on the other hand, are the second-hand processed data from the computing analysis of direct data, such as a map contour, recommended road, and buffer zone. Indirect data is a relative concept because additional indirect data can be extracted from indirect data. Spatial data are stored as analog data and digital data (Chen and Gong 1998).

During the processing of spatial data, in addition to general databases, other different types of databases with distinguishable natures and requests are involved. Some of these databases focus on the characteristics of time and location, some concentrate on the mathematical and physical characteristics of the data, and some pay more attention to the data characteristics of data. It is difficult to illustrate the spatial relationships between objects in a general database, but a spatial database can produce a detailed description of the length, width, and height of objects and the distance between one object and another object as well as the boundaries of the objects. Time is also very important for such databases because users want to know when the imagery data are obtained, such as when certain buildings were demolished and the various changes over time. Other significant concepts include adjacency and inclusion. For example, a specific building may belong to a park, and its adjacent buildings may include a government office building. Here, the adjacency indicates the nearest distance, such as the school or fire station nearest to a house. All of these special requirements constitute the characteristics of spatial databases.

For the integration of a spatial database system, a live data acquisition system focusing on the collection, organization, and management of environmental parameters is an important component. When addressing environmental data, there are a number of professional data, such as the noise level, temperature, and harmful gas content in the atmosphere, which are often acquired from experiments and observations. These values often comply with the scientific law of formula, which is used as the binding for such databases; however, the constraints and function dependence in relational databases are different. In the case of function dependence, the name of each symbol is defined either equivalently or directly through samples. In other words, when defining the names or symbols of a database, we should set a concrete or abstract object and mark or imply its compatibility with symbols. The relationship between the symbol and the name is purely external and cannot be produced by some process. However, the binding can be generated by a particular process, but it also has accuracy and similarity issues that may make it unable to match the binding with real values. In such databases, there still exists the conversion between the object and unit as well as value changes in experiments over time.

3.1.5 Spatial Data Fusion

As the spatial data collected through diverse sensors or methods are easily affected by various factors, their uncertainty is ubiquitous and everywhere. In data analysis and data identification, it is inadequate to process the data one by one without considering their connection. With an eye to the problem, it is necessary to integrate the spatial data so that the representation of spatial object can maintain the consistency.

3.1.5.1 Fusion of Complementary Data and Cooperative Data

Spatial data can be divided into at least two categories: complementary data and cooperative data. Complementary data refers to spatial data derived from different data sources. The environment features are independent from one another. It reflects the different sides of a spatial object. The integration of complementary data reduces the misunderstanding of features resulting from a lack of some object characteristics so that the representation of the spatial object can be improved to be consistent and correct. As complementary data are collected by dissimilar sensors, the data differ a lot from the aspect of measurement accuracy, scope, and output form. Thus, it is extremely important to unify the representation of data extracted from different sensors before the integration of multiple data. In spatial data processing, if the processing of certain data depends on other data processing, they are called cooperative data. The integration of cooperative data is largely related to the time and order of vicarious data processing.

One approach to complementary data fusion is that, guided by evidence theory, the resulting representation can be synthesized from the evidence of spatial objects provided by multiple data resources. As a result, the essence of multi-sensor data fusion is that different evidence bodies are merged together to form a new one under one recognition framework. In image recognition, another method is to fuse images based on the product rule. When the results of an experiment show that the product of integration of the rule is not very effective because fusion based on posterior probability is inadequate as a product rule is derived from the channel data, the hypothesis is supported. If the data of a channel denies the hypothesis, the posterior probability will remain almost at zero, which will affect the fusion. In addition, considering the fact that a product rule is sensitive to the parameter error, Killer et al. (1998) came up with a fusion method based on an additive rule and posterior probability.

3.1.5.2 Fusion with Quotient Structure

Usually, we can hardly recognize an object when observing the details of an image without viewing the whole image. As a result, it is difficult to interpret. Sometimes, it is difficult to recognize a phenomenon in the image. In the interpretation of remote sensing images, there are two principles:

1. Interpretation from macro to micro view. Observing from a general view first is much easier because details are very likely to confuse the observers.
2. Multiple perspective study. As a resolution, it is necessary to compare the images considering factors such as seasons, colors, or solar elevation angles or to observe them in multi-perspective images.

These principles reflect a universal phenomenon of human thinking: from the abstract to the concrete, from the general to the individual, and from the part to the whole. It can be interpreted as "when multiple quotient space has been known, how can we compose its original space?" in terms of fusion under quotient structure. The fusion contains three branches: universal synthesis, topological structure synthesis, and characteristic function. In the quotient structure synthesis, the study on the connection and regularity between automatic searching of integrated terrain units and multispectral images shows the promising future of quotient space application, and attention needs to be paid to the reasoning so that it can be regarded as the synthesis processing.

3.1.5.3 Multi-source Remote Sensing Data Integration

Remote sensing images of multi-platform, multi-faceted, multi-sensor, multi-phase, multi-spectral, multi-perspective, and multi-resolution come like a flood at an amazing speed and establish a multi-source image pyramid within an area. Thus, it is extremely urgent to do research on the subject of useful images aggregation within one region. Fortunately, remote sensing image fusion, as a resolution, not only can enhance the capability in data analysis of multi-source and dynamic environment monitoring, but it also will perfect the timeliness and reliability of remote sensing data extraction. In addition, it will improve the data utilization rate, which lays a good foundation for large-scale remote sensing application research.

The schemes of multi-source remote sensing data fusion may be weighted fusion, HSI (hue-saturation-intensity) transform fusion, feature fusion with wavelet technique, classification fusion with Bayesian rule, and image fusion with local histogram match filtering (Li and Guan 2000). To test the quality of integrated spectral imaging data content, a comparison between a fused image and a spectral imaging radiation value can be used to define it. If two images match with each other, their global statistical parameters should be approximate. Among those parameters, the average grade and bias index are relatively important. Average grade reflects all the contrast of the small details and changing texture features and also shows the image definition, while the bias index refers to the specific value between the absolute value and the low-resolution image.

Traditional database technology is a single data source contained in a database employed for data processing, such as transaction processing and decision-making. The rapid development of computer applications in recent years has extended in two directions. The first is breadth calculation, which aims to expand computer applications and to realize extensive data exchange. The internet is one feature of breadth calculation. The second direction is depth calculation, which enables computers to participate more in data analysis, decision-making, etc. Apart from that, database processing can be classified into operative processing and analytical processing. Such a division draws a clear distinction between them as far as laying the groundwork for a systematic environment (Codd 1995).

Therefore, research efforts are needed to reach the requirements of SDM and to improve its analysis and decision-making efficiency within the foundation of network and data processing technology. In addition, analytical processing, as well as its data, need to be separated from operative processing. A method also is needed to extract analytical data from transaction processing and then reorganize it in accordance with SDM requirements. To establish such a new data memory, organization, analysis, processing environment, and powerful decision-making technology, data warehouses are emerging as the need arises.

3.1.6 Seamless Organization of Spatial Data

Complete representation and storage of spatial relationships and semantics among objects is an important task in the organization of spatial databases (Shekhar and Chawla 2003). A seamless organization based on logic and physics requires the consistent storage of features. Thus, each of the complete features can only own one object identifier (OID), regardless of the size. However, the integrity of objects is determined by the requirements of users and the application purposes.

For example, the Yangtze River can be regarded as a polygon or linear object or can be constructed into a complex geographical entity for management and description.

To meet the need of seamless organization in logic and physics, all spatial attribute data should be under engineering management. Adopting RDBMS to manage and save the attribute data in form of tables is suggested. Taking the rules as reference, users can classify the features. For example, the features can be divided into control points, roads, residential area, water system, and vegetation of various levels; then, an exclusive classification code of features can be assigned. Any complete feature usually corresponds to a piece of an attribute data record. The spatial features are connected with their corresponding attribute records. To effectively organize and describe spatial entities, the features can be extracted into multi-grades based on the size of a feature (the largest coverage area).

The selection and representation of a spatial object and its attributes is very important. For example, many attributes exist in urban areas, such as age, green area, educational level of the citizens, and main roads. The selection and representation of those attributes have a direct influence on geometric operation and predicate evaluation in a position network. Considering the fact that a position network can combine its geometric and analog data with a semantic network, it can be brought into the spatial database to realize a seamless combined organization.

A position network helps fulfill the logically seamless organization of spatial objects (Tan 1998). It is a set of geometric point network representations. The set of geometric points is linked together by set theory and certain set operations corresponding to position limitations determined by a way of thinking or physical factors. The internal contacts of a position network contain geometric operations, variables, and results. For example, a node represents the union set of two variables and the result is a point set. The reasoning is on the basis of elevation on the network. The expected relative positions of object features are stored in the network, in which a model is established in order to operate the fundamental calculations of the geometric relationships between objects:

1. Computing direction (left, right, south, north, upper, lower, reflection, etc.): The point set is given by the relative position and direction of other point-groups.
2. Region operations (close to, within the quadrangle, inside the circle, etc.): The establishment of a point set of non-direction related with other point-groups.
3. Set operations: Union, intersection, and subtract.
4. Predicate evaluation: An operation to remove certain point sets by computing the features of the objects.

3.2 Spatial Data Acquisition

Spatial data play an important role in SDM. For example, during the process of the construction of a durable and practical GIS, spatial data accounts for 50–70 % of the cost. In developing countries, the cost is higher but may reach only about 50 %

for spatial data accounts due to the cheap labor, However, the quality of the data would be affected were the cost proportion were to further decrease (Gong 1999).

Spatial data are collected via various macro- and micro-sensors or equipment (e.g., radar, infrared, photoelectric, satellite, multi-spectral scanners, digital cameras, imaging spectrometer, total station, telescopes, television cameras, electronic imaging, computed tomography imaging) or by accessing spatial data collected by conventional field surveying, population census, land investigation, scanning maps, digitizing maps, and statistical charts. Technology applications and analysis of spatial data include other sources, such as computers, networks, GPS, RS, and GIS. These sources contain the origin of the spatial data as well as the step, format, conversion, date, time, location, staff, environment, transport, and history to collect, edit, and store spatial data.

Spatial data can be obtained by utilizing data acquisition methods, which include point acquisition, area acquisition, and mobility acquisition. Point acquisition refers to the use of total station, GPS receivers, and other conventional surface measurements to collect the coordinates and attributes of surface points on Earth point by point. Area acquisition refers to the use of aviation and aerospace remote sensing to acquire large areas of images from which geometrical and physical features can be extracted. Mobility acquisition integrates GPS, RS, and GIS into Earth's observation system to obtain, store, manage, update, and apply spatial data.

3.2.1 Point Acquisition

Point acquisition refers mainly to total station and GPS. The theodolite, level instrument, distance measuring instrument, total station, and GPS receivers used in surface measurement can acquire geospatial data point by point. Two of the trendiest types—GPS and total station—are introduced here.

Total station is another name for total electronic station. All of the data collected can be automatically transmitted to card records, e-books by hand, directly mapped to the indoor computer, or transmitted to an e-pad, which maps automatically on-site in the field. This method can be applied in cities with numerous skyscrapers or in situations where high accuracy is required to obtain and update spatial data as supplementary information for aviation and space remote sensing.

GPS is a satellite navigation system that is able to regulate time and measure space. GPS consists of three parts: the *space*, which is made up of GPS satellites; *the control*, which consists of a number of ground stations; and *the user*, which has a receiver as the main component. The three parts have independent functions but act as a whole when GPS is operating. As technology has advanced, GPS has expanded from static to dynamic, from post-processing to real-time/quasi-real-time positioning and navigation, which has greatly broadened its scope of application. In the 1980s, China introduced GPS technology and receivers on a large scale for the measurement of controls at all levels. China also developed a number of GPS data processing software programs, such as GPSADJ, which has measured the accuracy of thousands of GPS points, including the national GPS A-level and B-level, the control networks of many cities, and various types of engineering control networks.

3.2.2 Area Acquisition

Area acquisition refers to aviation and aerospace remote sensing, which is the main rapid, large-scale access to spatial data, including aviation and space photogrammetry. Because they make use of a variety of sensors to scan spatial objects on Earth's surface for taking photos, the obtained images contain a large amount of geometrical and physical data that indicate the real and present situation. Compared to GPS and other ground measurements that acquire data point by point, this area access to geospatial data is effective.

Aerial photogrammetry and satellite photogrammetry acquire color or full-color images, from which highly accurate geometric data can be extracted due to their high-resolution and geometrical stability. On the other hand, aerial and aerospace remote sensing is characterized by multi-spectrum and hyper-spectrum or multi-polarization and multiband, from which more physical data can be extracted, which therefore means that they are complementary tools.

1. The combination of photogrammetry and remote sensing.
   - Remote sensing can greatly impact photogrammetry by breaking its limit on focusing too much on the geometrical data of observed objects, such as their shape and size; in particular, aerial photogrammetry for a long time only emphasized the mapping portion of a terrain diagram. Remote sensing technology, in addition to the use of a visible light framework black-and-white camera, offers color infrared photography, panoramic photography, infrared scanners, multi-spectral scanners, as well as imaging spectrometer, CCD array scanners, and matrix synthetic aperture cameras for detecting radar. Remote sensing was further developed after the 1980s, once again demonstrating its tremendous influence on photogrammetry. For example, the space shuttle, as a remote sensing platform or means of launch, can return to the ground and be reused, which has greatly enhanced the performance of remote sensing and is cost-effective. In addition, remote sensing's ground resolution (spatial resolution), radiometric resolution resolution (number of grayscales), spectral resolution (spectral bands), and time resolution (repetition cycle) of many new sensors have been enhanced significantly.
   Remote sensing's improved spectral resolution spawned the imaging spectrometer and hyper-spectrum remote sensing. The so-called hyper-spectrum remote sensing identifies multiband images at nanometer-level spectral resolution through the breakdown of the spectrum, which forms a cube image and is characterized by the unity of the images and spectrum. Thus, it has greatly enhanced a user's ability to study surface objects, recognize the object's type, identify the object's material composition, and analyze its conditions and trends.

- Photogrammetry has contributed greatly as well, especially digital photogrammetry, which utilizes remote sensing. It is well known that the high-precision geometry positioning and correctness of remote sensing images are an important application of modern photogrammetry theory. Now, a modern digital photogrammetry system is no different than a modern processing system of remote sensing images.

2. Analytical mapping instrument to obtain basic GIS data.
   - GIS is a computer-aided system made up of hardware, software, data, and applications, which gathers, edits, addresses, stores, organizes, simulates, analyzes, and presents spatial data digitally. Its tasks include the input, management, analysis, and presentation of data. GIS data can be classified into basic data and thematic data according to its content. GIS can access basic data, such as ground-based surveys, photogrammetry, map digitization, and data extraction from statistical charts. For spatial data that collect and reflect topographical characteristics, photogrammetry is the most advantageous method compared to other means. Digital photogrammetry obtains GIS data automatically, and the ways that photogrammetry can be used to acquire GIS basic data can be divided into two categories: (1) the machine-aided and machine-controlled method and (2) the full digital automatic acquisition method.
   - There are three approaches when using a machine-aided/controlled system togather basic spatial data:
     - Create a conventional simulated diagram and then obtain the digital form of the data through the process of digitization. Obviously, this approach is irrational because it would not only be a waste of time, but also a loss in accuracy.
     - Acquire the digital data directly without considering a GIS request for the data structure first, and then edit the data gathered according to the GIS data structure requirements.
     - Collect the data based on the GIS requirements so as to provide directly the structured digital data for the application of GIS.

3. Digital automatic approach to GIS basic data acquisition.
   There are c+urrently two ways to obtain GIS basic data with the digital automatic approach: (1) automatic recognition and vectorization of the digital maps and (2) automatic identification and positioning of digital photographs.

Digital maps are automatically recognized and vectorized. Full digital automation of a map is based on the digital scan of the map, which would be processed by computers automatically so as to achieve the identification and vectorization of the elements on the map. The study of automatic map processing technology has gone through two stages: the first stage addressed only the automatic processing of a single element map, and the second stage studied the direct method to process total factor maps automatically. The automatic computer process of a total factor map can be divided into the following six steps: map digitalization, image segmentation, element separation, vectorization, feature extraction, and factor recognition.

GIS basic data is automatically acquired with full digital photogrammetry. The birth of full digital photogrammetry enabled computers to obtain GIS basic data from aerial photographs or space remote sensing images. The process starts from digital images or digitalized images, which are then processed on computers to obtain the necessary data to establish GIS maps. GIS basic data can be simply divided into the terrain data (three-dimensional coordinates of any point), graphic data (thematic elements and basic graphics; i.e., point, line, and polygon) and attribute data. As a result, it is possible to provide terrain data by generating a digital elevation model (DEM), graphic data by producing an orthophoto and extracting its structure data, and attribute data by doing image interpretation and thematic classification.

Automatic interpretation and thematic classification of images, also known as the automatic recognition pattern of images, refer to the recognition and classification of the attributes of images by using computers, certain mathematical methods, and geographic elements (i.e., certain features of the subjects), so as to classify the real geographic elements corresponding to the image data. Through the interpretation and thematic classification of images, the attributes of basic graphics and images of thematic elements can be acquired, which are necessary for establishing GIS. To improve the accuracy and reliability of the automatic interpretation and classification of images, texture analysis of the images and use of the nearby landscape should be considered apart from the spectral characteristics of the image.

With the introduction of a large number of high-resolution satellite images, remote sensing images will become the main source of GIS data. To obtain data quickly and effectively from remote sensing images, a video-based GIS should be considered in the future to combine the analysis system of remote sensing images and the standard geographic data system.

3.2.3 Mobility Acquisition

Mobile access to spatial data refers to the integration of global positioning system (GPS), remote sensing (RS), and geographic information system (GIS). GPS, RS, and GIS are the three major technical supports (hereinafter referred to as *3S*) in the Earth Observation System to acquire, store, manage, update, analyze, and apply spatial data. 3S is an important technical way to realize sustainable development of modern society, reasonable use of resources, smart planning and management of urban and rural areas, and dynamic supervision and prevention of natural disasters. At the same time, 3S is one of the scientific methods that is moving geoinformation science toward quantitative science.

To achieve an authentic 3S, it is necessary to study and resolve some common basic problems in the process of design, implementation, and application of the integrated 3S system in order to further design and develop practical theories, methods, and tools, such as real-time system positioning, integrated data management, semantic and non-semantic data extraction, automatic data updates, real-time data communications, design of an integrated system, as well as spatial visualization of graphics and images.

1. Partial integrated "3S"
   - GIS can be integrated with GPS. The combination of the maps of GIS and the real-time differential positioning technology of a GPS receiver may consist of various GPS+GIS navigation systems that can be used for transportation, police detection, and automatic driving of automobiles and ships. GPS+GIS also can be used directly for GIS real-time data updates, which is the most practical, simplest, and least expensive method of integration.
   - GIS can be integrated with RS. RS is an important data source and the update method for GIS. At the same time, GIS provides the auxiliary data in the RS data processing system for the automatic extraction of semantic and non-semantic data. GIS and RS may be combined in a variety of ways— a separate but parallel combination (different user interfaces, different tool libraries and different databases), a surface seamless combination (the same user interface, different tool libraries and different databases) and an overall integration (the same user interface, tool library, and database). The integration of RS and GIS is mainly for monitoring changes and real-time updates, which involves computerized pattern recognition and image interpretation.
   - RS can be integrated with GPS/INS, which refers to the use of GPS satellite signals to correct or calibrate a solution from an inertial navigation system (INS). Target positioning in RS has been dependent on ground control points. If real-time RS target positioning without ground control points is to be realized, the location and attitudes of RS sensors for access to instant images must be recorded with a GPS/INS method. For low-precision positioning, a pseudo-range method is not necessary; however, for high-precision positioning, a phase difference method must be used.

2. Overall integrated 3S
   The integration of GPS, RS, and GIS as a whole system is undoubtedly the goal. 3S not only has the automatic, real-time function of collection, processing, and updating but is also able to analyze and apply data smartly to provide scientific decision-making advice and to answer a variety of complex questions put forward by users. For example, in LiDAR, four CCD cameras on the front of the car represent a remote-sensing camera system. GPS/INS compensates for loss of lock and other system errors. A GIS system is installed inside the car. GPS/INS provides the elements of exterior orientation, image processing to calculate the real-time parameters of point, line, and polygon of targets by comparing them with the data in GIS, in order to monitor changes, update data, and navigate automatically.

3. The integration of GPS and aerophotogrammetry
   With the rapid development of computer technology, photogrammetry has been able to use geodetic observation in a more rational sense, such as photographing the simultaneous adjustment of the data and the original geodetic observations, thereby taking advantage of the dynamic GPS positioning technology to obtain the location of the aerial photographic cameras to carry out aerial triangulation.

GPS-assisted aerial triangulation uses a GPS signal receiver on the plane and a GPS receiver at one or more base stations on the ground to observe GPS satellite signals constantly at the same time. It can obtain 3D coordinates of the station at the exposure time of the aerial surveying camera through the GPS carrier phase measurement with differential positioning technology, and then can introduce them as added observations into the photogrammetric area network adjustment to position the point location and evaluate its quality through a unified method of mathematical models and algorithms. If the ground control station is replaced by an onboard GPS aerial photo instrument and the artificial measurement of point coordinates for GPS auxiliary automatic aerial triangulation is replaced by a full digital photogrammetry system for automatic matching of the point changes of multi-chip images, the time and manual labor necessary will be dramatically less than conventional aerial triangulation. On this basis, construction and automatic updating will be realized by generating a digital elevation model (DEM) and orthophotos through full digital photogrammetry work stations and providing the necessary geometrical and thematic data for GIS. For example, for the construction of an industrial area road that is needed by different customers who are highly integrated, photogrammetry can be completed quickly and the results of spatial databases can be addressed efficiently using a system that is not only in accordance with the trends of aerial photogrammetry but also aligned with the sustainable development of society and a solid foundation for the realization of Digital Earth.

3.3 Spatial Data Formats

In computerized spatial information systems, spatial objects are mainly described by using two depictive formats: raster and vector. Figure 3.1 illustrates a curve; the sketch on the left is the vector while the right-hand sketch is the raster.

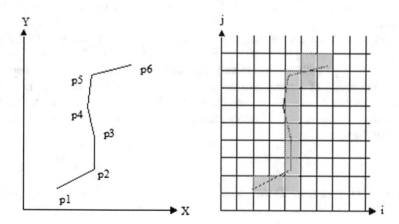

Fig. 3.1 Vector data (*left*) and raster data (*right*)

3.3.1 Vector Data

Vector data represent spatial objects as precisely as possible with points, lines, and polygons by continuously recording the coordinates. The point is located by the coordinates. The location and shape of the line are represented by the coordinate strings of the sampling points on its central axis. The location and range of the polygon are indicated by the coordinate strings of the sampling points on the outlines of their range. The vector format can store complex data by minimizing the data redundancy. Limited by the accuracy of the digital equipment and the length of the data record, the vector records the coordinates of the sampling points, which are taken directly from the spatial coordinates so that the objects can be represented in an accurate way. There are two vector methods that depict the characteristics of point, line, and polygon: path topology and network topology. The vector format features "explicit positions but implicit attributes."

3.3.2 Raster Data

Raster data represent spatial objects by dividing Earth's surface into a regular array that is uniform in size (i.e., grid, cell, pixel). They implicitly depict the relationship between spatial objects and a simple data format. Respectively, point, line, and polygon are indicated by a raster, a series of neighboring rasters along with lines, and neighboring rasters that share the same properties in an area. The basic unit of raster data is generally shaped like a square, a triangle, or a hexagon. While they share similarities, they also differ from each other in some geometrical features, such as directivity, subdivision, and symmetry. The raster data may be plane or surface, which are organized by the raster unit-oriented method, the instance variable-oriented method, and the homogeneous location-oriented method. The raster format is featured with "explicit attributes but implicit location."

Each raster has a unique pair of rows and columns and a code that determines the identity of a spatial location, and the code represents the pointer that correlates the attribute type with its pixel record. The number of rows and columns depends on the resolution of a raster and the attributes of a spatial object. Generally, the more complicated the attributes are, the smaller the raster size is and the bigger the resolution is. The amount of raster data increases in accordance with the square index of the resolution.

3.3.3 Vector-Raster Data

Vectors and rasters are compared in Table 3.1; along with Fig. 3.1, the table shows that the vector curve, represented by a series of points with coordinates, is able to

Table 3.1 Format comparison between raster and vector

Content	Raster format	Vector format
Suitability	Photographs, photo-realistic images	Typesetting and graphic design
Data sharing	Difficult to achieve	Easy to achieve
Topological relations	Difficult to represent topological relations, but easy to realize the overlays operations	Offer effective topological codes, easy to realize network analysis
Data volume	Huge	Small
Graphics quality	Cannot scale up to an arbitrary resolution without loss of apparent quality	Easily scale up to the quality of the device rendering them
Graphics operation	Simple with low efficiency	Complicated with high efficiency
Output display	Direct, cheap, lines with aliasing	Indirect, expensive, smooth
Editors	Revolve around editing pixels; e.g., Painter, Photoshop, MS Paint, and GIMP	Revolve around editing lines and shapes; e.g., Xfig, CorelDRAW, Adobe Illustrator, or Inkscape

be reproduced by the connection between two neighboring points. On the other hand, the raster curve, which is represented by the number of grids divided in accordance with the matrix form in the raster, has lines through it. It is clear that the precision of the vector is higher than that of the raster. To design a system for multiple purposes without additional storage, an integral format should combine the features of both vector data and raster data. The foundation of an integrated format on vector-raster data (Fig. 3.2) is to fill the linear object path and the areal object space. Every linear object records not only the original sampling points but also the raster that the path goes through. Apart from recording its polygons' boundaries, every areal feature also records its region's raster in the interior. A point, linear, and area adopt object-oriented depiction, which depicts data by following the location and explaining the topological relationships; as a result, its vector features are kept.

The integrated format of vector-raster data is hierarchical grids (i.e., fine grids and coarse grids). Fine grids improve the accuracy of the integrated format by the subdivision of grids, and coarse grids improve the speed of manipulating objects. In order to remain consistent with the format of the overall database, it is feasible to adopt the coding method of the linear quad-tree in coarse basic grids and fine subdivision grids by representing the sampling point and the intersection point of the linear objects and the boundaries of the basic grids with Morton codes. The integrated format is indexed with linear quad-tree coding. A basic grid constitutes a coarse grid, each of which is encoded by decimal Morton codes. A coarse grid will form an index table, whose order is obviously arranged by the linear quad-tree address codes. Every record in the index table has an indicator that points to the starting address of its first record in the linear quad-tree. For the sake of uniformity, every leaf node cannot surpass one coarse grid. In this way, the relationship between the index record and linear quad-tree is established. According to the

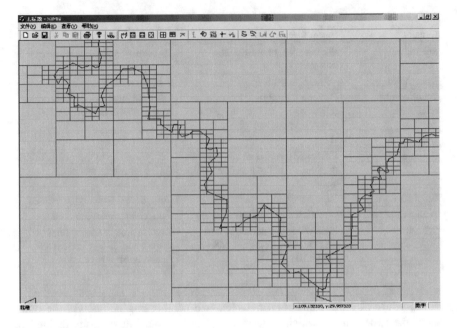

Fig. 3.2 Vector-raster data

location of a certain point, it can be directly entered into the index record, and the corresponding record in the linear quad-tree then can be determined. Thereafter, with the help of the indicators, its starting record number can be found in the linear quad-tree as well as the attribute value of the leaf node. In addition, due to the direct link between the coarse grid Morton code and the record number, the Morton code in the index documents also can be omitted and replaced by implicit record numbers.

When designing the integrated format on vector-raster data, the fundamental focus is the point, linear, and areal objects. The use of a sub-grid means that vector data may not have stored original samples that remain of good precision with the converted data format. Furthermore, all location data adopt the linear quad-tree address code as the basic data format, which ensures the direct counterparts of manifold geometric objects.

Point objects possess location without shape and area. It is unnecessary to regard the point-like objects as cover to decompose the quad-tree. Rather, they are only used to convert the coordinates of the points into Morton address codes, regardless of whether or not the entire configuration is quad-tree.

Linear objects have no area, and their shapes contain the whole path. We only need to express the path of each linear object by a series of numbers instead of decomposing the whole quad-tree. When it comes to the entire path, all the raster addresses through which the linear object has gone are supposed to be recorded.

A linear object may be composed of a few arcs with a starting point and ending point. The string of the mid-points contains not only the original sampling points but also all the intersection points of the grid's boundaries through which the path of the arc has gone, whose codes fill the entire path. Such a format also takes full account of the spatial characteristics that linear objects have on the ground. If a linear object passes over rough terrain, only by recording the grid elevation value of the DEM boundaries through which the curve has gone can its spatial shape and length be better expressed. Although this data format increases a certain amount of storage capacity compared to simple vector formats, it solves the quad-tree representation problem of the linear objects and enables it to establish an integrated data format based on the linear quad-tree code together with the point and area objects. This makes the query issue quite simple and fast in terms of the intersection between a point and linear objects, the mutual intersection between linear objects, and the intersection between linear and area objects. With the data files of the arc, the data format of linear objects is represented as a collection of arcs.

Areal objects should contain borders and the whole region. The borders are composed of arcs that surround regional data. Manifold objects may form multiple covering layers. For example, visual objects such as buildings, squares, farmland, and lakes can form one covering layer while the administrative divisions and the soil types can form another two covering layers. If every covering layer is single-valued, each grid only has the attribute value of one area object. One covering layer represents a space, and even an island has its corresponding attributes. In object-oriented data, the grid value of the leaf nodes is based on the identification number of the object instead of the attributes of the objects. By using the cycle indicators, the leaf nodes belonging to the same object are linked together to be object-oriented. A rounding method is used to determine the values of boundary grids in the areal objects. The value of the interacted grid of two ground objects is based on which one occupies a larger area in the grid. In order to carry out precise area or overlay calculations of the areal feature, it is feasible to further cite the boundary data of the arcs. These object-oriented data are featured with both vector and raster. The identification numbers of the areal object make it easy to determine its boundary arc and extract all the middle blocks along the arc.

3.4   Spatial Data Model

The spatial data of various types with public geographical coordinates are intrinsically linked, reflecting the logic organization of spatial data when they are placed in a spatial database structure. A spatial data model mathematically depicts spatial data content and presents the interrelation between two entities. At present, the hierarchical model, network model, relational model, and object-oriented model are in common use.

3.4.1 Hierarchical Model and Network Model

The basic relationship that a hierarchical model expresses is subordinative. When this "father and son" type of relationship is clear and defined, the hierarchical model owning the tree structure is appropriate for its depiction. Take, for example, soil, animals, and plants. They have a root node and many leaf nodes. Besides one root node and leaf nodes, only one father node and some sub-nodes exist. This hierarchical database is highly efficient for the user to store and access a website under the name of a certain chief element, such as street name. However, if the user changes the entry format, it will be inconvenient; for example, when the user searches for all the buildings in a certain limited area by inputting a hierarchical model with the keyword of street name, it will be difficult.

When the relationship between the data records is not subordinative, a network model enables an arbitrary record to be connected with any other records, which may greatly reduce the amount of data storage. Meanwhile, one node can have multiple father nodes, as well as a number of son nodes. A network model is an undirected graph with good versatility, but it is irregular in terms of structure.

When designing hierarchical models and network models, the paths of data storage and access should be taken into consideration. The paths are fixed once they are determined. In fact, as far as the complex relationships between data, there is no way to describe them in a fixed hierarchical model or network model. Thus, a relational model can be used to construct a large-scale spatial database.

3.4.2 Relational Model

A relational model depicts spatial data with their entities and relationships by using the connected two-dimensional table in certain conditions. Such a table is an entity, in which a row is an entity record while a column is an entity attribute. There may be many columns as the entity is depicted with many attributes. A unique keyword (single attribute or composited attributes) is selected to recognize the entity. A foreign keyword connects one table to another table, showing the natural relationships between entities (i.e., one to one, one to many, many to many). Based on Boolean logic and arithmetic rules, a relational model allows definition and manipulation of the data by using structured query language (SQL). The data description is of high consistency and independent, and complicated data become clear in terms of structure. Figure 3.3 shows a digital map and its relational database.

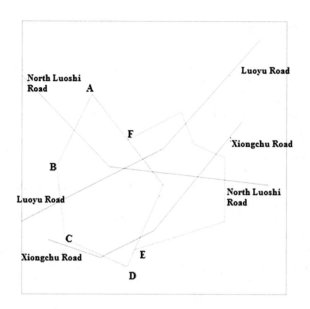

CITIES

Frame	Clid	X_1	X_2	Y_1	Y_2
WuHan	1	120	60	330	210
WuHan	1	60	75	210	90
WuHan	1	...	...	...	...
WuHan	1	180	186	45	50
WuHan	1	186	241	73	180
WuHan	2	273	345	305	150
WuHan	2	345	187	150	76
WuHan	2		⋮		
WuHan	2		⋮		
WuHan	2	183	345	309	150
WuHan	2	312	231	315	183

ROADS

Frame	Roid	X_1	X_2	Y_1	Y_2
WuHan	1	0	224	402	281
WuHan	1	224	467	281	227
WuHan	2	0	299	182	317
WuHan	2	299	466	317	479
WuHan	3	107	197	152	122
WuHan	3	197	287	122	167
WuHan	3	287	439	167	334

POS

Frame	Xsize	Ysize	Xcen	Ycen	LOC
WuHan	512	512	1792	256	/Pix/310

CINAME

Frame	Clid	Name
WuHan	1	WuChang
WuHan	2	HongShan

Frame	Roid	Name
WuHan	1	North Luoshi Road
WuHan	2	Luoyu Road
WuHan	3	Xiongchu Road

Fig. 3.3   Spatial relational database

3.4.3 Object-Oriented Model

The object-oriented model was the product of combining the object-oriented technique and the database technique. Since the late 1980s and early 1990s, application of the object-oriented technique in GIS has been highly valued. In Fig. 3.4,
three kinds of spatial objects—point-like, linear, and areal—are oriented along
with their annotations.

1. Object-oriented contents
   - Spatial objects may be single or composite in GIS. If various spatial objects
     are merged appropriately, there will be 13 classes of spatial objects and a
     data structure: node, point object, arc, linear object, area object, digital terrain
     model, fracture surface, image pixel, cubic object, digital 3D model, voxel,
     columnar objects, complex objects, and location coordinate. Each object has
     an identifier in the class; however, there is no identifier in location coordi-
     nates (e.g., a tuple of a floating point number in 2D GIS, a tuple of a floating
     number in 3D GIS). The obvious advantage of an object-oriented data model
     is that each class corresponds with one data structure and each object corre-
     sponds with one record. No matter how complicated the objects are or how
     many nested object relationships exist, there is still a structure table used for
     representation. It is easily understandable as each spatial object has one class,
     making it available to express the relationship by generalization, union, aggre-
     gation, etc.

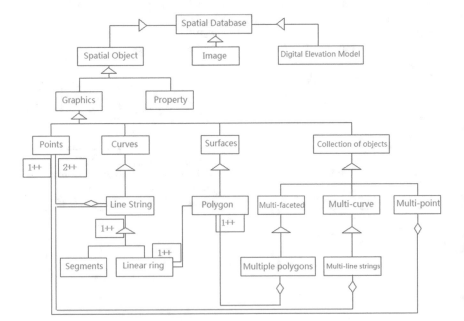

Fig. 3.4 Object-oriented data model

- Spatial objects mainly contain topology, object composition, and hierarchical structure, which chiefly carry the semantic concept of aggregation and union. An object-oriented data model involves a number of object-nested relationships and variable-length records. If it is converted into a relational data model, many relationship tables are appended to meet the database normalization. At present, most object-oriented database management systems support not only variable-length records but also the union of multiple nested objects and complex objects. Of course, the object records may not meet the requirements of a relational model. That is, in a piece of a record, an attribute may have multiple attribute values or object identifiers. Each spatial object corresponds with an object identifier given by the system, and the identifier contains the class identifier so that the geometric type to which the object belongs can be determined.

2. Object-oriented dimensions
   Spatial objects have two distinct characteristics: (1) the geometric features (i.e., size, volume, shape, and location) and (2) the physical features of object attributes, which can define whether it is a river or a road, for example. Concerned about the physical features, generally, we code spatial objects. Also, the state has classification and coding standards regarding spatial elements. In terms of geometric features, the spatial object in the abstract two-dimensional GIS can be transferred into a zero-dimensional object, one-dimensional object, two-dimensional object, and three-dimensional object (Gong 1999).

   - A *zero-dimensional object* can be abstracted as a point. A point-like object is totally a geometric type and also a spatial object, corresponding to the property code and the attribute table. A geometric node is a geometric topology element not belonging to any object, which is used only to express the geometry location and arc relationship. A node object belongs to the geometric topology as well as the spatial objects. For example, the node of a power plant is usually a power distribution station. A notation point is for a referenced location, and a polygon identifier point is auxiliary for polygons.
   - A *one-dimensional object* can be abstracted as a line. The point-to-point strings or arc strings can be reticulated, but they should be interconnected like rivers, a river network, and a road network. A topological arc is a kind of geometric topology that has no branch of the arc, but it does have start and end points. It may be part of the linear objects and the boundary of area objects. It can be directly coded and connected to the attribute tables. A nontopological arc is a geometric object, which also is called "spaghetti"; for example, it is not necessary to consider a contour line in terms of start node, end node, or left or right polygons. Arc-free topology is simpler than the arc topology. However, in terms of shape, arc-free topology is classified according to smoothness. A linear object features a line from one or several sections of an arc, allowing cross-branches to expand it, such as to a river basin or to deal with traffic problems. It must have encoded attributes and be connected to the table.

- *Two-dimensional (2D) objects.* An area object is constituted of the arcs around it with its own attribute codes and tables. Different from a planar polygon, it is necessary to consider the topographical change of the curved surface. An areal element is of no terrain significance and unconnected to the attribute tables. It only represents the random surface of a 3D object, while the projection of a surface model is not always a horizontal plane. A digital terrain model (DTM) is used to show the land surface and the topographical change of the layers underground or the surface configuration of a 3D body. At present, there are the grid method and the triangulation net method. A raster image is a digital image from remote sensing, aerial photography, or a scanned map, which can express the thematic data of areal objects. Usually, a pixel is called a spatial object when it is obtained with the digital number in a raster image.
- *Three-dimensional (3D) objects.* A homogeneous object consists of many areal elements with the same internal compositions. When it is internally empty, it becomes a hollow object. A non-homogeneous object contains many areal elements, but the internal components are different. It can fully express a three-dimensional object only when an object is connected to the raster formats. Based on digital models, an inhomogeneous 3D object divides into voxels with the same size and value, and the voxel positioned is by using spatial coordinates or 3D array, as well as the third series of numbers. A cylindrical object is made up of a series of fractures with a hollow center. For example, in a high-rise building, each layer is considered a fracture, the deployment presents its shape, and the attribute table is connected.

3. Object-oriented themes

In general, a GIS map is divided into multiple thematic layers according to the features. The whole map is the result of all the superimposed layers, which makes it easy for the user to find the target thematic layer in less time. Figure 3.5a shows spatial data organized with a thematic layer that is correspondingly mature in GIS. This organization is frequently used in the field of construction, road work, water systems, vegetation, and underground pipelines. The unified geographical foundation of all the layers is the public spatial coordinate system.

4. Object-oriented GeoStar

Object-oriented GeoStar is applied to manage 1/1,000,000 and 1/250,000 basic data in China. A demonstration project containing images, vector data, and digital elevation model (DEM) data was established in Guangdong, Shanxi, and other provinces in China with a 1/250,000 coverage areas of TM remote sensing images. The images covered 1/250,000 SPOT and were integrated with TM, 1/10,000 airborne digital orthoimages in Pearl River Delta, DEM, and vector data form a multi-scale, multi-sources database. The system can automatically employ different levels of data according to the scale the user has chosen. Object-oriented technology also can be used to construct 3D spatial

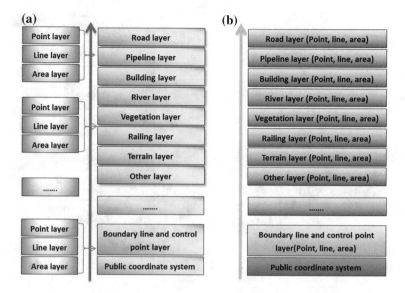

Fig. 3.5   Spatial data organization: **a** thematic layers and **b** object-oriented thematic layers

data models. In many cases of 3D spatial phenomena, using vector data alone cannot solve all the problems, such as the problem of non-uniform density within the entity or the irregular problem of a 3D surface, nor will using only a grid approach because the measurement and representation are of high accuracy. However, complete vector or grid data models are obtained by putting the above two methods into the three-dimensional data model construction.

3.5   Spatial Databases

When a spatial database is properly managed and is equipped with the appropriate tools, it performs organization, management, and input/output tasks (Ester et al. 2000). Different from the common transactional data which only have a few fixed data models with simple data conversion, spatial data are distinct in the understandings, definitions, representation, and storage needs of spatial phenomenon. As a result, spatial data sharing is very complicated.

3.5.1   Surveying and Mapping Database

A surveying and mapping database is the system responsible for collecting, managing, processing, and using digital surveying and mapping data (Fig. 3.6).

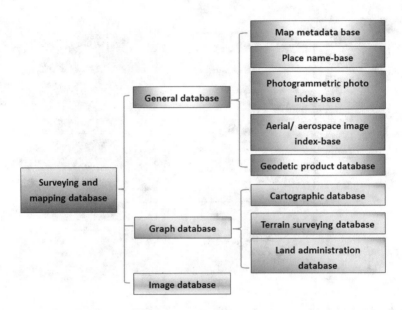

Fig. 3.6 Surveying and mapping database

In Fig. 3.6, it can be seen that a common database to store and retrieve observed data is very easy to construct. Databases are more popular now than ever before and are quite practical. There are about 2,000,000 place names in American geographical databases, and more than 10,000 names can be accessed by users in a few seconds. An image database is more complicated in terms of acquisition, storage, management, processing, and input–output. It contains digital maps or orthophotos observed in the field, photogrammetry, remote sensing, and image processing. To fulfill different purposes, graphic databases also can be divided into cartographic databases, terrain survey databases, and land administration databases. A cartographic database, which ranges from 1:10,000,000 to 1:1,000,000, contains the inventory data engaged in establishing thematic maps of various tasks for a wide range of services (national or regional) such as physical planning, economic planning, environmental sciences, ecological research, and national defense. Application software should contain the functions of projection transform, element extraction, and automatic synthesis. A terrain surveying database considers the national basic map (1:1,000,000 in China, 1:5,000 in Germany, and 1:24,000 in the United States) as the inventory data. It stores ground elevation models and digital topographic maps to serve the design work of the national economy departments, such as irrigation and water conservancy, irrigation-drainage networks, railways, highways, land-use in small range, economic planning, and environmental protection. The software should be qualified for the DEM production and output functions and content updating of the map to satisfy user needs. A land administration database is derived from the digitized map of a larger scale. An urban data system of 1:500~1:1,000 serves municipal management and town planning needs; in rural areas, the data system of 1:2,000~1:5,000 serves for land consolidation, agribusiness conglomerate, farmland planning, and irrigation works. As for cadastral management, data related to real estate and cadastral registration should be added. Also, in order to solve the problems of diverse and changeable data, the system should be interactive. The success of establishing image databases is due to the available real contents of various digital images and the development of digital sensors that are part of the measurement database and national spatial data infrastructure.

3.5.2   DEM Database with Hierarchy

A DEM (digital elevation model) database aims to organize related data efficiently and build a unified spatial index according to spatial distribution so that the user can find the data of any realm quickly (i.e., seamlessly roaming terrain). Usually, data display and computational efficiency are contradictory to each other in terms of the performance of large-scale databases. Due to the limitations of both computer display resolution and human eye sensory, the content capacity presentation is always restricted. In other words, a computer display remains close to zero as the computer reaches a certain complex level. At present, 2D GIS focuses on a great number of vector data, along with the management and operations of attribute data. However, establishing a DEM system with many details is very difficult compared to representing the 2D hierarchical relations of vector data. Fortunately, DEM helps not only to improve the function of 3D surface analysis, but also to add supplementary 3D representation data for the understanding of various spatial distributions. Vector data are identified by object identification without duplicate storage. However, a DEM of different levels belongs to different databases and the DEM of the lowest level belonging to the original data is the fundamental database of multi-scale, multi-resolution, and multi-source while the DEM located in the other layers is derived from the fundamental data. After the process of data fusion, data layers of the same scale are of one spatial resolution. To improve the efficiency of multi-scale data query, display, and analysis, a multi-scale DEM with hierarchy is of great importance in making it an alternative roaming operation and highly efficient interaction.

A hierarchical structure may enhance the browsing productivity of DEM databases. The spatial index of DEM data is constructed by hierarchical structure composed of "a project-workplace-layer-block-array." By virtue of this hierarchical structure, it is the only sure way to find the elevation within a DEM database. This spatial index opens access to the data and is seamless. Users can both view the whole and the part. They take advantage of the data from the sub-array of the same layer to build a mutually referable mechanism among the different layers. It becomes easy to automatically retrieve and transfer data of different levels according to the display range.

3.5.3 Image Pyramid

With the rapid development of remote sensing technology, the number of imagery data observed is growing geometrically. It is extremely urgent to establish a large spatial database to manage and integrate the multi-source, multi-scale, and multi-temporal images as a whole. However, due to the huge volume of images, it is necessary to distribute them in different segmentations. By watching the pointer during the process of indexing the recorded blocks, the data segmentations can be retrieved directly. Another problem is retrieving a more abstract image, which is transferred and extracted from the bottom layer, without any speed issues. As a solution, an image pyramid with hierarchical scales should be given so that data can be transferred in every layer.

The image pyramid stores and manages the remotely sensed images according to the resolution. In light of all the map scale levels, as well as the workplace and project division, the bottom layer is of the finest resolution and the largest scale. The top layer is of the coarsest resolution and the smallest scale—that is, the lower the layer, the larger the amount of data. As a result, an image database in the image pyramid structure makes it easy to organize, index, and browse the multi-source, multi-scale, and cross-resolution images (Fig. 3.7). When the image pyramid is used, there are three issues of concern:

1. The query is very likely associated with multiple maps and themes; for example, a road may exist in several maps.
2. Challenges are posed to maintaining the consistency and completeness of spatial objects.
3. Individual management threatens the exchange of data and the preliminary safety of spatial objects.

As a resolution, researchers of GIS companies (e.g., MG of Intergraph, ArcInfo of ESRI) have been engaged in establishing a seamless geographic database to complete the snap of the geometric edge or logical edge. As the physical storage is based on the frame of the maps, the correlations of all geometric identifications on the same spatial objects are processed invisibly for the users, which is called "logical seamless organization." To realize seamless organization of geographical data both logically and physically in terms of large geographical database, Li et al. (1998) proposed a better image display in different maps of various scales. The main organization form of GIS databases is the basic working unit established on the average map size; if the workplace identification is affected by image division, it is considered as the image workplace.

For the sake of saving time in storage and image compression (or decompression), the workplace can be reorganized when it is put in storage. As for the image database, a workspace file index is established so that the system can support the internal effective image files. The system can name the workspace files with sheet numbers. Within a row of spatial area of certain scale, the image database established on the basis of the unit of at least a piece of a digital image of the same plotting scale placed in the image workplace is called an *image project*.

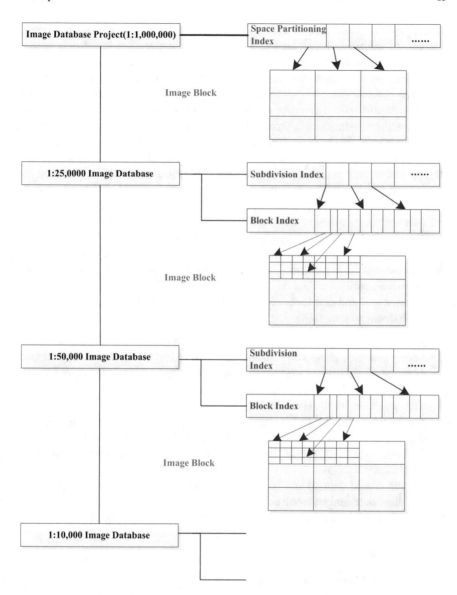

Fig. 3.7 Image pyramid with hierarchical scales

To accomplish fast retrieval and historical data archive management, several image projects of the same plotting scale, called *image sub-projects*, can be established. As an image sub-project is independent in the spatial index data, it becomes more convenient in the management of the data and the data product distribution. Contributing to the application of vector data and DEM data, this method stimulates the development and sharing of new image products and standardizes the database system application.

Suppose that we define an average size sheet as the workplace and take multi-source and multi-resolution remote sensing image data as the target data objects (mainly orthographic images) so that the image pyramid and multi-hierarchical file system can be combined to control image geometrical space and tonal space. At the same time, a seamless spatial database system can be established by stitching because this spatial database is cross-scale, cross-projection zone, cross-image project, and cross-image workspace. Then, it becomes easy and quick in terms of spatial indexing, dispatching, seamless browsing, distribution of image products, and sharing national imagery spatial data. By virtue of distributed management objects technology, which regards an image sub-project as an object, not only can the system safety be improved but the video distribution and data access can be fully controlled. Under this system, the maintenance and management of topological data are not confined to one server or client. It can both transfer data files into an image (sub-) project registry that records the topological data from the distributed image databases' file systems dynamically and transform the topological data in the servers of local networks. Network operation by synchronization replication establishes a seamless spatial database system for real-time query and roaming.

To better use and distribute geospatial data, a spatial database should be seamless with distinct features:

1. Retrieval, roaming, and inquiries should be conducted in the entire database.
2. Vector graphics should be translucently covered in raster images and capable of updating traditional graphic data with new images, including amendments and supplements, as well as deleting it.
3. The adaptive, multi-level resolution display based on the image pyramid should be realizable; that is, reading images from the corresponding layer on the image pyramid according to the chosen display scales automatically.
4. Its basic functions should include 3D display, query, roaming, analysis of images, and a DEM.
5. All the spatial data should be transferable to other data systems supported by GIS hardware and software according to the needs of users.
6. Visual products can be output according to map sheets or the scopes given by the user.

3.6 Spatial Data Warehouse

A spatial data warehouse is a subject-oriented, integrated, time-varying, and relatively stable collection of spatial data (Inmon 2005; Aji et al. 2013). As a kind of data warehouse, it integrates meta-data, thematic data, data engines, and data tools (extraction, transformation, load, storage, analysis, index, mining, etc.). When characterizing spatial data warehouses from spatial databases, spatial dimension and temporal dimension are basic. A spatial data warehouse is a tool to effectively manage spatial data and distribute them to the public to enhance spatial data-referenced decision-making.

3.6.1 Data Warehouse

The essence of a data warehouse is to assemble the data of daily business according to certain design principles, change them at any time as needed when refining to data products, and finally store them in the database. The basic idea of a data warehouse is to pick up the required data for applications from multiple operational databases and transform them into a uniform format; with the use of a multidimensional classification mechanism, the large amount of operational data and historical data can be organized so as to speed up the inquiry. Usually, the data stored in the operating databases are too detailed for decision-making, so generally a data warehouse will characterize and aggregate them before storage. In fact, a data warehouse contains a large amount of data from a variety of data sources, each of which is in charge of their respective databases. In general, transactional databases serve the online transactional process (OLTP), with standard consistency and recoverability while dealing with rather small transactions and a small amount of data. In addition, they are mainly used for the management of the current version of the data. On the other hand, a data warehouse stores historical, comprehensive, and stable data sets for online analysis and processing (OLAP) (Codd 1995). At present, there are two ways to store data in a data warehouse: relational database storage structure and multi-dimensional database storage structure. A star schema, snowflake schema, or their hybrids are often used to organize data in the relational database structure, whose products include Oracle, Sybase IQ, Redbrick, and DB2. In a multi-dimensional database storage structure, the data cube is employed; these products include Pilot, Essbase, and Gentia.

3.6.2 Spatial Data Cubes

A spatial data cube organizes the multi-dimensional data from different fields (i.e., geospatial data and other thematic data from a variety of spheres) into an accessible data cube or hypercube according to dimensions, using three or more dimensions to describe an object. These dimensions are perpendicular to each other because the analysis result the user needs is just in the cross-points (Inmon 1996). A data cube is a kind of multi-dimensional data structure with a hierarchy of multi-dimensions to identify and gather data representing the requirements of making a multi-faceted, multi-angle analysis for processing data in management and decision-making. For example, the time dimension can express a hierarchy of current, daily, weekly, monthly, quarterly, biannually, and annually.

In statistical analysis, a variety of complex statistical charts are often used, the units of which sort sub-totals, monthly sub-totals, annual totals, etc. This example is essentially a two-dimensional representation of a multi-dimensional data cube. Long before the appearance and application of computers, this way of thinking and method of calculation prevailed in China. This concept of data analysis is

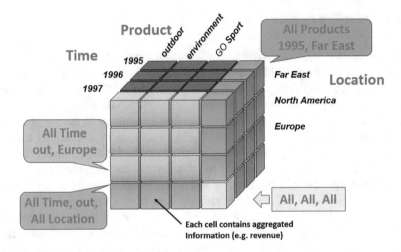

Fig. 3.8   Data cube

worth applying to today's data warehouse technology. In addition, the Microsoft Excel spreadsheet software also has a pivot table function, which is based on a similar idea of design.

The materialization of a data cube refers to the prior implementation of certain calculations, the results of which are stored and used directly in data analysis in order to improve efficiency. Figure 3.8 is a schematic diagram of a data cube, divided into the three dimensions of product, time, and sales location to describe a manufacturer's sales and income. Each dimension includes the hierarchical discrete values of its corresponding attributes and each cell of the cube stores a total value, such as the income in this case. At the end of each dimension, there is also a unit of the total of its own. Of course, in practical use there are many dimensions and the question of how to calculate efficiently to achieve a data cube; that is, the focus is the materialization of data cube (Wang 2002).

The idea of a spatial data cube came from three different fields—data warehouses, visualization, and GIS—but it is also relevant to spatial data warehouse, OLAP analysis, and map. The extension of the data warehouse is the spatial data warehouse. Visualization is a process of mapping from computerized representation to perceptual representation, from the choice of encoding technology to maximum understanding and communication between people. Its basic goal is to describe and provide insight into the data space through means of graphs, images, tables, etc., and its function is to produce visual graphics and images from complex multi-dimensional data.

A map is the graphic representation of spatial relations and forms. Cartography is an art, a science, and the technology of map-making. The function of GIS is to collect, manage, analyze and distribute data.

The spatial data cube is the core of OLAP. If the relevant data are selected and collected from different operational databases and stored in the cube beforehand, they can be directly observed and analyzed when making decisions, from different perspectives and to different extents, which can greatly enhance the efficiency of data analysis. The typical operations of a data cube are drill down, roll up, slice and dice, and pivot. All of these operations are called OLAP methods. OLAP is the main data processing and analysis technology of data warehouses, whose main functions are to analyze multi-dimensional data and generate reports. It generalizes data and can be seen as a simple data mining technique. However, data mining techniques such as induction, relevance, classification, and trend analysis are more powerful analysis tools than OLAP because the data in OLAP are aggregated through simple statistics while those in data mining require more complex techniques (Srivastava and Cheng 1999).

3.6.3   Spatial Data Warehouse for Data Mining

A spatial data warehouse reorganizes heterogeneous data under thematic subjects (Li et al. 2006). Spatial data are refined and stable and are updated over a period of time. Due to the complexity of the structure, relationship, and computation in spatial data, it is more complicated and difficult to study and establish spatial data warehouses. The development of spatial data warehouses lags behind that of business data warehouses, such as IBM DB2 OLAP Server, Oracle Express Server, Microsoft OLAP Services, Informix's MetaCube, Sybase IQ. At present, for spatial data that are not very large in size, a spatial data warehouse can be developed on an existing GIS platform and is considered a GIS tool. As for large-scale diverse spatial data, a spatial data warehouse is developed in a spatial database engine by using a common relational database management system to manage spatial data, with the mature technologies in data security and data consistency and ease of maintenance, such as ESRI's Spatial Database Engine and MapInfo's Spatialware. At present, there is no uniform standard on spatial data warehouses.

SDM may be implemented on spatial data warehouses or embedded as one of the data tools. OLAP is a basis of SDM in spatial data warehouses, and SDM is the deepening of spatial OLAP. SDM's efficiency is improved when it is combined with spatial OLAP. In fact, it is more efficient to uncover knowledge in data warehouses than in original databases because the data in spatial data warehouses—after a process of selection, cleaning, and integration—provide a good foundation for SDM. When extracting patterns from data warehouses, data mining is often driven by validation and discovery. Validation-driven data mining is supervised by a user hypothesis and is used to verify or deny the assumption through the use of the tools of recursive query at a relatively low level; discovery-driven data mining automatically finds out unknown but useful patterns from a large amount of data. In principle, SDM methods can be used in spatial databases as well as spatial data warehouses, despite some differences in implementation details and efficiency. As this book focuses on the theories and methods of SDM, spatial databases are generally the source for the purposes here. In spatial data warehouses, these theories and methods can be applied directly or even after minor improvement.

3.7 National Spatial Data Infrastructure

The National Spatial Data Infrastructure (NSDI) is the in-depth development of the National Information Infrastructure (NII) in the field of geomatics and is the basic data platform for SDM. The NSDI is being established for social and economic development and to coordinate the acquisition and use of geographic data. It is one of the basic ways to solve the serious problems in sustainable development related to population, resources, environment, and disasters, while guaranteeing the people's health and ensuring their safety and happiness. It may assure national security, social stability, and human health and benefit both the public and the private departments of a nation, such as public transportation, precision agriculture, farmland consolidation, urban planning, emergency relief, environmental protection, energy supply, forestry management, hydrological infrastructures, disaster reduction and prediction, and socioeconomic statistics. NSDI includes the system and structure of the coordination and distribution of spatial data, spatial data clearinghouses, spatial data exchange standards, and digital global data frameworks.

3.7.1 American National Spatial Data Infrastructure

In the United States, fully sharing high-quality geographic data is a bottleneck. Geographic data acquisition is a business worth billions of dollars. In many cases, the data have been repeated; for a given geographical area, such as a state or a city, there may be a number of organizations and individuals who are acquiring the same data. In theory, remote network communication technology allows data sharing, but it is very difficult to share data because a goal of making it not so easy to transfer the data of certain application into those of others has been set, which makes technical instructions unsuitable for common work. It is possible that the data collected locally are the best, but they may not be applicable to the planners of state and federal governments. Second, the common entry of data is another problem worth considering, as in the marine areas where there are little relevant digital products and the locations of the valuable geographic data occasionally provided are not easy to find for the public. Third, digitalization may be incomplete or incompatible, but it is impossible for the users to know the reason due to the lack of information about many data groups. The database is in want of the data and metadata of "who, what, when, where, why, and how (4W1H)" inhibiting the ability to find and apply data, and making data-sharing among agencies

more difficult. Thus, better entry standards and completely compatible digitalization will improve the well-being of our society.

The solution to these problems is the establishment of a NSDI. The Federal
Geographical Data Committee (FGDC) of the United States proposed a strategy
for NSDI in April 1997. In Executive Order 12906, former President William
Clinton recognized the urgent need for the country to find methods to build and
share geographic data. The document called for the establishment of a coordinated
NSDI to support the application of geographic data. Hence, NSDI is seen as a part
of the developing NII, in order to provide citizens access to important government
data and strengthen the democratic process.

To develop NSDI, federal agencies and various nongovernmental organizations
undertook major activities to establish the National Geospatial Data Clearinghouse
(NGDC) between data producers and users, as well as standards on the collection
and exchange of spatial data. A national digital spatial data network was set up
by NGDC for data producers and cooperative groups, which included the important data classes effectual to a wide range of user types. A new relationship may
allow the institutions and individuals from all fields to work together and share
geospatial data. Their mission is to make it easy for spatial data to be utilized for
economic growth and improvement and protection of environmental quality at the
local, national, and global levels, as well as contribute to social progress. There are
four goals.

The first goal is to enhance the knowledge and understanding of NSDI's
wishes, concepts, and benefits through education and to highlight its position. The
tasks are: (1) to prove the benefits gained through participation in the existing and
expected NSDI; (2) to develop the principles and practical use of NSDI through
formal or informal education and training; and (3) to improve the state of development of NSDI and promote the helpful activities.

The second goal is to find common solutions for the discovery, storage, and
application of geospatial data to meet the needs of a wide range of groups. The
tasks are: (1) to continue to develop seamless national geospatial data interchange
stations; (2) to support the reform of the common methods used to describe geospatial data sets; (3) to support the study of the tools for making applications
easier to use and to exchange data and results; and (4) to research, develop, and
implement the structures and technologies of data supply.

The third goal is to resort to the approaches based on community to develop
and maintain the common collection of geospatial data by selecting the right factors. Its tasks are: (1) to continue to develop the national network of geospatial
data; (2) to provide citizens, government, and industry the required additional geospatial data; (3) to accelerate the development of general classification systems,
content standards, and data models, as well as other common models in order to
facilitate the improvement, sharing, and use of data; and (4) to provide the mechanism and stimulus that bring the multi-resolution data from many different organizations into the NSDI.

The fourth goal is to establish a relationship between organizations to support the continuous development of NSDI. The tasks include: (1) to establish a

procedure allowing the stakeholder groups to formulate the logic and complementation that support NSDI; (2) to establish a network and make contacts with the public who are interested in NSDI through conferences; (3) to eliminate the obstacles in management and administration to unify a format; (4) to search for new resources for data production, collection, and maintenance; (5) to recognize and support the political and legal organizations, technologies, individuals, schools and economics which promote the development of NSDI; and (6) to participate in the international geospatial data groups that develop global geospatial data systems.

1. FGDC and NSDI of the United States
   - FGDC is the leader in the coordination and development of national spatial data infrastructure. It was organized according to Circular No: A-16 of the United States Office of Management and Budget (OMB), whose responsibilities are to develop, maintain, and manage the national distributed database of geospatial data, to encourage the study and use of relevant standards and exchange formats, to promote the development of spatial data-related technologies and transfer them into production, to communicate with other federal coordination agencies working on processing geospatial data, and to publish relevant technical and management reports.
   - FGDC and its working groups provide the basic structure for every research institute and private communities, cooperating to discuss the various issues related to the installation of NSDI. The standard working groups of FGDC, who study and set standards through structural procedures, make great efforts to unify their standards as much as possible. They get support from experts through public channels but have nothing to do with some expertise. The reference model of FGDC standards set the principles to guide the standard projects of FGDC. It defines the expectations of FGDC standards, describes different types of geospatial standards, and explains the FGDC standard process.
   - FGDC's mission is to formulate geospatial data standards so as to achieve the data sharing between manufacturers and users, as well as to support NSDI. According to Executive Order 12906, NSDI stresses the collection, processing, storage, distribution, and improvement of the technologies, policies, standards and human resources needed for using geospatial data; while FGDC focuses on how to strengthen its coordinative role between NSDI and state and county governments, academic organizations and private groups. Through its committees and working groups, FGDC supports the activities of the NSDI in four ways: to develop a national spatial data exchange network, to formulate spatial data-sharing standards, to create a national geospatial data framework composed by basic thematic data, and to promote geospatial data protocol of cooperative investment and cost-sharing among partners outside federal organizations.
   The formulation and application of FGDC standards go through five stages: application, planning, drafting, public review, and final approval. These five stages ensure the generation of standards in an open or consistent way and that the nonfederal organizations' participation in the formulation should be as wide as possible and their standards should be in accordance with other

FGDC standards. At present, there are 4 geospatial data standards formally signed by FGDC: Spatial Data Transform Standards (SDTS), Digital Geospatial Metadata Content Standard, Cadastral Data Content Standards and Classification of Wetlands, and Deepwater Habitats of the United States. SDTS is a tool used for data transformation between different computer systems, ruling the data transfer protocol and addressing format, structure, and content, as applied both to vector data and to raster data. SDTS consists of the conceptual model, standards of the quality report and the conversion model, and the definition of spatial features and properties. The advantages of SDTS include the sharing of data and cost, flexibility, and high quality without any loss of data.

2. NGDC and DGDF
   - The National Geospatial Data Clearinghouse (NGDC) is the most economical way to find geospatial data, determine its applicability, and acquire and order it. The clearinghouse is an electronic network linking the producers, managers, and users of geospatial data, from which metadata can be obtained. As the rapid changes that are happening are a sign of our new data society, every data producer and defender can describe the acquirable data in electronic form with a variety of software tools and upload the description (metadata) to the internet, by which the sites of the clearinghouse are connected. Thus, one main requirement for the clearinghouse is that it can search geospatial data on the internet.
   - The National Digital Global Data Framework (DGDF) is a basic and common set of digital global data, whose goal is to effectively organize and strengthen activities in the geospatial data industry in order to meet all the requirements. DGDF should be a trustworthy and reliable data provider and the requirements of techniques and other kinds of tools for every unit should be minimal and stable, which can be complemented quickly and generated according to the needs and ability of users. The cost of data access should be as low as possible, aiming to provide data at a price lower than transmission. In DGDF, the data are a sound and complete data collection to represent the elements in the real world (not the graphic symbols). Users can integrate the data into practical applications and maintain their properties and other data without increasing investment. When designing DGDF, the needs of federal, state, and local government users must be taken into account, as well as those of private enterprises. DGDF should be extended in accordance with the requirements and ability of the contributing organizations, accepting a large number of the widely distributed geographical organizations, and making good arrangements for their tasks, objectives, resources, and plans.
   - DGDF will benefit society as a whole. For an enterprise, DGDF can be helpful in focusing on reducing the cost in data acquisition and integration, simplifying and speeding up the application development, and attracting users for other data products and services. The enterprise may benefit from the data collected by other enterprises being faster and easier. By improving the practicality of geospatial data, efforts should be made to spread the use of

the data in the framework to all organizations, not only a society or a group of customers. DGDF will take full advantage of the superiority of the geospatial data that are establishing by local and regional governments, public utilities, non-governmental organizations, and the offices of state and federal agencies. Most of these geospatial data are built to solve a specific problem and meet the needs of the local, so gathering, maintaining, and distributing the data will involve many organizations and departments. To make the data consistent and easy to integrate, many works should be done to collect, synthesize, and validate the local data. To achieve this goal, 6 institutional responsibilities have been identified: policy-making, project evaluation, framework management, regional integration, data producers, and distributors. These tasks can be assigned to a number of different organizations to undertake them; the organizations with policy, mission, and authority will be the most successful participants.

3. OGC

- The Open GIS Forum (OGF) was set up in the United States in 1992. It deals with the problems encountered by the governmental and industrial geodata users and their expenses. These users tend to share and distribute spatial data; however, due to technical or organizational reasons, they are prevented from this goal. OGF has changed its name to the Open GIS Consortium (OGC).

   OGC is a non-profit organization whose purpose is to promote the use of new technologies and professional methods to improve the interoperability of geographic data and then reduce the negative effects of the non-interoperability on industry, education, and arts. The membership of OGC includes strategic membership, principal membership, application integration membership, technical committee membership, and associate membership. They share a common goal to establish a national and global data infrastructure in order to facilitate people's free use of geographic data and the resources they deal with—even to open a new market for communities that have not yet become involved the geographic data processing, bringing a new business model for the whole community.

- To share data, the core of OGC is an open GIS data model. There are 6 kinds of geographic data: digital map, raster image data, point vector data, vector data, 3D data, and spatiotemporal serial data. Because of growing environmental problems, government and commercial organizations have increasingly high demand for effective operation, making the need to integrate geographic data of different sources more and more important. However, it is troublesome, trivial, full of errors, and even completely impossible to share geospatial data. Due to the extensiveness of the content and scope of geographic data, the formatting is more complicated than those of any other digital data. Moreover, it becomes much more complicated as software platform, methods of data acquisition, data representation and transformation, and users (individuals and organizations) vary. OGC's software specification is the Open Geodata Interoperability Specification (OGIS), which is mainly composed of an open geographic data model, service model, and the model of the information society. OGIS is a software model used to acquire geographic data and geographic data resources through comprehensive distribution. It offers a framework for software developers in the world to produce software, which allows users to obtain and process data through a common software interface on the basis of open data technology.

- Compared with traditional geographic data processing technology, OGIS establishes a common technical basis for processing open geographic data. Its features are interoperability, support for the data society, universality, credibility, and ease of use, portability, cooperation, scalability, extensibility, compatibility, and enforceability. It should be noted that the standard OGC is still in the concept stage. Whether or not the final standard will be able to meet these mentioned objectives remains to be seen.

4. USGS and its geographic data products

- The United States Geological Survey (USGS) is a federal agency in federal government under the U.S. Department of the Interior. In the geospatial data infrastructure of the United States, USGS systematically carries out the classification of public land and the investigation of mineral resources and geological structures. It provides geological, topographical, and hydrological data so as to reasonably manage the nation's natural resources. The data provided include maps, databases, description, and analysis of water resources, energy, mineral resources, land surface, geological structure, and dynamic processes of Earth. It further provides users with access to the database of the U.S. GeoData through the internet.

- USGS has been responsible for America's topographic survey and geological survey, creating simulation maps and digital products in many forms and of many places from the surface of Earth to the planets and the moon:

  - Simulation maps: topographic maps, county maps, state maps, national park maps, shaded topographic maps, contour maps, and a map of the Antarctic.
  - Image maps: orthophoto maps, orthophoto quads, border maps, and satellite photo maps.
  - Geological maps: oil and gas survey maps, geological square maps, multi-purpose geologic maps, multi-purpose maps for outdoor survey, coal maps, hydro-geological maps, geological maps of states, maps of the moon and planets, and a variety of atlases.
  - Digital geographic data products: four-dimensional products of USGS geographic information make up the data basis of NSDI. These digital products are Digital Orthophoto quad (DOQ), Digital Elevation Model (DEM), Digital Line Graphs (DLG) and Digital Raster Graphics (DRG). USGS also has Geographic Names Information System (GNIS) and Global Land Information System (GLIS).

3.7.2 Geospatial Data System of Great Britain Ordnance Survey

The Ordnance Survey (OS) is the national agency authorized by the Queen of England to take charge of surveying and mapping in England, Scotland, and Wales. The OS headquarters are located in Southampton, while the base of Northern Ireland's Ordnance Survey is in Belfast and the Republic of Ireland's OS is in Dublin. All three are completely independent but are using the same coordinate's origin.

Britain attaches great importance to the comprehensive application of geographic data. To make the most of the advantages of GIS and to promote its application, the Department of the Environment (DOE) conducted an extensive investigation in 1987 and published a Chorley Report "Handling of Geographic Data: Report of the Government Committee of Inquiry." Among government departments, there are more than 40 major geographic data producers and users, but a low level of data sharing exists and there is a lack of uniform national standards in all communities. Stimulated by the American NSDI, the British government proposed the National Spatial Data Framework (NSDF) in 1995 to organize and coordinate cooperation in all aspects and to use an acceptable way to share spatial data.

The NSDF encourages the collaboration of collecting, providing, and applying geospatial data; promotes the use of standards and the practice use of geospatial data in collection, provision, and application; and promotes access to geospatial data. It is a superset of NSDI, and they share some similarities. The difference is that NSDF does not demand that different databases share the same public data, although it is very difficult to exercise.

3.7.3 German Authoritative Topographic-Cartographic Information System

The German Interior Ministry Survey Office decided in 1989 to establish the Amtliches Topographisch-Kartographisches Information System (ATKIS). ATKIS is a national authoritative topographic-cartographic information system that is basically for spatial distribution-related information systems.

ATKIS consists of a number of Digitale Landschafts Modelle (DLM) and Digitale Kartographischen Modelle (DKM). DLM is a digital landscape model described by geometry properties, while DKM is a digital mapping model described by visualization. In ATKIS, DLM and DKM have clear specifications. DLM includes a catalogue of ground objects, ATKIS-OK, and a DLM data model, whereas DKM includes a catalogue of ground symbols, ATKIS-SK, and a DKM data model.

As Britain and Germany have rather small territories, the mapping and updating task had been performed well. The basic topographic maps at 1:5,000-scale had been completed; thus, they did not stress or formulate that digital orthophoto maps must be included in the NSDF. As the high-resolution orthophotos maps are important, the two countries also have made orthophoto maps at a 1:5,000 scale.

Just like the DRG products of the American USGS, Germany also has digital raster data of topographic maps, which are shared by each state survey office. In addition, it has a total of 91 topographic maps at a 1:50,000-scale (TK50). The digital raster data of maps at a 1:25,000-scale (TK25) also was completed later in 1991. The raster data are mainly used as the background data of thematic maps to build the database systems for the environment, transportation, disaster prevention, and nature protection. Combined with DEMs and remote sensing images, it can generate 3D landscape maps as well as provide statistical data.

3.7.4 Canadian National Topographic Data Base (NTDB)

In 1963, R. Tomlinson developed the first GIS in the world, the "Canada Geographic Data System (CGIS)." CGIS is a thematic GIS in practice that has brought Canada a good reputation in the GIS field and extends the development of GIS into Canadian private companies. After the 1990s, Canada's spatial information technology clearly pursued the integration of RS, GPS, and GIS, whose scope of application is expanding.

Canada is entering the era of digital products and data sharing, including the emergence and standardization of digital data and the dissemination and provision of data. Canada also has a wide range of GIS data products, including street network files, digital borderline files, post data exchange files, and a variety of property files including census documents. Due to the complexity of the data, among most Canadian digitalized maps, it is still often difficult to accept the data for a certain project because the 1:50,000-scale index image of the spatial database in NTDB of EMR shows that the data are still not conforming, not uniform, and incomplete—far from the standards of the 4D products by USGS. For DEMs, Canada only has the national contours at a 1:250,000–scale recorded by discrete points, rather than the raster DEM data.

The lack of an acceptable standard is the main reason why Canada has no unified digital spatial data. Thus, Canada is working to resolve this issue and hopes to achieve the transformation between different GIS data through a common data transformation standard. The Inter Agency Committee of Geomatics (IACG) provides national initial data products and being widely representative, works on the development of national geographic standards. The Gulf Intracoastal Canal Association (GICA) is a national business organization whose members include advanced departments in GIS, remote sensing, and mapping. The main Canadian departments who are engaged in GIS are the Canadian Institute of Surveying and Mapping (CISM), the Canadian Committee of Land Survey (CCLS), the Association of Canadian Land Surveyors (ACLS), the Canadian Hydrographic Association (CHA), the Canadian Cartographic Association (CCA), the Canadian

Remote Sensing Society (CRSS), the Urban and Regional Data System Association (URISA) in Canada, the Urban Data Systems Association of Ontario, the Association of Geography in Municipal Quebec (AGMQ), and the Geographic Society in the province of Nova Scotia.

3.7.5 Australian Land and Geographic Information System

Australia began working on a digitalized map in the early 1970s. The basic surveying, mapping, and land management of Western Australia is in the charge of the Department of Land Administration (DOLA), who started the digitalization of topographic map and the establishment of Land Information System (LIS). In order to let the whole society share the state's geospatial data, a number of departments are involved in the establishment of the Western Australia Land Information System (WALIS). The digitalized product is known as the map data, which can be used in GIS after transforming structure. Later, the reproduction of a GIS data model was used, known as the Geodata. Then, with the development of the NSDI policy in the United States, Australia also adopted the same method to create a spatial data framework. There are many kinds of AUSLIG digital maps, the production of which has the following characteristics: compatibility for GIS, national unity, guarantee for quality, comprehensive documents, and regular maintenance.

The Australian government has focused on the application of GIS in support of the land bureau and its land management operations. To coordinate the land-related activities between the national and regional levels, the Australian Land Data Council (ALIC) was established to deal with land issues and to deliberate related policies at the national level, support the formulation and use of the national guidelines and standards in land data management, provide a forum at the national level where experience and data of land data management policies can be exchanged, and publish the annual report of the development of the Australian Land Information System. ALIC is committed to building a national land data management strategy to encourage all communities of the Australian economy to effectively obtain land data in order to provide a solid basis for all levels of government and private organizations to make effective decisions when using land, as well as to develop and provide an effective mechanism for data transformation. As a result of the wide use of GIS and the rapid development of web technologies, the land data system announced by ALIC is actually a service system for a spatial data warehouse and data distribution center.

3.7.6 Japanese Geographic Information System

The construction of Japan's NSDI was in response to the earthquake in Kobe in 1995. The earthquake awakened the government to its need for emergency management services and access to the related data. This need stimulated the establishment of a joint ministerial committee on GIS under the supervision of the cabinet, whose members are the 21 representatives of the Japanese government agencies, including the Ministry of International Trade and Industry. The Office of the Cabinet Secretariat was responsible for the committee, with assistance from the National Mapping Bureau and National Land Bureau. In 1996, the joint committee published its implementation plan through the beginning of the 21st century. The first phase of the plan, which ended in 1999, included the standardization of metadata and clarification of the roles of all levels of government and private companies to promote the building of NSDI, for which a federation for the advancement of NSDI was established to support these activities and whose members include more than 80 private companies.

3.7.7 Asia-Pacific Spatial Data Infrastructure

The Asia-Pacific Spatial Data Infrastructure (APSDI) project of the Permanent Committee on GIS Infrastructure for Asia and the Pacific (PCGIAP) is a database network. The databases are distributed in the whole region but work together to provide the needed basic data for the development of the region's economic, social, and human resources needs and the realization of environmental goals. PCGIAP spawned the idea of a database network distributed in various regions, in which the databases connect and are compatible with each other. Each database will have the technology and technical driving force required by all countries to maintain a database and ensure they abide by the principles of the regulatory agencies. APSDI will provide the organizations, technology, and system to ensure that they meet the level of unity, content, and coverage that the region needs. The infrastructure also will guarantee that each country's efforts will be intensified and coordinated to maximize their investment in data collection and maintenance from the perspective of the region. In the end, it will be an infrastructure to help the region achieve more through better support for economic, social, and environmental policy-making.

The overriding objective of APSDI is to ensure that users have access to the complete and unified spatial data they need, even if the data are collected and maintained by different countries. As a result, the problem lies in how the countries and their data can meet the needs of the whole region. To solve this problem, PCGIAP developed a model for spatial data infrastructure consisting of four parts: the institutional system, technical standards, basic data, and network interface. The institutional system sets rules for and manages the establishment, maintenance, access, and application of standards and data. The technical standards mold the technical features of the basic data to make a successful integration between them and other environmental, social, and economic data. As for the basic data, they are produced in the institutional system and are in full compliance with technical standards. The network interface is also in accordance with the policies and technical standards set by the institutional system.

APSDI will be based on the NSDI of all the countries in the region, but it will be closely connected to other international projects, such as Agenda 21, the global map, and global spatial data infrastructure. The establishment of this infrastructure and its standards and data management policies will help maximize the whole region's investment returns on spatial data and build a viable geographic data industry. The establishment of APSDI is a huge project, but PCGIAP believes that the project has enormous potential benefits and all efforts possible must be made to complete it. The determination, goodwill, and cooperation shown by PCGIAP will ensure that this goal is achieved; over time, a detailed implementation plan will continue to develop and mature and be adaptable to the ever-changing technological environment and the technical and administrative obstacles will be overcome. Due to the increasing awareness of the importance of spatial data resources, investment in data collection, management, and development will increase as well. A vibrant spatial data industry thus will appear, serving the government, industry, and society. The sharing, synergy, and benefits of knowledge and experience brought by the APSDI project will have a much greater effect than that of an individual country in the Asia-Pacific region.

3.7.8 European Spatial Data Infrastructure

European Spatial Data Infrastructure (ESDI) integrates the diversity of spatial data available for a multitude of European organizations. The European Union (EU) project Humboldt contributes to its implementation. Its vision is that geographic information (GI) with all its aspects should become a fully integrated component of the European knowledge-based society (i.e., EUROGI). The EUROGI/eSDI-Net initiative offers a sustainable network as a platform to exchange experiences in the field of spatial data infrastructure (SDI) and GI as well as international contacts to SDIs, GI and SDI key players, GI associations, and other related networks and projects.

Although many people think that the coordination of GIS in Europe and even around the globe is a good thing, in reality, political, historical, and cultural factors affect the design and implementation of this project. Because the level of economic development and training facilities is imbalanced, the skill level and the awareness of geospatial information among the European countries are also different, partly due to the different status of geography and other disciplines play in schools and universities and the technical differences between the countries, such as different applications for coordinates and benchmarks.

The greatest obstacle to implementing SDI in Europe and the whole world arises from a large number of local issues, rather than the opposition to it. The resistance stems from the oppression of the other organizations working on geographic information from the government, which restricts their activities from overseas; the lack of working capital for oversea activities; the lack of interest and awareness; the impact from politics and other factors; the limitations of the links between subjects; the lack of clear leadership from the EU-DG (Directorate-General) and other EU institutions; and the doubts and difficulties in copyright and other legal matters.

OGC came up with the idea of OGIS to address the technical problems of the data exchange between different spatial databases. In a very short period of time, OGC has convinced geographic information software vendors and others that there will be better business prospects for the geographic information industry if spatial data can be shared and exchanged without the restrictions of professional standards. The OGIS of OGC aims to provide a comprehensive set of open interface specifications that will allow software developers to write interoperable components and provide clear access to geographic data and geographic resources in the network. What is more worthy of attention is that the business members of OGC in North America are very active in Europe, such as ESRI, ERDAS, Intergraph, MapInfo, Autodesk, Genasys, PCI, Trimble, Oracle, Informix, Microsoft, Digital, Hewlett Packard, Silicon Graphics, Sun, and IBM. These companies provide the vast majority of the required geographic information software, hardware, and database technology for European countries, and there are several thousands of European employers who obviously have decision rights in their companies.

Before OGIS, the interoperable method in the ESDI was considered a top-down approach. The national geographic information organizations in EUROGI have indirect links with the EU Council and the European Commission so that when the senior leadership has approved certain treaty on ESDI, the relevant standards, orders, and agreements will be handed to the national geographic information organization through EUROGI, who will distribute them to its government and business members. However, in essence, the OGC model is a bottom-up approach based on commercial interests. Because Europeans buy most of their geographic information products from vendors in the United States, any interoperation standards and procedures passed by the American business community are most likely to become the Vivendi in Europe despite their own hopes. It also shows that the international market has far-reaching influences on the success and the goals of the infrastructure described by GI2000.

All of these show that European standards and interoperability specifications cannot be divorced from those of the world and they are especially dependent on the geographic information business in the United States. If the market forces can push European users to accept the interoperability specifications, it is still not necessary for the EU to invest a lot on their geographic information standards; instead, the funds will go to GI2000. The participation of the Europeans is needed in the OGIS, and they will influence the discussion on interoperation through EUROGI, the European business community, and the various activities of the EU. In view of the interests of European businesses and geographic information, the European people should be encouraged to become involved in the OGIS.

3.8 China's National Spatial Data Infrastructure

China's National Spatial Data Infrastructure (CNSDI) is included in China's National Information Infrastructure, as one of the key projects of the national plan. It is comprehensively planned and is gradually being implemented under the leadership of the National Administration of Surveying, Mapping, and Geoinformation (NASMG), referring to the Surveying and Mapping Law of the People's Republic of China, current reality, and experience around the world (Li 1999; Chen 1999). CNSDI provides geospatial data services to the whole country.

3.8.1 CNSDI Necessity and Possibility

CNSDI is necessary. In the process of building GIS, the collection of geospatial data, the manpower, and the costs to create databases were considerable. To avoid duplication of data collection, the state will set up NSDI all over the country. It is a fundamental project that will not only meet the social, economic, and sustainable development requirements in such areas as population census, resource utilization, environmental protection, and disaster prevention, which are conducted by the central government and the provinces (cities), but also to meet the demands of the market. NSDI will assure the authority, consistency, and safety of China's geospatial data and avoid the re-establishment of data projects during the process of Digital Earth in China. China's industrialization lags behind the developed countries; however, with reform policies opening up to the world, China is stable and unified. Under these circumstances, China's informatization needs to be improved as soon as possible in order to keep up with global development.

At present, China is qualified to build NSDI. The State Council approved NASMG to oversee the work of national surveying and mapping and the standardization of technology and industry. The National Fundamental Geographic Information System (NFGIS) database will be designed to provide the state's other professional information systems with a basic geospatial data framework. China started the digital transformation of traditional surveying and mapping technology in the early 1980s, achieving a number of modern techniques such as GPS, RS, GIS, and their integration. China is a leader in digital photogrammetry and digital orthophoto production technology around the world.

3.8.2 CNSDI Contents

CNSDI includes four main aspects—the coordination organization, the NFGIS database, the National Geospatial Data Switched Network (CNGDSN), and the geospatial data standardization. It will be constructed gradually from an initial experimental stage to the complementary stage (Fig. 3.9).

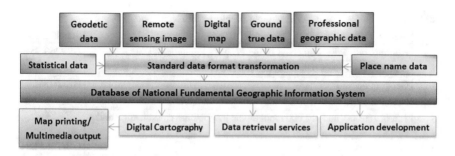

Fig. 3.9 Technical structure of NSDI

Coordination organization, including high-level macroscopic organization and project implementation organization, are gradually becoming responsible for CNSDI construction. China's National Geographic Data Committee (CNGDC) is responsible for coordinating and organizing the establishment and revision of geographic data technology. It is a leading organization for coordinating the extraction and usage of China's geographic data. CNGDC is set up under the leadership of the State Economic Informatization Joint Meeting of State Council. It consists of the leaders and experts in more than 30 national departments that are engaged in the production and application of geospatial data.

NFGIS consists of basic databases (coordinate database), thematic databases (topographic, place names, geodetic, gravimetric, images, raster graphic, DEM, etc.), and new databases from the development of technology. The project is to build 1:250,000 topographic databases and place-name databases, establishing the Land Resources Data System of Hainan Province, and carrying out the design and trial of 1:50,000 terrain databases and place-name databases. The first objective is to finish the topographic database, place-name database, digital raster graphic database, and DEM (with an accuracy of 1, 5, and 20 m) with a vector form of 1:1000,000, 1:250,000, and 1:50,000 on a national scale. The second objective is to build 1:10,000 digital orthoimage databases in economically developed eastern China and areas along the river and the railway line in order to extract various thematic data. After completion, the databases will be updated annually. The third objective is to build large-scale spatial databases with 1:500 to 1:2,000 and 1:2,000 orthoimage databases in major Chinese cities and special economic zones, which would be updated once or twice every year.

CNGDSN is based on the National Public Data Network to realize the electronic interconnection of the central and relevant ministers with the unified standard data files. It was formed within the internal network and external network of NFGIS.

Geospatial data standards should be established under the guidance of CNGDC. Their mission statement includes setting the CNSDI-related standards for various data and their compression, transmission, and exchange; as well as guiding the establishment of a variety of professional GIS data standards in order

Table 3.2 China's standards for geographical information techniques

No	Standard name	Standard code
1	Administrative division codes of the People's Republic of China	GB2260-95
2	National Geomatics data classification and code	GB/T13923-92
3	Geographic grid	GB/24090-90
4	National basic scale topographic maps and numbering	GB13984.92
5	1:500, 1:1,000, and 1:2,000 topographic map feature classification and codes	
6	World regions name code	GB/T2654.94
7	Highway route naming number and coding rules	GB9171-89
8	Highway classification code	GB914.94
9	Administrative divisions below county code compiled reason	GB10114-94
10	People's Republic of China railway stations code	GB10302-94
11	1:250,000 topographic map compilation of specifications and drawings	CHIV-302-85

to make them in accordance with national geospatial data. NASMG is a member of ISO/TC211 and has undertaken China's domestic work related to ISO/TC211, which will accelerate the connection of the geospatial data standards of China with the world. Based on the actual needs and the experimented results, China has issued ten national standards and one industry standard for geographic information technology (Table 3.2). GIS standards include a data dictionary for fundamental geographical databases; digital surveying and mapping product models; digital surveying and mapping product quality standards; spatial data exchange format standards, basic geographic data terms; name codes of country rivers, lakes, and reservoirs; GPS and GIS data interface standards; basic geographic data updated regulations; guidelines for the standardization of urban GIS; provisions for communication security and confidentiality of basic geographic data; basic geographic data description standards; and basic standards for geographic data network technology.

The completed major experimental projects include the construction of the Demonstration Bases for Digital Surveying and Mapping Production, the experiments of the National and Regional NFGIS databases, GIS applications, and the development of the Digital Surveying and Mapping Technology Standard. The successful model and building experience of the Demonstration Bases for Digital Surveying and Mapping Production will continue to the other provincial departments and promote the technology of basic surveying and mapping production.

3.8.3 CNGDF of CNSDI

China's National Geospatial Data Framework (CNGDF) was established with the combination of point, line, and area and multi-resolution due to the vast size of Chinese territory, different development degrees between the east and the west, different natural and social conditions in different regions, and the national power restrictions. Producing a geospatial data framework is more efficient than producing total factor map vector quantization because the storage capacity and computing speed are able to handle a large capacity of geospatial data.

First of all, CNGDF should follow the principles of speediness and efficiency and also take into account the integrity, importance, machinability, and expandability of the data. At present, it is time-consuming to establish a database with all the elements, which cannot meet the demands of the rapid updating speed of basic geographic data. As different devices can use different techniques to meet the demands of product quality, a variety of data acquisition programs are presented below to generate basic geographical data—digital elevation model (DEM), digital orthophoto model (DOM), digital line graphs (DLG).

1. *DEM*. Fully digital automatic photogrammetry, interactive digital photogrammetry, analytic photogrammetry, scanning vector contour, and DEM interpolation.
2. *DOM*. Digital photogrammetry, single chip digital differential correct, Orthophoto map scanning, digital aviation orthophoto, and remote sensing image processing (TM images, SPOT images, control point images).
3. *DLG*. 3D digital photogrammetry, computer-aided digital mapping, map scanning vectorization, and artificial semi-automatic tracking feature elements on DOM. For example, road, contour, settlement, water system, and administrative boundary.

DEM, DOM, and DLG constitute a geospatial data framework with a scale from 1:10,000 to 1:150,000. The database can display the images in different image resolutions and scales. When the scale enlarges, the details of the ground can still be seen clearly. The database can also use dynamic window, zoom, and roam in any direction. For example, 1:50,000 spatial data infrastructure of the National Geographic Information System includes seven databases: DEM, DOM, DLG, digital raster graphic (DRG), place name, land cover, and meta-data.

3.8.4   CSDTS of CNSDI

China's Spatial Data Transform Standards (CSDTS) is an essential issue to sharing data in CNSDI. Because a variety of professional GISs become a management tool for governmental departments and enterprises, spatial data are not disposed of or used by one small community alone; rather, they are to be shared with others. Each professional GIS has its own internal data formats and data storage methods. For example, a city's planning bureau has several departments, each of which purchases different software according to their own preferences and needs. GeoStar and MGE are used for surveying, ARC/INFO is used for a pipeline network, and MAP/INFO is used for district planning and building construction. If a worker in district planning does not exchange or manage spatial data in a database management system every minute, the data he or she uses may be a few days old or a few months old. Currently, GIS software cannot directly manipulate the data of other GIS software, and it is necessary to go through a data exchange. According to statistics, data exchange costs a great deal of manpower and capital resources. The cost of GIS spatial data exchange in developed countries has reached 30 %. Thus, it is important to set spatial data standards in order to produce spatial data compliant with the standards and allow other industry communities to share data and avoid duplicate collection of basic spatial data as well. The exchange standards address the conversion issues of spatial data among different GIS software.

There are currently three data exchange methods for sharing data: external data exchange, data interoperability specification, and data sharing platform.

An external data exchange usually is defined in ASCII code files in order to convert the data into other software because many GIS software do not allow the users to read and write internal data directly. However, external data exchanges are defined by software vendors, and the contents and representation methods are not the same, such as AutoCAD DXF, MGE ASC Loader format, and ARC/INFO EOO format. It is difficult to update spatial data in a timely manner and maintain data consistency through an external data exchange. For the sake of standardization and consistency, many countries and industries set their own external data exchange standard to call for using public data exchange formats in a country or a department, such as STDS in the United States.

Data interoperability specification consists of drawing up a set of spatial data with manipulation function API, which can be accepted by all parties. All of the software vendors follow this standard to provide driver software in accordance with the API function. In this way, different software is able to manipulate each other's data. At present, the work has progressed smoothly as Open Geodata Interoperability Specification (OGIS) in several GIS companies in the United States. For example, Geo-Media, which was launched by Intergraph Corporation, can call ARC/INFO directly. This method is more convenient than the external data exchange. However, the data provided by the defined API function might be the smallest for the overall situation due to the spatial data stored by different GIS software. The data also may be inconsistent and not new data because each of the software programs mainly manages its own system.

The data sharing platform makes it possible for all spatial data and each application software module of a community to share a platform; all of the data exist on the server and all of the software are the programs of one client. Data can be accessed from the server through this platform. Any application's updated data are reflected in the database in a timely manner so as to avoid inconsistency of the data. However, this approach is more difficult to achieve. Right now the owners of a number of GIS software programs are unwilling to discard the software's base to adopt a public platform. Only when the base server of the software proves to be absolutely superior to other systems and has shown that it can manage a large number of basic spatial data will the data sharing platform be possible. A spatial data sharing platform also can adopt general data management software, such as Oracle SDO.

The principles of CSDTS may include the completeness, simplicity, scalability, and expandability between internal data and external data. Completeness means that spatial data in CSDTS are as plentiful as possible and also cannot be lost because of data exchange. Simplicity enables users to read and understand CSDTS easily, which has made CSDTS popular. Scalability means that spatial data are heterogeneous from a variety of sources. Expandability means that CSDTS will be open in order to append data as people further understand the real world in the future. The development of CSDTS following the principles may create a simple and practical plan as soon as possible. For example, define spatial data models first, then define the related data structures, and finally develop relevant data exchange formats and simultaneously leave some further expansion in the future.

Thus, it is necessary to construct CNSDI in the context of NSDI. The overall planning and construction of CNSDI by the state would avoid the waste made by the repetitive collection of basic spatial data, realize data sharing in public, and form an effective mechanism to update data in a relatively short period. In CNSDI, organization coordination, equipment, and standardization of data and data-sharing policies have great importance, as should cultivating and training GIS talent.

3.9   From GGDI to Big Data

When NSDI is implemented in the world, Global Geospatial Data Infrastructure (GGDI) may appear. An expansion and extension of GGDI is Digital Earth. Digital Earth puts the real Earth into the networked computers. By describing the characteristics of human existence, work, study, and life on Earth in the data era, Digital Earth has been implemented throughout the world from which the digital areas, digital cities, digital towns, digital enterprises, and institutions have been derived. The Internet of Things integrated human societies and physical systems. Cloud computing is the provision of dynamically scalable and virtualized resources as a service over the internet, which is accessed from a web browser while the software and data are stored on servers. Both the Internet of Things and cloud computing accelerate the Smart Planet.

3.9.1   GGDI

Global Geospatial Data Infrastructure (GGDI) became an ongoing initiative on an international scale when NSDI aroused interest among data providers, vendors, and users in public organizations in North America, Europe, and many Asia-Pacific countries. NSDI's achievements in the world have played a major role in data collection, packaging, and distribution. GGDI shows the diversity and similarity, consistencies and differences, and needs and bounds at the same time. The development of GGDI was affected by the local interests, demands, and constraints of each involved country or organization. Understanding the different characteristics between global work and local, regional, and national work is important in order to develop the skills to change from regionalization to globalization, which guides the GGDI project.

In the 1960s, the proponents of integrated mapping advocated to register, overlay, interpret, and analyze the different layers and thematic data of spatial-relevant data sets in order to solve the key issues of land-use planning and resource investigation. In the 1970s, a report entitled "Information Society" was published in France to discuss global information infrastructure along with telecommunication infrastructure as technical sovereignty. Canada accelerated the development of its electronic highway. In the early 1980s, the concept of data being used as a common resource and a newly formed data resource management movement encouraged individual organizations to adopt a common approach to collect, manage, and share the specified paper copy and computer-based data that were in the interest of all legal entities. The form of these data-sharing methods changed from the early dreamy "centralized land information database" to a more complex network of distributed land information.

The "information infrastructure" introduced by Anne Branscombe in 1982 referred to a variety of media, carriers, and physical infrastructures for information distribution. In 1983, the Teletopia plan was announced in Japan to ensure good information communication. The internet appeared to give a set of global data resources in 1984. In 1987, a green book on the telecommunication industry in Europe concluded that a united, universal, and advanced telecommunication infrastructure would act as the neural network of socioeconomic development. In 1989, the World Wide Web (WWW) accelerated socioeconomic growth throughout the world. In the early 1990s, the development of the spatial data infrastructure (SDI) concept was originally to support and accelerate the work related to geographic data exchange standards. In 1990, McLaughlin and Nichols proposed that spatial data infrastructure should consist of spatial data sources, databases and metadata databases, data networks, technology (relating to data collection, management and representation technology), institutional settings, policies and standards, and end-users. In 1990, Neil Anderson suggested that data infrastructure should have three important characteristics: the data, network and control procedures in telecommunications should be standardized, resources and users networked, and the network open for third-parties to access them easily. In 1993, the National Information Infrastructure (information highway) was put forward in the United States to highlight the priority role of the infrastructure in communication and information. In March 1994, at the first international conference on telecommunication, some principles were proposed on the basis of global information infrastructure. In May 1994, the Bangemann Report on the European and global information society appeared to be fundamental to international or national plans.

These early intentions of globalization were put forth by several different groups. The military institutions of some North Atlantic Treaty Organization (NATO) countries have participated in the preparatory work of VMap data products and DIGSET tools in support of "global geospatial data and services initiatives." The International Hydrographic Organization has done a lot of work in the cooperation and development of defining global standards to address the nature and content of an electronic route map, which would be created by different countries. The first meeting of the budding GGDI was held near Bonn, Germany in September 1996. To carry out the agreement on multinational environments, the Global Mapping Regional Institute (GMRI) declared in Santa Barbara that there was a strong demand for accelerating the collection, update, and use of the products created by a national and global mapping plan and the harmonious development of GGDI.

Geospatial data management changes from integrated mapping to spatial data infrastructure. Kelley thinks that spatial data infrastructure should include basic spatial data sets with a wide range of applications; the planning, management, and technology to support the use of the spatial data, system, standards, and agreement that will allow timely access a large number of intensified spatial data. Therefore, it can be concluded that GGDI includes four elements (policies, techniques, standards, and human resources) for the effective collection, management, access, distribution, and use of geospatial data in the world. However, the assumption that a similar accord will happen in every country is childish in foreign affairs, problematic in operation, and impossible in economics. It is still a long-term goal to establish a real global infrastructure.

Interested governments, armed forces, and business departments recently began establishing transnational geospatial data infrastructure (TGDI) projects. The military, scientific research, and sea transportation industry are the three key groups of the global (or at least transnational) geospatial data infrastructure. Other groups—especially the organizations that locate, track, and orient people or things—will soon join them. The implementation of GGDI may be driven by data, technology, mechanisms, markets, and applications. Each of these entities has its own view, but they are interdependent and complementary.

3.9.2 Digital Earth

Digital Earth was presented by Al Gore, the former U.S. vice president (Al 1998). The digital geospatial data framework provides the basic and public data sets for researching and observing Earth and conducting geographical analysis. Data can be attached to the users based on the Digital Earth. There is a large amount of data in this framework along with the additional user data, which can be used for mining and the decision-making process.

Digital Earth is the unified digital reproduction and recognition of real Earth and its relevant phenomena on computer networks. It is the multi-dimensional description of Earth of multi-resolution, multi-scale, multi-space, and various types by applying massive Earth information through broadband networks on the base of computer technology, communication technology, multimedia technology, and big storage devices. Digital Earth can be used for supporting human activities and promoting the quality of life, such as for global warming, sustainable economic and social development, precision farming, intelligent transportation system, digitized battlefields, and so on (Grossner et al. 2008). To put it simply, Digital Earth is a multi-resolution virtual globe based on terrestrial coordinate system, which involves massive geographical information and can be indicated by 3D visualization.

Digital Earth is an inevitable outcome of global information as it depicts the features of human living, working, and learning on Earth in this data era and it can be applied to many fields involving politics, economy, military, culture, education, social life, and entertainment. Since it was proposed, research and applications of Digital Earth have been undertaken all around the world and stretched to multiple levels such as digital region, digital city, digital urban area, and digital enterprise. At the same time, the construction and development of Digital Earth in turn accelerated the pace of global information, and to a large extent, it has changed people's lifestyles.

Digital Earth is the digital record of the real Earth in a computer network system. Its realization in electronic computers needs to be supported by plenty of technologies, including information infrastructure, high-speed networks, high-resolution satellite images, spatial information infrastructures, massive data processing and high-capacity storage, scientific computation, visualization technology, and virtual reality. Among them, geospatial information technology is indispensable, such as GPS, GIS, RS, and their integration. Digital Earth provides the most basic data sets for researching and observing Earth. It requires a wide sharing of spatial information to which user data shall be attached. From the above, we can see that SDM is one of the key technologies for Digital Earth because it can recognize and analyze the massive data accumulated in Digital Earth to find out the law as well as new knowledge.

SDM is one of the key technologies of Digital Earth (Goodchild 2007). Earth is a complex giant system where the courses of events are mostly nonlinear and the span changes of time and space are different. Only by using data mining technology based on high-speed computing can massive data gathered in Digital Earth be found and the rules and its knowledge learned. The construction of Digital Earth requires sharing data on a large scale. The spatial data warehouse provides effective tools for managing and distributing spatial data effectively. In Digital Earth, the object of SDM generally is the spatial data warehouse. The next generation of the Great Globe Grid (GGG) and Spatial Data Grids based on network computing may create a better environment for SDM in Digital Earth. Therefore, Digital Earth would definitely bring great benefits to the use of framework data if SDM technology can be used in a digital geospatial data framework.

3.9.3 Smart Planet

It is a trend to build a "smart planet" through the comprehensive integration of Digital Earth and the Internet of Things (Li et al. 2006; Wang 2011). *Smart Planet* first refers to applying the new-generation information technology and Internet technology to all communities of society. Secondly, it means embedding and equipping sensors into global hospitals, power grids, railways, bridges, tunnels, roads, buildings, water supply systems, dams, and oil–gas pipelines. In addition, Smart Planet enables man to live and manage production in a more subtle and dynamic way in the world through interconnection to form the Internet of Things and then through supercomputers and cloud computing. American president Barack Obama praised the Smart Planet as a new productivity and a new growth point for solving economic crises.

Smart Planet is based on Digital Earth. Its neural network is the Internet of Things, and its thinking is implemented via cloud computing. The intelligence of the Smart Planet is not human collective intelligence quotient but systematic intelligence. The core of Smart Planet is changing the way governments, companies, and human beings are interacting by using new techniques in a smarter way and improving the specificity, efficiency, flexibility, and response speed of the interaction. There are three major intelligent factors of the Smart Planet: more thorough perception, broader interconnection and interworking, and deeper intelligentization, which come together to form a hierarchical structure. Deeper intelligentization obtains more intelligent insights and puts it into practice by using advanced technology and then creates new values. It refers to deeply analyzing collected data in order to acquire newer, more systematic, and more comprehensive insights for solving given problems. This calls for utilizing advanced technology, such as SDM tools, scientific models, and powerful computing systems to conduct complicated data analysis, collection, and computing in order to integrate and analyze a large number of cross-regional, cross-industrial, and cross-departmental data and information, while applying specific knowledge to specific industries, particular scenarios, and particular solutions in order to better support decision-making and action. SDM is and will continue to play an irreplaceable role in the Smart Planet.

3.9.4 Big Data

Big data are complicated data that is tremendous in volume, variety, velocity, and veracity (Wu et al. 2014). Big data offer an opportunity in the information world to observe the whole world completely instead of just partial samples. Before big data, statistics only could be sampled at random and conclusions were drawn from sampled data because of the limitation of spatial data collection, calculation, and transportation, which led to only partial information, such as the proverbial blind man grasping a part of an elephant and seeing the segment as a whole. Therefore, the deficiency of data sampling and the scattering of sample data contributed to the difficulty in recognizing the overall laws and extraordinary changes when they occurred.

In September 2008, a special issue on big data was published in *Nature*. In May 2009, the Global Pulse project within the United Nations released "Big Data for Development: Challenges & Opportunities," a report promoting the innovative ways to rapidly collect and analyze digital data (United Nations Global Pulse 2012). In 2010, the National Broadband Plan was issued in the United States. In February 2011, a special issue called "Dealing with Data" was published by *Science*, which included related topics on the importance of data for scientific research such as "Science: Signaling," "Science: Translational Medicine," and "Science: Careers." In May 2011, the McKinsey Global Institute released "Big Data: the Next Frontier for Innovation, Competition, and Productivity," a report that analyzed the potential application of big data in different industries from the economic and commercial dimensions and spelled out the development policy for government and industry decision-makers to deal with big data (McKinsey Global Institute 2011). In January 2012, the *Wall Street Journal* argued that big data, smart production, and wireless networks will lead to new economic prosperity (Mills and Ottino 2012). In March 2012, the U.S. government released the "Big Data Initiative," which moved the development and application of big data from a business activity to national strategic deployment in order to improve the ability to extract knowledge from large and complex data in order to help solve some of the nation's most pressing challenges (Office of Science and Technology Policy 2012). In April 2012, *Nature Biotechnology* invited eight biologists to discuss finding correlations in big data on the paper entitled "Detecting Novel Associations in Large Data Sets" in *Science* (Reshef et al. 2011).

Handling big data is also attracting some enterprises and industries (Wang and Yuan 2014). In 2011, Microsoft released Windows-compatible, Hadoop-based Big Data Solution as a component of SQL Server 2012. IBM released InfoSphere BigInsights in 2011, which combines DB2 with NoSQL for databases, and made a series of software vendor acquisitions in the years prior. In 2007, IBM acquired Cognos (business intelligence); in 2009, they acquired ILOG (business rules management), SPSS (data analysis and statistics), and Netezza (database analysis); and in 2010 Coremetrics (web analysis). In 2009, Elastic MapReduce by Amazon uses a hosted Hadoop framework that users can adjust to suit their needs for their data-intensive distributed programs work. Oracle combined the NoSQL Database and Big Data Appliance, which enables customers to handle unstructured massive data directly. Google uses Bigtable to distribute the storage of large-scale structured data and uses BigQuery SQL for data queries. In addition, Apple's iCloud, Facebook's Open Compute Project, and EMC's Greenplum HD are also committed to providing data solutions and applications.

Tackling big data has attracted as much attention in China as it has around the world. Baidu has used Hadoop to do offline processing since 2007. Currently, Baidu has over 20,000 Hadoop servers, which is more than Yahoo and Facebook. In these servers, 80 % of the Hadoop clusters are processing a total of 6 TB data every day on log analysis. Tencent, Taobao, and Alipay are also using Hadoop to establish data warehouses and to handle big data. In April 2010, Taobao issued a data mining platform "data cube," which resides on a 100 billion level database named OceanBase, which supports 4–5,000,000 times more update operations, including over 2 billion records, containing more than 2.5 TB data in one day. In

May 2010, China Mobile established a massive distributed system and structured massive data management system on the cloud. Huawei analyzes data based on mobile terminals and stores massive data through the cloud to obtain valuable information. Alibaba analyzes transactional data in business data through big data technology to conduct credit approval operations. In March 2012, when China's Ministry of Science and Technology released their list of key national alternative projects for science and technology in 2013, big data research was the first priority. According to a document released on September 5, 2015, "China Seeks to Promote Big Data," the State Council has called for a boost in the development and application of big data in restructuring the economy and improving governance.

Under the current condition of big data, it is possible to create, copy, and calculate data greatly to overcome a data sampling deficiency. The overall data can literally reproduce the original appearance of the real world, describe the whole appearance of spatial objects, and imply the overall laws and development trends to promote the efficiency of human beings in understanding the world and predicting the future. The United States employed professional knowledge and modern information technology to predict the influence of disasters accurately and in a timely manner to release early-warning information (Vatsavai et al. 2012). The high-resolution images of the tsunami around the railway station in Galle produced by the IKONOS 2 and QuickBird satellites explained the situation of the architecture. In the monitoring system of precipitation in the Google Earth, users now need only to open the Google Earth 3D terrain image, which is automatically accompanied by a satellite cloud picture, a precipitation chart, single station rainfall data, soil data, and onsite photos provided by the weather bureau; the stereo disaster effect could be demonstrated to undergo an inundation analysis and thereby provide the basis for decision analysis. ArcGIS can make any kind of disaster map; ArcGIS Mobile can meet the needs of quick reporting of disasters and collecting information for any kind of disaster.

Therefore, SDM focuses on the value of big data and using it efficiently by providing a process that extracts the information from data; discovers knowledge from the information and gains intelligence from the knowledge; improves self-learning, self-feedback, and self-adaption of this systems; and realizes human–computer interaction.

3.10 Spatial Data as a Service

GIS is being applied increasingly as the need for geo-information services grows. Since its introduction in the 1960s, GIS has gone through a long process of development and accomplished remarkable achievements in mapping sciences, resources management, environmental monitoring, transportation, urban planning, precision farming, etc. The promulgation, distribution, and publishing of geospatial information are growing rapidly. Michael F. Goodchild proposed "Citizens as

Voluntary Sensors" to describe the cooperative production of geospatial information and the transmission and sharing of geospatial knowledge in a more persuasive way. He projected the possibility of 6 billion residents of the world equipped with facilities that are capable of uploading what they consider to be important and effective sources of geospatial information (Goodchild 2007; Goodchild et al. 2007).

Network mapping is changing the way we see the world, including Tianditu, Google Earth, and Virtual Earth. The arrival of Web 2.0, the smart sensor web, and the Internet of Things have made geo-information so popular that not only professional users but also public users can do a variety of work on a uniform geospatial information service platform with increasingly more functions. With such a platform, spatial data is integrated with other data in such a way to answer actual questions. The variety of data sources includes statistical data, population data, social and economic data, and comprehensive services of measurement-on-demand for e-government services, e-commerce, public security, and transportation industries. The openness of this network has boosted its application and popularity greatly.

Web 2.0, as a human-oriented network, is revolutionary to the advocates of openness, anticipation, sharing, and creativity for geo-information services and thereby satisfies individual needs and transforms passive users into innovative and creative users. It also provides a variety of services with characteristics that include experience, communication, variation, creativity, and relation. For geospatial information services, visual services are the basis of experience, measurable services ensure variation, and creativity and minable services allow for relations among professional applications. Traditional measurement-by-specification may be transformed to measurement-on-demand by uploading the external orientation elements of Digital Measurement Images (DMIs) and a corresponding measurement software kit via the internet on Web 2.0. Any end user can measure objects as needed, which means that the public users are also data and information providers and geospatial data are changing from being outdated to active participatory services through the integration of a smart sensor web and Web GIS. Spatial data infrastructure with these new real-time geo-information services provide a better solution to users.

Mobile sensors on vehicles, fixed sensors at road intersections, captured video data, data in monitoring centers, road condition inspection data, and emergency data are used to a great degree in intelligent transportation system (ITS) on the sensor web for improving transportation, reducing traffic delays and traffic accidents, and decreasing gas consumption (Fig. 3.10). In ITS, traffic jams can be decreased by about 20 %, traffic delays by about 10–25 %, traffic accidents by about 50–80 %, and gas consumption by about 30 %. All of the aerial and aerospace sensors can be integrated together to build a large smart sensor web and thereby accomplish real-time data updating, information extraction, and services.

Different kinds of data and services can register through the registration center and build catalog services. For an end user, there are three ways to implement SDM based on the sensor web (Fig. 3.11). The first way is based on the sensor

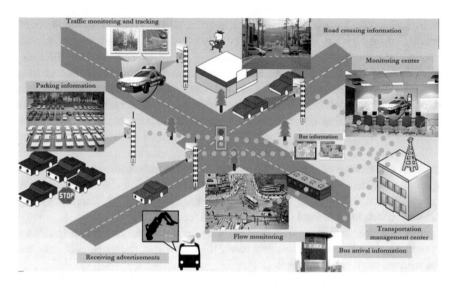

Fig. 3.10 Application system in ITS based on sensor Web Reprinted from Ref. (Li and Shao 2009), with kind permission from Springer Science+Business Media

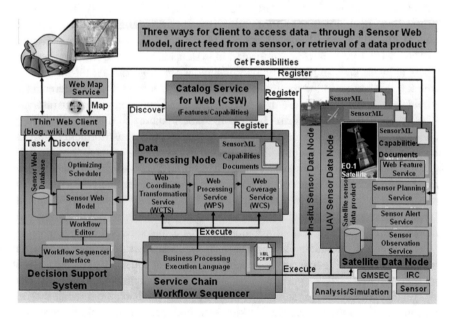

Fig. 3.11 Reference architecture for an interoperable sensor Web Reprinted from Ref. (Li and Shao 2009), with kind permission from Springer Science + Business Media

web model. When an end user sends a request on the client side, the decision-making support system (DSS) transfers the received request to a service chain and searches the corresponding registered services in the catalog services through a workflow sequence. Then, the registered services acquire the information of interest and send it to the end user. The second way is based on direct feedback from the sensor web, which is applied when the corresponding registered service cannot be found in the catalog services through the workflow sequence. In this case, new sensor web feedback is further searched; if the required service can be found, it will be sent to the end user and registered at the registration center. The third way is based on retrieval of the digital products, which is also applied when the corresponding registered service in catalog services cannot be found through the workflow sequence. In this case, the requested information is further searched through the sensor web node instead of the sensor web feedback.

The new spatial information systems may be distinguished from a traditional system on the view of the data providers, the data users, the geospatial data, and the measurement and data share.

The new geo-information systems provide services not only to professionals but also to all public users; a great many of these users require fundamental information regarding professional and individual applications. It is unfortunate that this information cannot be discovered from traditional 4D products directly, which cannot satisfy the need for integrity, richness, accuracy, and reality in geospatial information. In this new geo-information era, such information can be acquired from DMIs, which are released on the internet according to specific requirements.

The users of these new geo-information systems are data and information providers as well. Data users can upload or annotate new geospatial information in Web 2.0 in addition to the traditional downloading of information of interest. The provided data and services are transferred from fixed updating at regular intervals to more popular updating forms (i.e., from static updating to dynamic updating). The boundary between the data provider and the data user thus is quickly blurring thanks to this total open accessibility to spatial information.

Geospatial data from outdated sources live through a smart sensor web. Sensor web technology provides access to services that meet the specific needs of professionals and general users in a multimedia and dynamic service environment. All the sensors can be integrated together to build a large smart sensor web that provides real-time data updates, information extraction, and services.

Measurement includes measurement by specification to measurement on demand. Geospatial data in service are DMIs instead of simple image maps. DMIs are digital stereo images appended with six exterior orientation elements acquired by Mobile Measurement Systems (MMSs). By using DMIs on the internet that are accompanied by measuring software kits, measurement of a special object at centimeter level precision is available. Undoubtedly, the overlapping of DMIs with GIS data makes the representation of geographic objects more comprehensive and vivid and facilitates visible, searchable, measurable, and minable functions.

Share is from data-driven to application-driven. Service-Oriented Architecture (SOA) is a software structure that achieves interoperability by packaging the

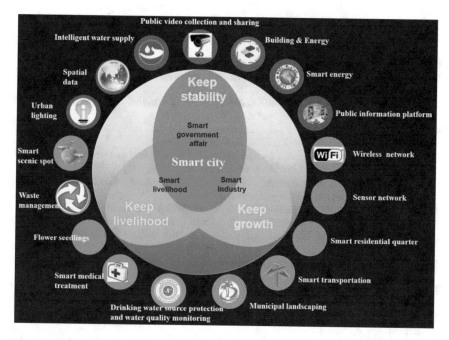

Fig. 3.12   A smart city mechanism

program units to accomplish a given task, which is a widely accepted as a standard. Service-Oriented Integration (SOI) integrates traditional objects with highly flexible web services. SOI provides an abstract interface for systems to communicate with one another, instead of using low-layer protocols and self-defined programming interfaces to prescribe communication with other systems. The system only needs to appear in the form of service to select the interactive system, make simple discovery, and bind with this service at runtime or in design.

Therefore, spatial data act as a service in the context of heterogeneous and distributed spatial information systems. The new geo-information era will boost the whole chain of the geospatial information industry to prosperity and will bring tremendous economic benefits to the whole world with the ultimate goal of providing geo-information for all aspects. Locally, it may bring people a smart city (Fig. 3.12).

References

Aji A et al (2013) Hadoop-GIS: a high performance spatial data warehousing system over MapReduce. In: Proceedings of the 39th international conference on very large data bases, VLDB endowment, vol 6(11), pp 1009–1020 (August 26–30th 2013, Riva del Garda, Trento, Italy)

Al G (1998) The digital earth: understanding our planet in the 21st century, speech at the california science center, Los Angeles, California, on January 31, 1998. url:http://www.isde5.org/al_gore_speech.htm

Codd E (1995) Twelve rules for on-line analytic processing. Computer world, April 1995

Chen J (1999) On the construction of NSDI in China. J Remote Sens 3(2):94–97

Craglia M, Bie K, Jackson D (2012) Digital Earth 2020: towards the vision for the next decade. Int J Digital Earth 5(1):4–21

Chen J, Gong P (1998) Practical GIS. Science Press, Beijing

Ester M et al (2000) Spatial data mining: databases primitives, algorithms and efficient DBMS support. Data Min Knowl Disc 4:193–216

Grossner KE, Goodchild MF, Clarke KC (2008) Defining a digital earth system. Trans GIS 12(1):145–160

Goodchild MF (2007) Citizens as voluntary sensors: spatial data infrastructure in the world of Web 2.0. Int J Spat Data Infrastruct Res 2:24–32

Goodchild MF, Fu P, Rich P (2007) Sharing geographic information: an assessment of the geospatial one-stop. Ann Assoc Am Geogr 97(2):249–265

Gong JY (1999) Theories and technologies on the contemporary GIS. Wuhan Technical University of Surveying and Mapping Press, Wuhan

Inmon WH (2005) Building the data warehouse, 4th edn. Wiley, New York

Killer J et al (1998) On combining classifier. IEEE Trans Pattern Anal Mach Intell 20(3):226–239

Li DR (1999) Information superhighway, spatial data infrastructure and the digital Earth. J Surveying Mapp 28(1):1–5

Li DR, Guan ZQ (2000) Integration and implementation of spatial information system. Wuhan University Press, Wuhan

Li DR, Shao ZF (2009) The new era for geo-information. Sci China Ser F-Inf Sci 52(7):1233–1242

Li DR, Wang SL, Li DY (2006) Theory and application of spatial data mining. Science Press, Beijing

McKinsey Global Institute (2011) Big data: the Next Frontier for Innovation, Competition, and Productivity, May 2011

Mills MP, Ottino JM (2012) The coming tech-led boom, 2012-10-12. www.wsj.com

Office of Science and Technology Policy (2012) Big data initiative: announces $200 Million In New R&D Investments, March 29, 2012. www.WhiteHouse.gov/OSTP

Reshef N et al (2011) Detecting novel associations in large data sets. Science 334:1518

Shekhar S, Chawla S (2003) Spatial databases: a tour. Prentice Hall Inc

Srivastava J, Cheng PY (1999) Warehouse creation-a potential roadblock to data warehousing.IEEE Trans Knowl Data Eng 11(1):118–126

Tan GX (1998) The integration of spatial data structure and its indexing mechanism. J Surveying Mapp 27(4):293–299

The Executive Order 12906 by the President of the United States, 1994, The Harmonization of Geographic Data Access and Storage: National Spatial Data Infrastructure (NSDI) the version published by the United States Federal Register on April 13, 1994, vol 59, 71, pp 17671–17176

United Nations Global Pulse, 2012, Big Data for Development: Challenges & Opportunities, May 2012

Vatsavai RR et al (2012) Spatiotemporal data mining in the era of big spatial data: algorithms and applications. In: Proceedings of the 1st ACM SIGSPATIAL international workshop on analytics for big geospatial data, 6–9 Nov 2012. Redondo Beach, CA, USA, pp 1–10

Wang SL (2002) Data field and cloud model based spatial data mining and knowledge discovery, PhD thesis, Wuhan University, Wuhan

Wang SL, Shi WZ (2012) Data mining and knowledge discovery. In: Kresse Wolfgang, Danko David (eds) Handbook of geographic information. Springer, Berlin

Wang SL (2011) Spatial data mining under smart earth. In: Proceedings of 2011 IEEE international conference on granular computing, pp 717–722

Wang SL, Yuan HN (2014) Spatial data mining: a perspective of big data. Int J Data Warehouse Min 10(4):50–70

Wu X, Zhu X, Wu G, Ding W (2014) Data mining with big data. IEEE Trans Knowl Data Eng 26(1):97–107

Chapter 4
Spatial Data Cleaning

There are always problems in spatial data, which makes spatial data cleaning the most important preparation for data mining. If spatial data are input without cleaning, the subsequent discovery may be unreliable and thus produce inaccurate output knowledge as well as erroneous decision-making results. This chapter discusses the problems that occur in spatial datasets and the various data cleaning techniques to remediate these problems. Spatial observation errors mainly include stochastic errors, systematic errors, and gross errors, such as incompleteness, inaccuracy, repetitiveness, inconsistency, and deformation in spatial datasets from multiple sources with heterogeneous characteristics. Classical stochastic error models are further categorized as indirect adjustment, condition adjustment, indirect adjustment with conditions, condition adjustment with parameters, and condition adjustment with conditions. All of these models are considered parameters as non-random variables in a generalized error model. By selecting weight for iteration, Li Deren made the assumption of two multi-dimensional alternates of Gauss-Markov when he established the distinction and reliability theory of adjustment. As a result, Li extended Baarda theory into multiple dimensions and realized the unification of robust estimation and least squares.

4.1 Problems in Spatial Data

Spatial data vary by type, unit, and form with different applications, and problems inevitably occur that affect the quality of spatial data. In recorded items, for example, incompleteness, repetition, mistakes, inaccuracy, and abnormalities may occur (Koperski 1999; Wang et al. 2002; Shi et al. 2002), which may be additionally influenced by gross, systematic, and stochastic errors. The process of data acquisition brings along with it the data's systematic errors, such as instrument system distortion, observed materials deformation, Earth's curvature, atmospheric refraction, surveying instruments, and behavior of observers. The gross errors also appear during spatial data acquisition. With the rapid accumulation of massive data, the long-standing, manual, empirical analysis approach to correcting gross errors has become invalid (Wang and Shi 2012). The manual approach, which is based on the experience and professional knowledge of the operator, also may not uncover all the faults and some undiscovered gross errors may enter the adjustment. Even the proficient experts at the renowned Schwaben Laboratory in Germany tried their best to uncover gross errors in the local area network, which were later discovered by using the automatic positioning system method. Obviously, if the errors cannot be found and eliminated, they will impact the adjustment results and the truth in spatial data becomes even farther away from discovery.

4.1.1   Polluted Spatial Data

In the real world, most spatial data are polluted and are revealed a variety of ways, such as disunity of scale and projection among different data sources, inconsistency of sampled data and their forms, mismatch among different data sheets, redundancy of data, discrepancy between different coordinate systems, deformation of graphics and images, inconformity between the map scale and the unit of length of digitizers, loss and repetition of spatial points and lines, missing identifiers for regional centers, too-long or too-short lines when taking input, and discordance with the consistency requirements of topology in nodal codes or attribute codes. The most common of these are incompleteness, inaccuracy, repetition, and inconsistency.

4.1.1.1   Incomplete Spatial Data

The completeness (or integrity) of spatial data reflects the level of the overview and abstraction of it. Spatial data may be missing or incomplete for various reasons (Goodchild 2007). First, the defect of incompleteness is caused by errors such as ellipsis (Smithson 1989). For example, the necessary domains and instances are not recorded in a spatial database when they are designed; the decision rule of collecting and editing spatial data does not take all the variables and impact factors into account; the data in the spatial database cannot fully present all the possible attributes and variable features of objects; and the spatial database does not include all possible objects. Second, not all the characteristics in the measuring standards are collected according to the criteria, definitions, and other rules; and some important characteristics for recognizing spatial objects are fused because of the evaluation standards. For example, the data will be incomplete if the boundary points of a land parcel are omitted. Third, some elements are lacking in a process that intends to help guarantee the continuity of data analysis and the future development of editing systematic documents and a spatial system. For example, lazy input habits or the different demands for the data from different users can lead to missing necessary data values. In most cases, the missing values must be entered by hand and some missing values can be deduced by the spatial data source and related data.

There are various methods to deal with the data noises caused by unknown attribute values (Clark 1997; Kim et al. 2003). The first method is ignorance, where the records whose attributes have unknown values are ignored. The second method is adding values, where the unknown value is treated as another value of the attribute. The third method is likelihood estimation, where the unknown value is replaced with the most possible value of the attribute. The fourth method is Bayesian estimation, where the unknown value is replaced with the maximum possible value of the attribute under its function distribution. The fifth method is a decision tree, where the unknown value is replaced with the classification of the object. The sixth method is rough set, where the rules are deduced from an incompatible decision tree for the unknown values. The seventh method is a binary model, where the vector of the attribute values is used to represent the data in the context of transition probability between symbolic attributes. The average value, maximum value, and minimum value are also used.

4.1.1.2 Inaccurate Spatial Data

The inaccuracy of spatial data measures the discrepancy between the observed data and its true value or the proximity of the measurement information and the actual information. Many inaccuracy problems occur in the corresponding environments that are related to the data types, such as processing methods, sorting algorithm, positional precision, time variation, image resolution, and spectral features of spatial objects. The inaccuracy may be quantitative and qualitative, such as incorrect data whose values are contrary to the attributes of real objects, outdated data that are not updated in a timely manner, data from inaccurate calculation or acquirement, inexactitude category data which are difficult or impossible to explain, and vague data with fake values or eccentric formats (Smets 1996). Topographic features can be acquired by accurate measurement while the accuracy of forest or soil boundaries may be low due to the influence of line identification errors and measurement errors. If the accuracy of an attribute is high, the classification should be expected to be rigorously in accordance with the real world. However, it is difficult to ensure the complete accuracy of classification, and cartographic generalization might further classify various attributes as one and the same attribute. A large amount of the spatial data is not well used because of its low accuracy. Generally, spatial databases are designed for a specific application or business step and are managed for different purposes, which makes it difficult to unify all the related data and therefore makes it easy to induce errors into the integrated spatial data.

The basic methods of dealing with inaccuracy include statistical analysis to detect the possible errors or abnormal values (deviation analysis to identify the value that does not follow the distributions or regression equations), spell checking in documents processing, and a simple rule base (e.g., common sense rules, business-specific rules to examine spatial data values, external data, and reorganizing spatial data). Spatial data reorganization refers to the process of extracting spatial data from the detached primitive data for a certain purpose, transforming spatial data into more meaningful and integrated information, and mapping data into a target spatial database.

4.1.1.3 Repetitive Spatial Data

The presence of repetitive spatial data indicates that duplicate records of the same real object exist in one data source or in several systems. The records of similar attribute values are considered duplicate records. Due to errors and expressions in spatial data, such as spelling errors, and different abbreviations, it is possible that records that do not exactly match might also be duplicate records. Duplicate records are more common in multi-source SDM. The spatial data provided by each source usually include identifiers or string spatial data, which may vary in different spatial data sources or might cause some errors for various reasons, such as errors caused by printed or typed-in information or aliases (Inmon 2005). Therefore, whether two values are similar or not, it is not a simple arithmetical solution but rather a set of defined equivalent rules and some fuzzy matching techniques. Identifying two or more records that relate to one object in the real world and eliminating the duplicate records can not only save storage and computing resources, but also improve the speed, accuracy, and validity of the SDM-based approaches. Merge or purge is the basic method.

Merge or purge is also called record linkage, semantic integration, instance identification, data cleansing, and match or merge. This method detects and eliminates duplicate records when integrating spatial data sources presented in heterogeneous information (Hernàndez and Stolfo 1998). The current main approach is to use the multi-attributes of the records and let the user define the equivalent rules as to whether or not the records correspond to one spatial entity, thereby enabling the computer to auto-match the possible records that correspond to the same entity with the defined rules. Arithmetic-sorted neighborhood and arithmetic fuzzy match/merge are currently popular for this process. Arithmetic-sorted neighborhood sorts the entire spatial data set according to the user-defined code and groups the possible matched records together. Repeated sorting can improve the accuracy of the matching results. Arithmetic fuzzy match/merge adopts certain fuzzy technology to compare all the records in pairs after normalizing the spatial data of all the attributes and finally merge the results for comparison.

4.1.1.4 Inconsistent Spatial Data

Inconsistency in spatial data values often happens between the internal data structure and the spatial data source during spatial data cleaning, which can include a variety of problems that can be classified in two types. The first type is context-related conflicts, such as different spatial data types, formats, patterns, granularities, synonyms, and coding schemes caused by systems and applications. There might be semantic conflicts of the inconsistency caused by spatial data integrated by different multi-spatial data sources. That is, computerized spatial entities cannot satisfy the topological consistency of the spatial entities, the internal consistency of the data structure, and the logical consistency of data normalization. For example, if the differences in the radiation signal between the soil with a forest and the soil without a forest are not taken into consideration when judging soil types, then the classification results may be incorrect. In spatial information systems, such as GIS, illogical spatial database information should be avoided and warning information should be distributed when users work with the data improperly. The second type is context-free conflicts, which includes the state change of spatial databases caused by wrong input due to stochastic factors, hardware or software failures, untimely updates, and external factors, as well as the inconsistency caused by the different places, units, and times of the spatial data in the same system. The inconsistency in expression and context caused by input denormalization makes it difficult to determine the transfer functions.

Identification of the entity and resolution of the semantic conflicts in multi-spatial database systems is important. The purposes of detecting and solving the inconsistency of spatial data are to merge the spatial data from different spatial data sources, detect the semantic conflicts, set transfer standards, and detect the possible incorrect data values. Integrity constraints can be defined in order to detect the inconsistencies, and the connection can be found through analyzing spatial data.

4.1.2 Observation Errors in Spatial Data

The errors in spatial data are equally as relevant as the data, the measurement, and the data structure. Errors mainly originate from the uncertainty of the attribute definitions, data sources, data modeling, and analysis process, in which the uncertainty of the data sources because the measuring depends on subjective human judgment and the hypothesis during data collection (Burrough and Frank 1996). The attribute uncertainty of land cover classification data obtained by remote sensing is from the uncertainty of the space, the spectrum, and the temporal characteristics (Wang et al. 2002). In 1992, a MIT report pointed out that the quality problems of data were not uncommon. In most of the 50 departments or institutions involved in the sample survey around the world, the accuracy of their spatial data was also below 95 %, mainly because of the inaccurate rate and the repetitive rate (Fig. 4.1).

In surveying, users spend a great deal of time dividing errors into gross errors, systematic errors, and stochastic errors, according to the size, characteristics, and reasons for the errors.

$$\varepsilon = \varepsilon_g + \varepsilon_S + \varepsilon_n \tag{4.1}$$

In Eq. (4.1), ε is the total observation error, ε_g is the gross error, ε_S is the systematic error, and ε_n is the stochastic error. In many years of empirical experience,

Fig. 4.1 Statistical figure on
the average of inaccurate and
repetitive data

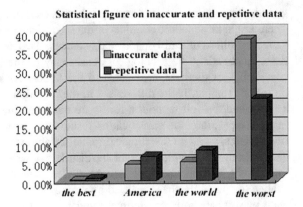

it has become a custom to classify errors into these three types. From the view
of statistical theory, however, there is no common and clear definition. The best
approach is to analyze and sort the errors from different perspectives.

First, all three types of errors can be considered as model errors and can be
described by the following mathematical model:

$$\varepsilon_S = H_S s \quad s \sim M(s_0, C_{SS}) \tag{4.2a}$$

$$\varepsilon_g = H_g \Delta l \quad \Delta l \sim M(\Delta \hat{i}, C_{gg}) \tag{4.2b}$$

$$\varepsilon_n = E_n \varepsilon_n \quad \varepsilon_n \sim M(0, C_{nn}) \tag{4.2c}$$

In Eqs. (4.2a–4.2c), $M(\mu, C)$ shows the expectation is μ, the variance-covariance
matrix is any one of the distributions of C, while the matrix is H_S, H_g, E_n and
decides the influence from the systematic errors, the gross errors, and the stochastic
errors on the observation values. The features of these three coefficient matrixes are
different (Fig. 4.2).

1. The elements in the coefficient matrix of the systematic errors, usually the
   position function and the time function, are universal or group by group. For
   instance, when additional parameters are considered as the regional invariants,
   the coefficient matrix H_s is occupied completely; when they are strip invariants,
   the coefficient matrix is occupied group by group and the systematic errors of
   the photo are the function of (x, y).
2. The gross errors coefficient matrix H_g is of sparse occupation, and normally
   there are only one or few nonzero elements in every column. For the p_1 different
   gross errors, there is Eq. (4.3):

$$H_g = \left[e_{i+1}, e_{i+2} \dots e_{i+p_1} \right] \tag{4.3}$$

where, $e_i = [0 \ 0 \cdots 0 \ 1 \ 0 \cdots 0]$ (only the ith element is 1).

3. The stochastic error is a diagonal matrix (usually an identity matrix).

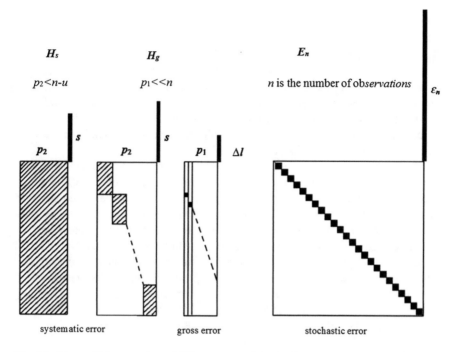

Fig. 4.2   The coefficient matrixes of different types of observation errors

It is well known that a stochastic error has a nonzero matrix. The systematic error only can be seen as the functional model error ($s = s_0$, $C_{SS} = 0$) or as the stochastic model error ($s = 0$, $C_{SS} \neq 0$), and of course can be both at the same time (Eq. 4.2a).

When determining its reliability, the gross error is always considered as the functional model error ($\Delta l \neq 0$, $C_{SS} = 0$) while to the position, it is better to be seen as the stochastic model error, which is beneficial for effective discovery and correction of the gross error.

In addition, the three types of errors also can be distinguished according to the reasons why the errors appear. The systematic errors brought out in data acquisition are due to certain physical, mechanical, technical, instrument, or operator errors. Normally, it has some rules or changes regularly. The gross errors happen because of irregular mistakes in data acquisition, data transmission, and data processing, which cannot be adopted by the assumed and estimated error model as the acceptable observation. As for stochastic errors, they are generated by the observation conditions (instruments, field environment, and observers). Different from the systematic errors, they show no regularity in size and symbol, and only the total of a large sum of errors is of certain statistical rules.

4.1.3 Model Errors on Spatial Data

The mathematical model for the observation errors in spatial data is described with a functional model and a stochastic model. Obtaining a set of observation values from the subject of interest and using it to estimate the relevant unknown parameters representing the subjects is called *parameter estimation* in mathematical statistics and adjustment in surveying. Realization of the errors first needs a mathematical model reflecting the relationship between the observation values and the unknown parameters (Li and Yuan 2002).

As a set of random variables, the observation vector can be described by its first moment (expectation) and its second center moment (variance-covariance). Here forward, the model used to describe observation expectation is called the functional model while the model for the precision characteristics of observation values is the stochastic model, and their combination is known as the mathematical model for adjustment.

The full-rank Gauss-Markov linear model is defined as follows. Assume that A is the known coefficient matrix $n \times u$ (usually known as the first design matrix), x is the unknown parameter vector $u \times 1$, l is the random observation vector $n \times 1$, its variance-covariance matrix is $D(l) = \sigma^2 P^{-1}$ (σ^2 is the variance of unit weight), and the matrix A is full column rank—that is, the weight matrix P is a positive definite matrix. Therefore, the full-rank Gauss-Markov linear model is:

$$E(l) = A\tilde{x}, \quad D(l) = \sigma_0^2 P^{-1} \tag{4.4}$$

This mathematical model is called the Gauss-Markov model because Gauss deduced the least square method from this model by using the likelihood function in 1809 (i.e., the optimum estimation method), and Markov used the best linear unbiased estimation to get the parameter of this model in 1912. This model is actually the adjustment of indirect observation values in survey adjustment. The goal of adjustment is to obtain the values of the unknown parameter and estimate its accuracy from the set of observation values.

4.1.3.1 Model Errors and Hypothesis Test

In statistics, the model errors can be defined as the difference between the models built (including the functional model and the stochastic model) and the objective reality, which in the form of an equation is

$$F_1 = M_0 - W \tag{4.5}$$

In Eq. (4.5), F_1 is the true model error, M_0 is the mathematical model used, W is the unknown reality, and $M_0 \neq W$.

If a mathematical model, according to verification theory in mathematical statistics, is considered, a hypothesis compared to the real world (the null hypothesis), the starting point when determining the model is to make the model errors zero for both the expectation and variance of the observation values. To test the null hypothesis, one or more standby hypotheses are needed. This kind of standby hypothesis tries to make a more precise extension of the model built in order to reduce model errors.

As the objective reality W is unknown, users must use a mathematical model extended and refined as much as possible to replace it. Thus, the definition of the difference between model M and the refined model M_0 as plausible model errors is meaningful for factual research:

$$F_2 = M_0 - M \tag{4.6}$$

The mathematical model M can be extended and refined as much as possible to make it very close to the objective reality $((M - W) \to 0)$. For instance, in terms of bundle adjustment with self-calibration, Schroth introduced a model in which both the functional model and the stochastic model are expanded. Under this premise, an equation is obtained:

$$F_2 = M_0 - M = (M_0 - W) - (M - W) \approx M_0 - W \tag{4.7}$$

Further discussion can derive from the model errors with this definition. In a hypothesis test, the difference between the model M_0 (the null hypothesis H_0) and an extended model M_1 (the standby hypothesis H_a) must be verified as palpable (Fig. 4.3). If $M_0 = M_1$, this model is untestable.

When making a choice between two standby hypotheses, whether or not the two extended models M_1 and M_2 proposed from the original model M_0 can be distinguished should be ensured. If they are identical to each other or one model is contained by the other, then they are indistinguishable (Fig. 4.4).

4.1.3.2   The Influence of Model Error on the Adjustment Results

In the Gauss-Markov model (Eq. 4.4), the observation expectation is represented as the linear function of a group of unknown parameters while its weight

Fig. 4.3 Relationship between an original assumption and its alternative assumption under a single alternative hypothesis

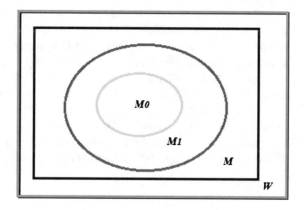

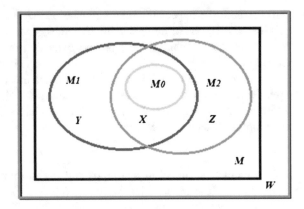

matrix is known. If there are some errors in the mathematical model, how would the adjustment results be affected? The following three situations may appear:

1. When too few unknown parameters are chosen in the functional model describing the observation expectation, there is a deviation in the estimation of the observation values, the covariance matrix of the unknowns is small, and the estimation of unit weight variance increases.
2. When too many unknown parameters are chosen, there is no deviation in the estimate of the observation values, the covariance matrix of the unknown increases, and the estimate of unit weight variance has no deviation.
3. When a wrong weight matrix is chosen, there is still no deviation in the estimate of the observation values. If the weight of observation values is small, the cofactor matrix of observation values decreases; otherwise, it increases. If the weight of the observation values in the model introduces errors, the estimation of the unit weight variance is deviated.

Therefore, the target of SDM is the dataset, which always involves a large amount of spatial data from many sources, which in reality may be polluted.

In the complicated designs and operations of spatial datasets, if the integrity of spatial data cannot be ensured, all the construction work and the following SDM will be meaningless. It is very important to remember that just a 4 % increase in data integrity could bring millions of dollars to any large enterprise.

If good data do not exist, it is not likely that SDM will be able to provide reliable knowledge, quality service, and good decision support. Moreover, spatial rules are very often hidden in large complex data, where some data may be redundant and others are totally irrelevant. Their existence may affect the discovery of valuable spatial rules. All of these reasons necessarily require SDM to provide a data cleaning tool that can select correct data items and transform data values (Dasu 2003).

4.2   The State of the Art

The traditional adjustments in geodetic surveying and photogrammetry are based on the fact that observation values contain only stochastic errors, while the current trends center on how to improve consideration of the possible gross errors and systematic errors in the adjustment process in order to guarantee the accuracy and reliability of the adjustment results. The more-specific term *spatial data cleaning* applies when data mining is employed.

4.2.1   Stages of Spatial Data Error Processing

In terms of adjustments in photogrammetry, given different types of observation errors, development of the adjustment model can be divided into four stages (Li and Yuan 2002).

4.2.1.1   The First Stage: $\varepsilon_S = 0$, $\varepsilon_g = 0$, $\varepsilon_n \neq 0$

The tasks for adjustment are as follows:

1. Find the most precise solution—that is, find the most common and rigorous functional model and program computer utility. For example, the rigorous mathematical model uses the independent model method and the bundle method, which is the combined adjustment in photogrammetry and non-photogrammetric data.
2. Study the theoretical accuracy of the adjustment results in order to well advise the design of the network.
3. Estimate the precision characteristics of the observation values, such as the variance-covariance component estimation.

Analytical photogrammetry was studied from 1957 to the early 1970s, during which it was determined that the gross errors could be avoided by conducting the necessary examination according to the observation values and extensively training the operators of measuring instruments and methods. However, in the adjustment, the gross errors were checked through some simple processes and selected by manual labor. In addition to the difficulty in operation, however, this method also was not ideal since the adjustment results often depended on the level of precision and theoretical knowledge of the operators. Using simple tests made it almost impossible to find small gross errors in observation values.

Moreover, in this stage, to compensate for the systematic errors, users could use only pre-calibrated equipment to provide the relevant parameters for the correction of systematic errors or to devise special measuring orders and methods to eliminate the influence of systematic errors.

4.2.1.2 The Second Stage: $\varepsilon_S \neq 0$, $\varepsilon_g = 0$, $\varepsilon_n \neq 0$

This stage was the simultaneous processing of the stochastic errors and the systematic errors. In the adjustment of photogrammetry, the unknown systematic errors in images need to be considered first. The processing method introduced them into the adjustment as additional unknown parameters and solved the additional parameters of the featured systematic errors at the same time as the unknown parameters were calculated. Filters and estimation also were used.

Most certainly, self-calibration adjustment with additional parameters can effectively compensate for the influence of systematic errors in images in order to make the unit weight of the mean square error of the analytic aero-triangulation reach the high precision level of 3–5 μm. Bundle block adjustment programs with additional parameters (e.g., PAT-B and WuCAPS) are effective in this stage. For further information about compensating for systematic errors, please refer to Li and Yuan (2002).

4.2.1.3 The Third Stage: $\varepsilon_S = 0$, $\varepsilon_g \neq 0$, $\varepsilon_n \neq 0$

This stage further processed the systematic errors and the gross errors at the same time in the adjustment from two aspects: (1) in theory, the reliability of the adjustment system was studied (i.e., its ability to find the gross errors and the influence of the undiscoverable gross errors); and (2) in reality, the importance of finding a practical, effective, and automatic gross error positioning method. Many useful requests and suggestions for optimal design in surveying areas are now available due to research on the reliability theory of the adjustment system; however, the exploration of automatic gross error positioning also introduced the automatic gross error positioning method in some computer utilities, which is achieving important results that are not possible for artificial methods.

The second stage and the third stage have generally paralleled since 1967.

4.2.1.4 The Fourth Stage: $\varepsilon_S \neq 0$, $\varepsilon_G \neq 0$, $\varepsilon_n \neq 0$

This stage, which began developing in the 1980s, deals with all of the possible observation errors simultaneously in the adjustment. This is a very practical approach because all three types of observation errors do exist at the same time in reality.

In this stage, the distinguishability theory to discriminate different model errors was raised theoretically. Here, the two types of model errors could be different gross errors, different systematic errors, or a combination of the two. While the theoretical distinguishability of the adjustment system was studied, in the meantime this work is informing the optimum design of surveying areas. In the real world, this period in the future should find the adjustment method that can process different model errors at the same time. For example, in the areas of analytic aero-triangulation and the data processing of LiDAR, efforts should be made to find a method which can simultaneously compensate for systematic errors and eliminate errors.

4.2.2 The Underdevelopment of Spatial Data Cleaning

In 1988, the National Science Foundation of the United States sponsored and founded the National Center for Geographic Information and Analysis. Among the 12 research subjects of the center, the first subject designated the accuracy of GIS as the first priority, while the twelfth subject designated the error problem in GIS as the first priority among its six topics. Meanwhile, also in 1988, the U.S. National Committee for Digital Cartographic Data Standards published spatial data standards for lineage, positional accuracy, attribute accuracy, logical consistency, and completeness. Later, currency (the level for spatial databases to meet the newest demands) and thematic accuracy (defined by the substance's attribute accuracy, class accuracy, and timing accuracy) were proposed, adding temporal accuracy to express time as the basis of the five standards. As lineage describes the procedure of data acquisition and processing, spatial data accuracy should contain spatial accuracy, temporal accuracy, attribute accuracy, topology consistency, and completeness.

Data cleanliness in SDM does not receive adequate attention from the public. Data access, query, and cleanliness are the three important topics in spatial data processing, but for a long time users have focused only on the solutions to the first two while seldom attending to the third. With the C/S structure, the dedicated spatial data warehouse hardware and software and a complete set of communication mechanisms link users with spatial data to solve the data access problem. When querying data, there are many choices in various impromptu query tools, report writers, application development environments, end-user tools of the second-generation spatial data warehouse, and multi-dimensional OLAP/ROLAP tools. Currently, only a small amount of basic theoretical research and application development of the second-generation spatial data has been conducted. The international academic magazine *Data Mining and Knowledge Discovery* observed that only a few articles had been available on data cleaning, such as "Real-world Data Is Dirty: Data Cleansing and the Merge/Purge Problem" (Hernàndez and Stolfo 1998). The development of data cleaning lags far behind that of data mining; even academic conferences lacked sessions addressing data cleaning. At the same time, far from the richness and completeness of spatial data access and query tools, there are only a few tools to solve the problem of spatial data cleanliness, mainly because of the enormous workload and costs. The present products providing limited spatial data cleaning functions are QDB/Analyze of QDB Solutions Inc., WizRules of WizSoft Inc., Integrity Data Re-Engineering Environment of Vitality Technology, Name and Address Data Integrity Software of Group ScrabMaster Software, Trillium Software System, i.d. Centric and La Crosse of Trillium

Software, MERGE/PURGE of Wisconsin, and Innovative Solutions of Innovative Systems. Moreover, it is all about the lack of English spatial data cleaning and the practical Chinese unstructured spatial data cleaning research, which may be due to a lack of fully understanding the influence of tainted spatial data and the features of spatial data cleaning. Fortunately, the *ICDM-02 First International Workshop on Data Cleaning and Preprocessing*, which was held in Maebashi, Japan in December 2002 offered three topics that addressed spatial data cleaning. The international journal *Applied Intelligence* published a special issue for this conference which stated, "We believe that the issue of data cleaning and preprocessing might become the next hot point of research in the KDD community" (Wang et al. 2002).

Next are the influential ways of choosing target data and judging their weight, which include the index method in the military, Delphi method of N. Dalkdy and O. Helmer's in the Rand Corporation, T. L. Saaty's Analytic Hierarchy Process (AHP) method from the United States, Deng Julong's grey system, the Availability + Dependability + Capability (ADC) method developed by the U.S. Weapons System Effectiveness Industry Advisory Committee (WSEIAC), the System Effective Analysis (SEA) method of A. H. Levis and others, and the Modular Command and Control Evaluation Structure (MCES) method by R. Sweet and others. However, the index method is similar to the traditional Lanchester equation method, which takes into account precise data calculation and comparison while ignoring the uncertainty. With regard to the grey system, a colored PETRI net theory is used in order to better reflect the information flow, order, and network structure in the system, but its rules of information movement are too harsh and the calculation is very difficult and complex. As for the ADC method, the final measurements obtained from its matrix operations have no corresponding physical meaning to the three elements. The SEA method is based on the belief that the efficiency index should reflect the degree to which a system can fulfill its tasks under a given environment and conditions, and the rate of the coincidence between the system trajectory and the trajectory demanded by mining tasks should be a probability, which is generally difficult to represent. The Delphi method and the AHP method are relatively more common. The Delphi method is good at choosing attributes but has too much qualitative analysis, and the AHP method controls the consistency among attributes layer by layer, but requires excessive quantitative calculations.

All of these methods do not take the features of spatial data into consideration and cannot be transplanted entirely into SDM. They also ignore the uncertainty and do not have a valid, convenient model for the exchange between the qualitative and quantitative analysis, whose results have no definite physical meaning. Thus, it is very hard to promote the use of the above data cleaning methods because they do take into account a user's initiative in the system.

As a result, when multi-source or heterogeneous spatial data are assembled together, there may be high-dimensional data, different data structures, different projection systems, and different measurement units, as well as the problem of abnormal data, noisy data, and even errors. Spatial data cleaning, or spatial data scrubbing in advance, therefore becomes an important procedure in the data preparation stage of SDM. Further study of efficient spatial data cleaning algorithms and the examination and correction of data abnormalities are of realistic significance.

4.3 Characteristics and Contents of Spatial Data Cleaning

Because a variety of errors and abnormalities exist in spatial data, in a broad sense, every process undertaken that contributes to the quality improvement of spatial data is spatial data cleaning. In a narrow sense, spatial data cleaning is understanding the meaning of a field in a database and its relationship with other fields; to examine the integrity and consistency of spatial data; setting cleaning rules according to practical tasks; and with the use of query tools, employing statistical methods and artificial intelligence, filling up the lost data, disposing of noisy data, calibrating spatial data, and improving the accuracy and overall availability so as to ensure spatial data cleanliness and to make the data suitable for the spatial data processing that follows (Faryyad et al. 1996).

4.3.1 Fundamental Characteristics

Spatial data cleaning is definitely not just about simply updating previous records to correct spatial data. A serious spatial data cleaning process involves the analysis and re-distribution of spatial data to solve spatial data duplication, both inside a single spatial data source and among multiple spatial data sources, as well as the inconsistency of the data content itself, which is not just inconsistency in form. For example, the inconsistency of models and codes can be handled in combination with the spatial data extraction process in which the conversion of models and codes can be completed. It is a relatively simple and mechanical process. Because SDM is a case-by-case application, it is difficult to build mathematical models. Also, because the cleaning methods are closely related to spatial data samples, even the same method may produce very different experimental results in various contexts of application. It is therefore difficult to conclude that a general procedure and method is best. The typical data cleaning process models include Enterprise/Integrator, Trillium, Bohn, and Kimball.

4.3.2 Essential Contents

Spatial data cleaning mainly consists of confirming data input, eliminating null values errors, making sure spatial data values are set in the range of definition, removing excessive spatial data, solving the conflicts in spatial data, ensuring the reasonable definition and use of spatial data values, and formulating and employing standards.

The available tools fall into three classifications: (1) data migration, which allows simple transformation rules, such as Warehouse Manager from the company Prism; (2) data scrubbing, which uses domain-specific knowledge to clean spatial data and employs parsing and fuzzy matching technology to carry out multi-source spatial data cleaning (e.g., Integrity and Trillium can specify the "relative cleanliness" of sources); and (3) data auditing, which finds rules and relations through statistical analysis of spatial data.

Choosing the right target spatial data is also a necessary part of spatial data cleaning; in fact, it is the top priority. A feature of SDM is its case-by-case analysis. For a given specific task, not all the attributes for the entities in spatial databases have impacts; rather, it varies with different attributes. Consequently, it is necessary to consider SDM as system engineering and to choose target data and to ascertain the weight of the role played by the different attributes of target spatial data. For example, for mining land prices, the point, line, and area factors have different impacts on the results.

4.4 Systematic Error Cleaning

Systematic errors can be understood as the reproducible errors caused by some objective reasons and are generally described by a certain rule. For example, image systematic errors can be shown and included by some functional relationships of the pixel position. The changing parts of systematic errors can be regarded as a time series in order to achieve adjustment in a stochastic model or can be simply regarded as part of the stochastic errors.

The relevant parameters of the systematic errors can be obtained, such as by laboratory calibration, so that the effects of the systematic errors are removed in advance before surveying adjustment. However, the adjusted results may show that there are still some certain systematic errors after adjustment. Thus, a strict adjustment method, such as the bundle method, cannot achieve the most accurate results. Also, there is a certain discrepancy between the actual accuracy and the expected accuracy in theory. Static laboratory calibration therefore cannot exactly represent the properties of an actual dynamic photogrammetric system and there must be some systematic errors, which are always difficult to estimate and measure.

In the decade from 1970 to 1980, a great deal of photogrammetry research was conducted around the world to address this problem. The third committee of the International Society for Photogrammetry and Remote Sensing (ISPRS) set up a working group to organize an international cooperative experimental study of this topic, which resulted in several methods to compensate for systematic errors. Systematic errors can be well compensated with these methods in order to make regional network adjustment achieve or close to the expected theoretic accuracy. At the International Photogrammetry Conference in Hamburg, Germany in 1980, Ackermann and Kilpelä summarized the compensation of systematic errors. Thereafter, the research on this topic focused on automatically compensating the additional parameters of systematic errors, overcoming excessive parameterization, the internal and external reliability of additional parameters measurement, and compensation of the relevant, non-constant systematic errors.

Systematic errors can be reduced by appropriate observation methods or regarded as additional parameters to be plugged into the adjustment in order to compensate them effectively. In principle, the methods of compensating systematic errors can be classified into the direct compensation method and the indirect compensation method.

4.4.1 Direct Compensation Method

The proving ground calibration method, introduced by G. Kupfer, is a direct compensation method. Considering that conventional laboratory calibration cannot completely represent the actual process of accessing photo data, Kupfer proposed determining the parameter value of compensating systematic errors with proving ground calibration under real photographic flight conditions. Assuming that the circumstances of the photogrammetric conditions are basically unchanged (i.e., camera, photographic time, atmospheric conditions, photographic material, photo processing conditions, measuring instruments, and observers), using this group of parameters will compensate for the actual systematic errors in regional adjustment. Due to the imitativeness of this method, its advantages are that it can correct systematic errors in a relatively accurate and reliable manner and will not increase the calculation workload of regional adjustment. Its disadvantages are that it increases proving ground calibration periodically and the photogrammetry system must remain unchanged in each cycle.

There is a simple direct compensation method—a self-canceling method— where aerial photographs are taken of two mutually perpendicular courses in which the side overlaps are 60 % in one testing zone in order to obtain the four photogrammetric data (four overlapped testing zones) in the testing zone; regional adjustment of these four groups of data occur at the same time. Although the deformation rules of photography systems are similar in the internal of each group of data, the systematic deformation of each group is random. Thus, adjusting four groups of data simultaneously can self-cancel or reduce some systematic errors. Testing showed that it can remove the effects of systematic errors reasonably well. The disadvantage of this method is that the workload of photogrammetry rapidly multiplies. Therefore, it would be used only in high-encrypted small areas, such as geodetic surveying networks with encryption three or four. This method also should cooperate with other compensation methods.

4.4.2 Indirect Compensation Method

There are two types of indirect compensation methods, which often cooperate with other compensation methods.

The self-calibration method uses the concurrent overall adjustment of additional parameters. It chooses a systematic error model consisting of several parameters and regards these additional parameters as unknown numbers or addresses them as weighted observation data to calculate them with other parameters in order to self-calibrate and self-cancel the effects of systematic errors in the adjustment process. This method does not increase any actual workload of photography and measurement; it also could avoid some inconvenient laboratory test work. Because the compensation is conducted by itself in the adjustment process, the accuracy would obviously increase if it is addressed properly. Its disadvantages are that it can compensate only the systematic errors reflected by the available connection points and control points; the selection of additional parameters is man-made or experiential and the results therefore would be different according to different selections; there is the possibility of worsening the calculation results due to the strong correlation of the additional parameters and the strong correlation between the additional parameters and other unknown parameters; and the calculation work obviously increases. Because both the advantages and disadvantages of this method are salient, there are many researchers who work on this method.

Another indirect compensation method is the post-test compensation method, which was first introduced by Masson D'Autume. This method aims at the residual errors of the observation value of the original photos or model coordinates and does not change the original adjustment program. It analytically processes the residual errors of the picture point (or model point) after several iterative computations and calculates the systematic errors' correction values of several sub-block interior nodes of the picture point (or model point). Then, the systematic errors' correction values of all the picture points (or model points) are calculated by the two-dimensional interpolation method and its next calculation is performed after correction. This post-test method of determining the systematic errors' correction value by repeatedly analyzing the residual errors can reliably modify the results and is easier to insert in all kinds of existing adjustment systems. It is called post-processing if least square filter and estimation of the coordinate residual errors of the ground control points are conducted after adjustment. Therefore, in a broad sense, this method also is a post-test method. This post-processing requires the ground control to have a certain density, which mainly removes the stress produced in the ground control network in order to better represent the photogrammetric coordinates in the geodetic coordinate system.

All of the above-mentioned methods also can be combined. For example, the post-test compensation method can be used in a self-calibration regional network and self-calibration adjustment can be used in duplicate photographic areas; these two methods can be combined with the proving ground calibration method. Through such combinations, the best strategy results can be achieved and the accuracy can be increased as much as possible while the workload is increased accordingly.

4.5 Stochastic Error Cleaning

The classic stochastic error processing model includes indirect adjustment and condition adjustment, indirect adjustment with conditions, condition adjustment with parameters, and condition adjustment with conditions (summary models). All of the above models treat the parameters as non-random variables. If the prior information of the parameters is considered, many models can be used, such as the least square collocation model, the least square filter estimation model, and the Bayesian estimation model. Wang (2002) summarized the above-mentioned models by using generalized linear models. Its principles are described in the following sections.

4.5.1 Function Model

The function model of the generalized linear summary model is defined as

$$
\left.
\begin{array}{l}
L = BX + AY + \Delta \\
CX + C_0 = 0
\end{array}
\right\}
\tag{4.8}
$$

where L refers to the observation vector of n; A is the known coefficient matrix of $n \times m$, in which $A = (A_1, 0)$; A_1 is the known coefficient matrix of $n \times m_1$; 0 is the known coefficient matrix of $n \times m_2$; B is the known coefficient matrix of $n \times u$; C is the known coefficient matrix of $d \times u$; C_0 is the constant vector of $d \times 1$; X is the non-random parameter vector of $u \times 1$; Y is the random parameter vector of $m \times 1$; $Y = \begin{pmatrix} S \\ S^w \end{pmatrix}$; S is the observed point random parameter of $m_1 \times 1$; S^w is the point not observed random parameter of $m_2 \times 1$; Δ is the observation error vector of $n \times 1$; n is the number of observations; u is the number of non-random parameters; $u \geq t$; m is the number of random parameters; and $m_1 + m_2 = m$; $d = u - t$ is the number of non-random parameters that are not independent.

4.5.2 Random Model

Let the unit weight variance $\sigma^2 = 1$, then

$$
E(\Delta) = 0, E(Y) = \begin{pmatrix} E(S) \\ E(S^w) \end{pmatrix} = \begin{pmatrix} \mu_S \\ \mu_{S^w} \end{pmatrix}
$$

$$
\mathrm{Var}(\Delta) = Q_{\Delta\Delta} = P_\Delta^{-1}
$$

$$
\mathrm{Var}(Y) = Q_{yy} = \begin{pmatrix} Q_{SS} & Q_{SS^w} \\ Q_{S^wS} & Q_{S^wS^w} \end{pmatrix} = P_y^{-1}
$$

$$
\mathrm{Var}(L) = Q_{LL} = P^{-1} = \left(Q_{\Delta\Delta} + B_1 Q_{SS} B_1' \right)
$$

$$
\mathrm{Cov}(\Delta, Y) = 0, \quad \mathrm{Cov}(Y, \Delta) = 0
$$

4.5.3 Estimation Equation

Let

$$L_y = \begin{pmatrix} L_S \\ L_{S^W} \end{pmatrix} = E(Y) = \begin{pmatrix} \mu_S \\ \mu_{S^W} \end{pmatrix} \tag{4.9}$$

Then, the virtual observation equation is

$$L_y = Y + \Delta_y = \begin{pmatrix} S \\ S^W \end{pmatrix} + \begin{pmatrix} \Delta_S \\ \Delta_{S^W} \end{pmatrix} \tag{4.10}$$

The error equation and condition equation are obtained from Eqs. (4.9) and (4.10):

$$\left. \begin{array}{l} V_y = \hat{Y} - L_y \\ V = B\hat{X} + A\hat{Y} - L \\ C\hat{X} + C_0 = 0 \end{array} \right\} \tag{4.11}$$

Let

$$\bar{L} = \begin{pmatrix} L_y \\ L \end{pmatrix}, \ \bar{V} = \begin{pmatrix} V_y \\ V \end{pmatrix}, \ \hat{Z} = \begin{pmatrix} \hat{X} \\ \hat{Y} \end{pmatrix} = \begin{pmatrix} \hat{X} \\ \hat{S} \\ \hat{S}^W \end{pmatrix},$$

$$\bar{\Delta} = \begin{pmatrix} \Delta_y \\ \Delta \end{pmatrix}, \ \bar{B} = \begin{pmatrix} 0 & E \\ B & A \end{pmatrix}, \ \bar{C} = \begin{pmatrix} C & 0 \end{pmatrix}$$

We obtain

$$\text{Var}(\bar{\Delta}) = \begin{pmatrix} Q_{yy} & 0 \\ 0 & Q_{\Delta\Delta} \end{pmatrix}, \quad \bar{P} = \begin{pmatrix} Q_{yy}^{-1} & 0 \\ 0 & Q_{\Delta\Delta}^{-1} \end{pmatrix} = \begin{pmatrix} P_y & 0 \\ 0 & P_\Delta \end{pmatrix} \tag{4.12}$$

$$\left. \begin{array}{l} \bar{V} = \bar{B}\hat{Z} - \bar{L} \\ \bar{C}\hat{Z} + C_0 = 0 \end{array} \right\} \tag{4.13}$$

Based on the principle of generalized least square

$$\bar{V}'\bar{P}\bar{V} = V'P_\Delta V + V_y'P_y V_y = \min \tag{4.14}$$

the function is established and solved for the parameter $\hat{Z}$

$$\Phi = \bar{V}'\bar{P}\bar{V} + 2K'\left(\bar{C}\hat{Z} + C_0\right) \tag{4.15}$$

where K is the vector of the connection number with $d \times 1$.

Equation (4.15) is used to compute the partial derivative of $\hat{Z}$ and to make it 0. After conversion, the following equation is obtained:

$$\bar{B}'\bar{P}\bar{V} + \bar{C}'K = 0 \tag{4.16}$$

From Eqs. (4.13) and (4.15), the normal equation can be obtained:

$$\left.\begin{array}{l} \bar{B}'\bar{P}\bar{B}\hat{Z} + \bar{C}'K - \bar{B}'\bar{P}\bar{L} = 0 \\ \bar{C}\hat{Z} + C_0 = 0 \end{array}\right\} \tag{4.17}$$

Let

$$\bar{N} = \bar{B}'\bar{P}\bar{B}, \quad \bar{U} = \bar{B}'\bar{P}\bar{L} = \begin{pmatrix} B'P_\Delta L \\ P_y L_y + A'P_\Delta L \end{pmatrix}$$

The following equation is obtained:

$$\begin{pmatrix} \bar{N} & \bar{C}' \\ \bar{C} & 0 \end{pmatrix}\begin{pmatrix} \hat{Z} \\ K \end{pmatrix} - \begin{pmatrix} \bar{U} \\ -C_0 \end{pmatrix} = 0 \tag{4.18}$$

Let

$$\bar{B}'\bar{P}\bar{B} = \begin{pmatrix} N_{11} & N_{12} \\ N_{21} & N_{22} \end{pmatrix} = \begin{pmatrix} B'P_\Delta B & B'P_\Delta A \\ B'P_\Delta A & P_y + A'P_\Delta A \end{pmatrix}$$

Then, Eq. (4.18) can be noted as follows:

$$\begin{pmatrix} N_{11} & N_{12} & C' \\ N_{21} & N_{22} & 0 \\ C & 0 & 0 \end{pmatrix}\begin{pmatrix} \hat{X} \\ \hat{Y} \\ K \end{pmatrix} = \begin{pmatrix} B'P_\Delta L \\ P_y L_y + A'P_\Delta L \\ -C_0 \end{pmatrix} \tag{4.19}$$

4.5.4 Various Special Circumstances

4.5.4.1 The Least Square Collocation

When $u = t$—that is, there are only t independent non-random parameters in Eq. (4.8)—$d = u - t = 0$, $u = t$ is the result. At this moment, $C = 0$, $C_0 = 0$, then Eq. (4.8) changes to

$$L = BX + AY + \Delta \tag{4.20}$$

Equation (4.20) is the function model of the least square collocation.

Because $C = 0$, $C_0 = 0$, Eq. (4.19) changes to

$$\begin{pmatrix} N_{11} & N_{12} \\ N_{21} & N_{22} \end{pmatrix}\begin{pmatrix} \hat{X} \\ \hat{Y} \end{pmatrix} = \begin{pmatrix} B'P_\Delta L \\ P_y L_y + A'P_\Delta L \end{pmatrix} \tag{4.21}$$

It is known that coefficient matrix B is full column rank from $u = t$. Therefore, the solution to Eq. (4.21) is

$$
\begin{pmatrix} \hat{X} \\ \hat{Y} \end{pmatrix} = \begin{pmatrix} N_{11}^{-1} + N_{11}^{-1} N_{12} R^{-1} N_{21} N_{11}^{-1} & -N_{11}^{-1} N_{12} R^{-1} \\ -R^{-1} N_{21} N_{11}^{-1} & R^{-1} \end{pmatrix} \begin{pmatrix} B' P_{\Delta} L \\ A' P_{\Delta} L + P_y L_y \end{pmatrix}
$$

$$(4.22)$$

where, $R = P_y + A' P_{\Delta} A - A' P_{\Delta} B (B' P_{\Delta} B)^{-1} B' P_{\Delta} A$

That is,

$$
\left.\begin{aligned}
\hat{X} &= \left[B' (Q_{\Delta\Delta} + A_1 Q_{SS} A_1')^{-1} B \right]^{-1} B' (Q_{\Delta\Delta} + A_1 Q_{SS} A_1')^{-1} (L - A_1 \mu_S) \\
\hat{Y} &= L_Y + Q_{YY} A' (Q_{\Delta\Delta} + A Q_{YY} A')^{-1} \left(L - B\hat{X} - A L_Y \right)
\end{aligned}\right\}
$$

$$(4.23)$$

Consider

$$
A = (A_1 \quad 0), \quad \hat{Y} = \begin{pmatrix} \hat{S} \\ \hat{S}^w \end{pmatrix}, \quad Q_{YY} = \begin{pmatrix} Q_{SS} & Q_{SS^w} \\ Q_{S^wS} & Q_{S^wS^w} \end{pmatrix}, \quad L_Y = \begin{pmatrix} \mu_S \\ \mu_{S^w} \end{pmatrix}
$$

We obtain

$$
\left.\begin{aligned}
\hat{S} &= \mu_S + Q_{SS} A_1' (Q_{\Delta\Delta} + A_1 Q_{SS} A_1')^{-1} \left(L - B\hat{X} - A_1 \mu_S \right) \\
\hat{S}^w &= \mu_{S^w} + Q_{S^wS} A_1' (Q_{\Delta\Delta} + A_1 Q_{SS} A_1')^{-1} \left(L - B\hat{X} - A_1 \mu_S \right)
\end{aligned}\right\}
$$

$$(4.24)$$

4.5.4.2   The Least Square Filter and Estimation

When $u = 0$—that is, Eq. (4.8) does not include non-random parameters—$B = 0$, $C = 0$, $C_0 = 0$ is the result. Therefore, Eq. (4.8) changes to

$$
L = AY + \Delta \tag{4.25}
$$

Because Y is the random parameter, Eq. (4.25) is the least square filter and estimation model (Wang 2002). Bringing $B = 0$ into Eqs. (4.12) and (4.13) makes it possible to obtain

$$
\left.\begin{aligned}
\hat{Y} &= L_Y + Q_{YY} A' (Q_{\Delta\Delta} + A Q_{YY} A')^{-1} (L - A L_Y) \\
\hat{S} &= \mu_S + Q_{SS} A_1' (Q_{\Delta\Delta} + A_1 Q_{SS} A_1')^{-1} (L - A_1 \mu_S) \\
\hat{S}^w &= \mu_{S^w} + Q_{S^wS} A_1' (Q_{\Delta\Delta} + A_1 D_{SS} A_1^T)^{-1} (L - A_1 \mu_S)
\end{aligned}\right\}
$$

$$(4.26)$$

According to the third line of Eq. (4.26), although it is the random parameter on the point that is not observed and there is no function relation between it and the observations, if we know the covariance matrix Q_{S^wS} of S^w and S, S^w can be estimated. This indicates that the understanding of the statistical correlation between the individual values is beneficial.

4.5.4.3 Bayesian Estimation

In Eq. (4.19), when $B = 0$, $C = 0$, $C_0 = 0$, then $N_{11} = 0$, $N_{12} = 0$, $N_{21} = 0$, considering $N_{22} = P_Y + A'P_\Delta A$, then Eq. (4.19) changes to

$$\left(A'P_\Delta A + P_y\right)\hat{Y} = \left(A'P_\Delta L + P_y L_y\right) \tag{4.27}$$

We obtain

$$\hat{Y} = \left(A'P_\Delta A + P_y\right)^{-1}\left(A'P_\Delta L + P_y L_y\right) \tag{4.28}$$

Equation (4.28) is the Bayesian estimation. It is clear that Bayesian estimation is a special case of the generalized linear summary model in parameter estimation.

4.5.4.4 Linear Model with Linear Constraints

When $m = 0$—that is, Eq. (4.8) does not include random parameters—then $A = 0$. At this moment, Eq. (4.8) changes to

$$\left.\begin{array}{l} L = BX + \Delta \\ CX + C_0 = 0 \end{array}\right\} \tag{4.29}$$

Equation (4.29) is the linear model with linear constraints. The corresponding error equation and condition equation is

$$\left.\begin{array}{l} V = B\hat{X} - L \\ C\hat{X} + C_0 = 0 \end{array}\right\}$$

As $m = 0$, Eq. (4.29) becomes

$$\begin{pmatrix} B'PB & C' \\ C & 0 \end{pmatrix}\begin{pmatrix} \hat{X} \\ K \end{pmatrix} = \begin{pmatrix} B'PL \\ -C_0 \end{pmatrix}$$

Its solution is

$$\begin{pmatrix} \hat{X} \\ K \end{pmatrix} = \begin{pmatrix} B'PB & C' \\ C & 0 \end{pmatrix}^{-1}\begin{pmatrix} B'PL \\ -C_0 \end{pmatrix}$$

It is clear that the linear model with linear constraints is also a special case of the generalized linear summary model.

4.5.4.5 The General Linear Model

When $m = 0$—that is, Eq. (4.8) does not include the random parameter $A = 0$—and if $u = t$—that is, there are only t independent parameters—then $d = u - t = 0$, $u = t$. Thus, $C = 0$, $C_0 = 0$. Then, Eq. (4.8) changes to

$$L = BX + \Delta$$

This is the general linear model, whose solution is

$$\hat{X}_{LS} = \left(B'PB\right)^{-1}B'PL$$

From the above derivation, Eq. (4.8) includes the least square collocation model, the least square filter estimation model, the Bayesian estimation, the linear model with linear constraints, and the generalized linear summary model of the general linear model. As for more complex nonlinear model parameter estimation theories and their applications, readers can refer to Wang (2002).

4.6 Gross Error Cleaning

Reliability gives the adjustment system the ability to detect gross errors, indicating the impact of the undetectable gross errors on the adjustment results as well as the statistical measurements in the detection and discovery of the gross errors. Internal reliability refers to the ability to discover the gross errors in detectable observations, which is commonly measured by the minimum or the lower limit of detectable gross errors. The smaller and lower the limit is, the stronger the reliability is. External reliability, on the other hand, is used to represent the impact of the undetectable gross errors on the adjustment results or the adjustment result's function due to its limited number of detectable gross errors (gross errors that are lower than the limits are undetectable). The smaller the impact of the undetectable gross errors on the results is, the stronger the external reliability is. The redundant observations also are key in the detection of gross errors. As for both internal and external reliability, the more redundant the observations are, the better.

Another demand of reliability research is to search for the location of gross errors. Locating gross errors is about finding a way of automatically discovering the existence of gross errors in the process of adjustments and pointing out their locations precisely so as to reject them from the adjustment. This is an algorithmic problem more than a theoretical issue because an algorithm is required to carry out the process of automatic detection controlled by programs based on different adjustment systems and manifold types of gross errors that are likely to arise. In measurements, there are two specific ways in which the rejection of gross errors and dealing with them can be generally categorized:

- *Mean-shift model.* Driven by the reliability theory introduced by Baarda, this model makes use of data snooping or distributed snooping so as to discover and reject the gross errors by incorporating them into the function model.
- *Variance inflation model.* By incorporating the gross errors into a random model, this model makes use of the testing functions (driven by the variance estimation theory or posterior variance estimation theory introduced by Robust) so as to realize the automatic rejection of gross errors after giving them an extremely small weight in successive iterative adjustments.

4.6.1 The Reliability of the Adjustment System

Reliability research has two major tasks. One is a capacity study of the adjustment system in discovering and distinguishing the gross errors of different models as well as the impact of the undetectable and indistinguishable model errors on the adjustment results in theory. The other task is the quest for methods that automatically discover and distinguish model errors while determining the location of model errors in the process of adjustments. The former task applies to the reliability analysis and optimal design method for a system that could design the best graphics in the surveying area and meets the requirements as an integration of precision, reliability, and economy. The latter task is able to perfect current manifold adjustment programs by lifting the adjustment calculation to a higher level of automation. As a result, the measurement results that are provided to the different sectors of the national economy will not be limited to the geodetic coordinates of the points but also their numerical precision and the numerical value of their reliability.

Reliability research is based on the mathematical statistical hypothesis test. The classical hypothesis testing theory was introduced by Neyman and Pearson in 1993. In the field of survey adjustment, reliability theory was proposed by Baarda in 1967–1968. Driven by a single one-dimensional alternative hypothesis, Baarda's reliability theory studies the capacity of the adjustment system in discovering single model errors and the impact of the undetectable model errors on the adjustment results. The former is called *internal reliability* and the latter is called *external reliability*. Here, the model errors refer to the gross errors and system errors. In addition, starting from the known unit weight variance, Baarda also derived data snooping to test gross errors, which considers residual standard errors that follow the normal distribution as statistical measurements.

Förstner and Koch later applied this theory to a single multidimensional alternative hypothesis so as to enable it to discover multiple model errors. From the perspective of the statistics of single gross error test, Förstner and Koch calculated the amount of testing needed for an unknown variance factor. Pope and Koch calculated the test variable. Among several gross error tests, the F test variable was calculated by Förstner and Koch. In 1983, Förstner introduced the possibility of

Table 4.1 Adjustment system reliability research methods

Researchers	Number of candidate hypothesis	Dimensions of the alternative hypothesis	Function index	Gross error test variable
Baarda	1	One dimension	Single error discoverability	Individual not related to gross errors in observations
Koch	1	Multi-dimension	Multi-error discoverability	Single gross error vector
Förstner	2	One dimension	Single error discoverability and distinguishability	Single gross error vector
Li Deren	2	Multi-dimension	Multi-error discoverability and distinguishability	Single gross error vector

distinguishing model errors and, from two one-dimensional alternative hypotheses, drew the conclusion that distinguishing the possibility depends on the relative coefficients of the variable of test. Li Deren, while completing his doctoral dissertation in Germany, introduced the theory that an adjustment results system has distinguishability and reliability on the basis that a Gauss-Markov model contains two multidimensional alternative hypotheses. Therefore, Li Deren extended Baard's theory from one dimension to multiple dimensions and unified least squares and robust estimation. Table 4.1 presents the research methods of the reliability of the adjustment result system.

The theory of distinguishability and reliability proposed by Li Deren, which can be used to study the ability of systems to find and distinguish model errors and the influence of model errors that are impossible to distinguish from other model errors in the adjustment results, provides the basic research theory and quantitative measurement for distinguishing and positioning model errors. If the unknown parameter vector ∇_s is decomposed into a unit vector S that indicates direction and a scalar $\nabla(S)$ that indicates the size—that is, $\nabla_S = S\nabla(S)$—and assumes that decentralized parameters are unrelated to the direction—that is, $\delta_0(S) = \delta_0$—then the method to determine the reliability of the survey adjustment system (Li and Yuan 2002) is as follows:

1. Give the basic designed matrix B and the weight matrix $P = Q_\mu^{-1}$ of observations
2. Give the model errors $H\nabla_s$ and its coefficient matrix H
3. Calculate $Q_{vv} = P^{-1} - B\left(B^T PB\right)^{-1} B^T$
4. Calculate $\underline{P}_{SS} = H^T PH, P_{SS} = H^T P Q_{vv} PH$
5. Calculate internal reliability and the eigenvalues and their eigenvectors

$$\nabla_0 S = \frac{\sigma_0 \delta_0 S}{\sqrt{S^T P_{SS} S}} \tag{4.30}$$

6. Calculate external reliability, the eigenvalues, and their eigenvectors

$$\left(\underline{P}_{SS} - P_{SS}\right)t = \frac{\underline{\delta}_0^2(S)}{\delta_0^2}P_{SS}t \tag{4.31}$$

In this way, the vector length $\underline{\delta}_0^2(S)$ can be evaluated in different directions.

In his book entitled *Sequel to Principle of Photogrammetry*, Wang Zhizhuo (2007), a member of the Chinese Academy of Science, wrote: "In the aspect of distinguishable theory, i.e., a reliability theory under double alternative hypothesis, German Förstner and Chinese Li Deren have made effective researches." He further specifically introduced the iteration method with variable weights, which is presented in Sect. 7 of Chap. 4 (Wang 2007), and named it the "Li Deren method."

4.6.2 Data Snooping

Data snooping, introduced by Baarda, consists of standard normal statistics by the observed correction v_i according to the adjustment result, namely:

$$w_i = \frac{v_i}{\sigma_{vi}} = \frac{v_i}{\sigma_0\sqrt{q_{ii} - B_i\left(B^{\mathrm{T}}PB\right)^{-1}B_i^{\mathrm{T}}}} \backsim N(0, 1)$$

where v_i is the correction of the ith observation, calculated by the error equation; σ_{v_i} is the middle error of v_i; σ_0 is the middle error with unit weight; q_{ii} is the ith element on the main diagonal of total inverse matrix P^{-1}; B_i is the ith row of the matrix of error equation; and $B^{\mathrm{T}}PB$ is the coefficient matrix of normal equation.

As for the statistic w_i, which is used to detect gross errors, according to the recognized significance level of Baarda in which $\alpha = 0.001$, it can be obtained that $w_i = 3.3$. With $N(0, 1)$ as the null hypothesis, if $|v_i| < 3.3\sigma_{v_i}$, the null hypothesis can be accepted and there are no gross errors under this condition. However, if $|v_i| \geq 3.3\sigma_{v_i}$, the null hypothesis must be rejected and there are gross errors.

4.6.3 The Iteration Method with Selected Weights

It can be seen from the discussion in the last section that it is difficult to position the gross error, especially several gross errors, when they are incorporated in a function model. If the observations containing gross errors are regarded as the sample of large variance of the same expectation, then it can lead to the iteration method with selected weights for positioning gross errors.

The basic theory of the method is that, due to unknown gross errors, the adjustment method is still least squares. However, after each adjustment, according to the residuals and other parameters and the weight function selected, the weight of the iterative adjustment in the next step is calculated and included in the adjustment calculation. If the weight function is chosen properly and the gross errors can be positioned, the observations containing gross errors become smaller and smaller until they reach zero. When the iteration is suspended, the relevant residuals point out the values of the gross errors; subsequently, the results of the adjustment will not be affected by gross errors. Thus, automatic positioning and correctness are realized. The iteration method with selected weights can be used to study the discoverability and measurability of the model errors (gross errors, system errors, or distortions) in any adjustment system, and the influence of the immeasurable model errors on the result of adjustment thus is calculated.

The method starts from the minimal condition $\sum p_i v_i^2 \to$ min, in which the weight function is $p_i^{(v+1)} = f\left(v_i^{(v)}, \dots\right)$, $(v = 1, 2, 3, \dots)$. Some known functions include the iterative method with minimum norm, the Danish method, the weighted data snooping method, the option iteration method proceeding from the principle of Robust, and the option iteration method deduced from the posteriori variance estimation principle. The various types of functions can be divided into the residual function, the standardized residual function, and the variance valuation function according to content, and into the power function and the exponential function according to the form.

To position gross errors effectively, the selection of a weight function should meet the following conditions:

1. The weight of the observation containing gross errors should be inclined to zero through iteration; that is, the redundant observation should be inclined to one.
2. The weight of the observation containing no gross errors should be equal to the weight of the group observation (given before the posteriori variance is tested or calculated) when the iteration suspends. The weight is one when there is only one group of observation values with equal precision.
3. The selection of the weight function should guarantee the contraction of iteration at a faster speed. As for the iteration method with selected weights, it is an ideal positioning gross errors weight function to designate the posteriori variance of all observations as the basic component in the form of an exponential function.

4.6.4 Iteration with the Selected Weights from Robust Estimation

The Robust estimation method, which is a special method of maximum likelihood estimation, can guarantee that the estimated parameters are not affected or minimally affected by model errors (gross errors). This method is mainly used to find gross errors and to position them. Because this method can resist the external disturbance brought by gross errors, it is named the Robust estimation method.

However, the usual least squares estimation and maximum likelihood estimation have a stronger possibility of gross errors; therefore, the estimates are likely to be unable to resist external disturbance. Also, the usual least squares estimation is not the Robust method.

The equation of Robust estimation is as follows (Wang 2002):

$$V = B\hat{X}_R - L = \begin{pmatrix} b_1 \\ b_2 \\ \vdots \\ b_n \end{pmatrix} \hat{X}_R - \begin{pmatrix} L_1 \\ L_2 \\ \vdots \\ L_n \end{pmatrix} \tag{4.32}$$

where b_i is the ith row of the designated matrix and $\hat{X}_R$ is the robust estimate for unknown parameter X. Suppose the weight of the ith observation is p_i; the calculation of the unknown parameter is to solve the optimization problem.

$$\sum_{i=1}^{n} p_i \rho(\upsilon_i) = \sum_{i=1}^{n} p_i \rho\left(b_i \hat{X}_R - L_i\right) = \min \tag{4.33}$$

In Eq. (4.33), take the derivative of $\hat{X}_R$, make it zero, and note $\varphi(v_i) = \partial_\rho / \partial_{vi}$. Then

$$\sum_{i=1}^{n} p_i \phi(\upsilon_i) b_i = 0 \tag{4.34}$$

Let

$$\varphi(\upsilon_i) / \upsilon_i = W_i, \bar{P}_{ii} = P_i W_i \tag{4.35}$$

In Eq. (4.35), W_i is called the weight factor and $\bar{P}_{ii}$ is the equivalent weight. Equation (4.34) can be noted as follows:

$$B'\bar{P}V = 0 \tag{4.36}$$

Taking Eq. (4.29) into Eq. (4.36) will lead to

$$B'\bar{P}B\hat{X}_R - B'\bar{P}L = 0 \tag{4.37}$$

Then,

$$\hat{X}_R = \left(B'\bar{P}B\right)^{-1} B'\bar{P}L \tag{4.38}$$

Because of the introduction of $\bar{P}$, Eq. (4.38) can both resist the disturbance of gross errors and keep the form of least squares estimation. Equation (4.38) was called the robust least squares estimation by Zhou Jiangwen.

It can be observed from Eq. (4.36) that the weight factor W_i is a nonlinear function of the residuals. To make the equivalent weight more practical, it is necessary to improve W_i through the iteration calculation. The maximum likelihood estimation can be made robust through the following actions (Fig. 4.5):

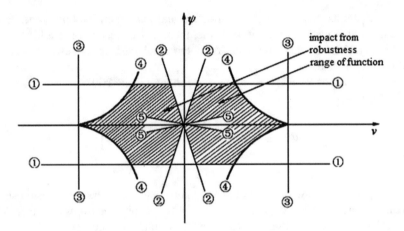

Fig. 4.5 Principles affecting the option of functions

1. The impact of gross errors on adjustment results must have an upper limit.
2. The impact of minor gross errors on the adjustment results should not reach the upper limit once; instead, it should increase gradually with the growth of errors. That is, the increase rate affecting the function (the influence of an added observation on the estimation) should have an upper limit.
3. The gross errors above a certain limit should not affect the adjustment result; that is, an influence value of null should be set.
4. The changes in minor gross errors should not bring about big changes in results; that is, the decline affecting the functions should not be too abrupt.
5. To make the estimate robust, the increase rate affecting the functions should also have a lower limit to guarantee fast contraction during calculation.

With the above five principles, we can obtain the optional robust scope, which affects the functions. If the options affecting the functions are in the scope, the estimate is called robust estimation.

It must be pointed out that these five principles are established for gross errors (i.e., genuine errors). As a matter of fact, users can use the observation residual value only and must change the influence function and the weight function into a correction function. This is a fatal deficiency in the robust estimation. The influence function in the classical least squares adjustment method can only meet robust conditions (2) and (5) and cannot meet the requirements of gross error proof, so it is not a robust adjustment method. The adjustment result of the least squares adjustment method will be seriously influenced in proportion to the gross errors. The small power exponent in the influence function in mini-norm can basically meet conditions (1) and (2); therefore, they can get part of the robust characteristics. The influence function in the Huber method already reaches the most important conditions of (1), (2), and (5), but large gross errors still have some impact on the adjustment result. The Hampel method meets all the requirements, and only condition

(4) is approximately met. The Danish method introduced by Krarup almost meets all the requirements except condition (3). The Stuttgart method satisfies all five robust conditions. In the beginning of the iteration, because of many gross errors, an observation with the large residual values is not abandoned casually; however, at the end of the iteration, even the smallest gross errors are removed strictly.

4.6.5 Iteration Supervised by Posteriori Variance Estimation

The above-mentioned weight functions are usually chosen from the empirical approach and the weight is represented by the function of correction. Because the correction is only the visible part of the genuine errors, the above-mentioned weight functions (except the Stuttgart method) do not take into account the geometrical conditions of adjustment. In fact, gross errors are thought to be a subsample of orthostatic, whose expectation is zero and the variance is large. The posteriori variance calculated through the posteriori variance of the least squares method can then be used to find the unusual large observation containing gross errors. Next, according to the classic definition that an observation posteriori variance is out of proportion to its weight, a relatively small weight is given to it to carry on the iteration adjustment, which makes it possible to position gross errors. This method was proposed by Li Deren in 1983.

$$\hat{\sigma}_i^2 = \frac{V_i^T V_i}{r_i} \quad (i = 1, 2, \ldots, k \text{ is the number of group}) \tag{4.39}$$

where, $r_i = \text{tr}(Q_{VV}P)_i$.

Then, the weight of each group's observation in the next iteration is

$$p_i^{(v+1)} = \left(\frac{\hat{\sigma}_0^2}{\hat{\sigma}_i^2} \right)^{(v)} \tag{4.40}$$

where, $\hat{\sigma}_0^2 = \frac{V^T PV}{r}$.

To find the gross errors in each group's observations, the variance estimate $\hat{\sigma}_{i,j}^2$ of any observation $l_{i,j}$ and the relative redundant observation r_{ij} are calculated. Equation (4.41) can be obtained from Eq. (4.39)

$$\hat{\sigma}_{i,j}^2 = \frac{v_{i,j}^2}{r_{i,j}} \tag{4.41}$$

where $r_{i,j} = q_{v_{i,jj}} p_{i,j}$.

Therefore the following statistics used for checking whether the variance is unusual—that is, whether the relative observation contains gross errors can be built.

H_0 hypothesis: $E\left(\hat{\sigma}_{i,j}^2\right) = E\left(\hat{\sigma}_i^2\right)$

Statistical parameter

$$T_{i,j} = \frac{\hat{\sigma}_{i,j}^2}{\hat{\sigma}_i^2} = \frac{v_{i,j}^2 p_i}{\hat{\sigma}_0^2 r_{i,j}} = \frac{v_{i,j}^2 p_i}{\hat{\sigma}_0^2 q_{v_i,jj} p_{i,j}} \tag{4.42}$$

Here, p_i can be considered the posteriori weight or a priori weight on test.

Suppose the observation $l_{i,j}$ does not contain gross errors (i.e., the hypothesis is correct), then, the statistic $T_{i,j}$ approximates the F distribution whose degree of freedom is 1 and r_i. If $T_{i,j} > F_{a,1,r_i}$, the variance of the observation is obviously different from the group's observation and probably contains gross errors. Then, the weight of the next iteration adjustment is calculated as the following weight function:

$$p_i^{(v+1)} = \begin{cases} \frac{\hat{\sigma}_0^2}{\hat{\sigma}_i^2} & \left(T_{i,j} < F_{a,1,r_i}\right) \\ \frac{\hat{\sigma}_0^2 r_{i,j}}{v_{i,j}^2} & \left(T_{i,j} \geq F_{a,1,r_i}\right) \end{cases} \tag{4.43}$$

As for the adjustment containing only one group of observations of the same precision, its statistic and weight functions are as follows:

$$T_i = \frac{v_i^2}{\hat{\sigma}_0^2 q_{v_{ii}} p_i} \tag{4.44}$$

$$p_i^{(v+1)} = \begin{cases} 1 & \left(T_i < F_{a,1,r}\right) \\ \frac{\hat{\sigma}_0^2 r_i}{v_i^2} & \left(T_i \geq F_{a,1,r}\right) \end{cases} \tag{4.45}$$

To compare this with Baarda's data snooping, the statistic parameters in Eq. (4.42) are taken. In the first iteration $p_i = p_{i,j}$. Therefore,

$$T_i^{\frac{1}{2}} = \frac{v_i}{\hat{\sigma}_0 \sqrt{q_{v_{ii}}}} = \tau_i \tag{4.46}$$

This is the variance τ during the data snooping of the unknown unit weight variance.

It can be observed that data snooping equals the first iteration of the method. As in the first iteration, the weight containing the gross error observation is incorrect and all the residual values and $\hat{\sigma}_0$ are affected; therefore, the estimate is not precise. However, if taking advantage of the method in which the posteriori variance estimate changes the observation weight during the process of iteration, the observation weight containing the gross errors will decline gradually to null and ultimately will have no influence on the adjustment results, making the estimate more precise.

4.7 Graphic and Image Cleaning

Spatial graphics and images from multiple sources may be derived from different time periods or sensors. During the imaging procedure, the spectral radiance and geometric distortion of spatial entities determine the quality of its graphics and images to a great extent. In order to make spatial graphics and images approach a primitive spatial entity to the utmost, it is necessary to correct and clean the radiation and geometric distortion (Wang et al. 2002).

4.7.1 The Correction of Radiation Deformation

The output of the sensors that generate graphics and images has a close relationship with the spectral radiance of the target. Cleaning the graphic and image data of radiation deformation mainly corrects the radiation factors that affect the quality of remote-sensing images, which include the spectral distribution characteristics of solar radiation, atmospheric transmission characteristics, solar altitudinal angle, position angle, spectral characteristics of surface features, altitude and position of sensors, and properties and record modes of sensors.

4.7.1.1 The Correction of Sensors' Radiation Intensity

These methods mainly correct the errors of the sensor systems and the missing data in accessing and transmitting. The manufacturing components and the properties (e.g., the spectral sensitivity and energy conversion sensitivity of every sensor) are different; therefore, the methods to correct systematic errors are different. For example, the systematic errors of sensor MSS are caused by the non-uniformity of gain and drift of every detecting element arrayed by the sensor's detector and the possible changes in work. Compensation for the noises of the radiation measurement value of sensor MSS can be performed according to the statistical analysis of the calibration grey wedge and image data generated from satellites. After the calibration parameters are obtained and the gain and drift of the voltage values are calculated, they can be corrected through each pixel of every scanning line, one by one. When the calibration data are absent or unreliable, statistical analysis can be used for calibration, which means using spatial entity data in the scanning field to calculate the total and average total of all the samples of every detecting element, determining the gain and drift of all the detecting elements in each band, and then correcting the data. Moreover, a signal transmission is needed before recording the signals received by the internal sensors. In the process of signal transmission, dark current is easy to generate, which decreases the signal-to-noise ratio. The sensitivity characteristics, response characteristics, position, altitude, and attitude of the sensors also can affect the quality of spatial graphics and images and need to be corrected.

4.7.1.2 The Correction of Atmospheric Scattering

The frequency low pass filter effect caused by the atmosphere absorbing and scattering electromagnetic waves and the atmospheric oscillation effects and changes in the radioactive properties of satellite remote sensing images. The main factor is atmospheric scattering. Scattering is the result of molecules and particles in the atmosphere, which can affect electromagnetic waves several times. The increasing intensity value caused by the scattering effect does not include the information of any target but will decrease the contrast ratios of images, lead to reductions in image resolution, and can have the following three serious effects on the radioactive properties of images: losing ground useful information of some short-wave bands; producing interference in radioactive properties between neighboring pixels; and forming sky light with cloud reflections. Therefore, the frequency low pass filter effect must be corrected; here, we present three methods to do so:

1. Correction by solving radiation equations entails substituting atmospheric data into radiation equations to calculate the gray value equivalent to the atmospheric scattering as the atmospheric approximate correction value. However, this method is not often used.
2. The field spectral measurement method considers the image intensity value that corresponds to the spectral data and images of measuring land features as the correction value of regression analysis calculating radiation.
3. Multi-band image comparison addresses the effects of the scattering that mainly occur in the short band and rarely influence the infrared band. This approach takes an infrared image as the standard image without any apparent influences and compares it with other band images using the histogram method and the regression analysis in certain fields. The difference is the scattering radioactive value needing correction.

4.7.1.3 The Correction of Illumination

The lighting conditions in photography (solar position, altitude angle, solar position angle, the irradiance of the light source caused by direct solar light and skylight, etc.) also can affect the quality of images. The solar altitudinal angle and the position angle are related to the irradiance and optical path length of the horizontal surface and affect the magnitude of directional reflectivity. When the solar altitude angle is 25–30°, the images acquired by photography can form the most stereoscopic shadow and appropriate images; however, such results are difficult to guarantee in actual photography. For satellite images under good atmospheric conditions, the changes in lighting conditions mainly refer to the changes of solar altitude while imaging solar altitude can be determined by the imaging time, season, and position and then corrected through these parameters. The correction of lighting conditions is realized by adjusting the average brightness in an image. The correction results can be obtained by determining the solar altitude angle, knowing the imaging season and position, calculating the correction constants, and then multiplying the constants by the value of each pixel. The solar light spot, which is the phenomenon where the solar border is brighter than the field around it when the sunlight is reflecting and diffusing on the surface can be corrected by reducing the border light. This can be achieved by calculating the shaded curve surfaces (the distortion parts caused by the solar light spot and the reduced border light in the shading change areas of images). Generally, the stable changing compositions extracted from images by Fourier analysis can be regarded as shaded curve surfaces.

4.7.1.4   Noise Removal

During the process of obtaining remote sensing images, the abnormal stripes (generally following a scanning stripe circle) and the spots caused by differences in the properties, interference, and breakdown of detectors could not only cause directly quotative error information but also can lead to bad results and need to be removed.

1. *Periodic noises*. The origin of periodic noises can be the periodic signals in raster scan and digital sampling institutions being coupled to the image electronic signals of electron-optical scanners or mechanical oscillation in the electronic scanner and magnetic tape recorder. The recorded images obstruct graphics stacking with the primitive scenery by periodicity with changing amplitude, frequency, and phase and produces periodic noises. The periodic noises reflected in two-dimensional images are along the appearance of scanning periodicity and distributed vertically with the scanning line and can be shown in two-dimensional Fourier spectrum. Therefore, the noises can be reduced by using Fourier transform in the frequency domain through band-pass or notch filter.
2. *Striping noises*. Striping noises are produced by equipment, such as changes in gain and drift of the sensors' detectors, data interrupt, and missing tape records. This type of noise presents obvious horizontal strips in the images. The correction of striping noises usually compares the average density of the stripe and the adjacent line parallel to the stripe and then chooses a gain factor, which can show the difference or can use a similar method to reduce periodic noises.
3. *Isolated noises*. Isolated noises are caused by error codes in the data transmission process or the temperature perturbation in analog circuits. Because they deviate from the image elements of the neighboring data in number, they are called isolated noises. They can be dealt with through median filtering or noise removing algorithm.
4. *Random noises*. Random noises are attached to the images and their values and positions are not fixed, such as the granularity noise of photo negatives. They can be restrained through averaging several images of sequence photography of one changing scene. Moreover, there are also other methods, such as bidirectional reflectance correction, emissivity correction, terrain correction, and inversion of remote sensing physical variable.

4.7.2 The Correction of Geometric Deformation

The graphic and image cleaning of geometric deformation corrects the image matching, which chiefly points to the relative displacement or graphics changing between images. It mainly refers to the direction x displacement parallax and direction y displacement parallax. If the left and right images are resampled according to the epipolar lines, there are no vertical parallaxes in the corresponding epipolar lines. Assuming the radiometric distortion has been corrected, the geometric deformation of a pixel mainly is the displacement p of direction x, and the gray-scale function $g_1(x, y)$, $g_2(x, y)$ of the left and right images should satisfy

$$g_1(x, y) + n_1(x, y) = g_2(x + p, y) + n_2(x, y) \qquad (4.47)$$

where, $n_1(x, y)$, $n_2(x, y)$ respectively are the random noises of the left and right images. Its error equation is

$$v(x, y) = g_2(x', y') - g_1(x, y) = g_2(x + p, y) - g_1(x, y) \qquad (4.48)$$

For any point $P(x_0, y_0)$ that falls in the parallactic grid (i, j) of column i and row j, according to the bilinear finite element interpolation method, the parallactic value p_0 of p can be calculated by bilinear interpolating the parallax $p_{ij}, p_{i+1,j}, p_{i,j+1}, p_{i+1,j+1}$ of the four vertexes $P(x_i, y_j)$, $P(x_{i+1}, y_j)$, $P(x_i, y_{j+1})$, $P(x_{i+1}, y_{j+1})$ in the grid. That is,

$$p_0 = \frac{\begin{aligned}&p_{ij}(x_{i+1} - x_0)(y_{j+1} - y_0) + p_{i+1,j}(x_0 - x_i)(y_{j+1} - y_0)\\ &+p_{i,j+1}(x_{i+1} - x_0)(y_0 - y_j) + p_{i+1,j+1}(x_0 - x_i)(y_0 - y_j)\end{aligned}}{(x_{i+1} - x_i)(y_{j+1} - y_j)} \qquad (4.49)$$

where, $x_i \le x_0 \le x_{i+1}, y_j \le y_0 \le y_{j+1}$. Substitute Eq. (4.49) in error Eq. (4.48), linearize it, and calculate it; the parallactic value of the regular grid point $P(i, j)$ can be obtained and the parallactic grid can be formed to correct the geometric deformation.

(a)   **(b)**

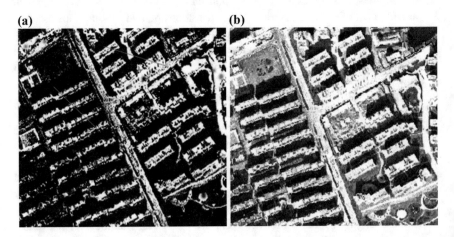

Fig. 4.6 The aerial images before and after cleaning. **a** Original image. **b** Cleaned image

4.7.3 A Case of Image Cleaning

To have a better understanding of spatial graphics and images cleaning, this section uses an aerial image of a district in Nanning city, Guangxi province as an example (Fig. 4.6a) and cleans and calculates it by the above-mentioned method. This image was shot in 1998 and its measuring scale is 1:2000. Figure 4.4b is the result of the image data cleaning.

Comparing Fig. 4.6a, b, it can be concluded that the original aerial image before cleaning was affected by radioactive distortion to a great extent, obscured with poor readability. It is also difficult to recognize the objects and geometric deformation occurring to the left; however, the aerial image after cleaning is a clearer image, the readability is enhanced, and the objects are easier to recognize. The most obvious example is the square in the right corner of the image where the fog that exists in the original aerial image was removed by the cleaning.

References

Burrough PA, Frank AU (eds) (1996) Geographic objects with indeterminate boundaries. Taylor and Francis, Basingstoke

Clark CF (1997) Evaluating the uncertainty of area estimates derived from fuzzy land-cover classification. Photogram Eng Remote Sens 63:403–414

Dasu T (2003) Exploratory data mining and data cleaning. Wiley, New York

Fayyad U, Piatetsky-Shapiro G, Smyth P, Uthurusamy R (eds) (1996) Advances in knowledge discovery and data mining. AAAI/MIT, Menlo Park, pp 1–30

Goodchild MF (2007) Citizens as voluntary sensors: spatial data infrastructure in the world of Web 2.0. Int J Spat Data Infrastruct Res 2:24–32

Hernàndez MA, Stolfo SJ (1998) Real-world data is dirty: data cleansing and the merge/purge problem. Data Min Knowl Disc 2:1–31

Inmon WH (2005) Building the data warehouse, 4th edn. Wiley, New York

Kim W et al (2003) A taxonomy of dirty data. Data Min Knowl Disc 7:81–99

Koperski K (1999) A progressive refinement approach to spatial data mining. Ph.D. Thesis, Simon Fraser University, British Columbia

Li DR, Yuan XX (2002) Error handling and reliability theory. Wuhan University Press, Wuhan

Shi WZ, Fisher PF, Goodchild MF (eds) (2002) Spatial data quality. Taylor & Francis, London

Smets P (1996) Imperfect information: imprecision and uncertainty. Uncertainty management in information systems. Kluwer Academic Publishers, London

Smithson MJ (1989) Ignorance and uncertainty: emerging paradigms. Springer, New York

Wang XZ (2002) Parameter estimation with nonlinear model. Wuhan University Press, Wuhan

Wang ZZ (2007) Sequence to principles of photogrammetry. Wuhan University Press, Wuhan

Wang SL, Shi WZ (2012) Chapter 5 data mining, knowledge discovery. In: Kresse W, Danko D (eds) Handbook of geographic information. Springer, Berlin, pp 123–142

Wang SL, Wang XZ, Shi WZ (2002) Spatial data cleaning. In: Zhang S, Yang Q, Zhang C (eds) Proceedings of the first international workshop on data cleaning and preprocessing, Maebashi TERRSA, Maebashi City, 9–12 Dec, pp 88–98

Chapter 5
Methods and Techniques in SDM

Many methods and techniques can be used in SDM. Their performance—good or bad—will have a direct impact on the quality of the discovered knowledge. In this chapter, usable methods and techniques are presented based on their characteristics in the context of SDM (Li et al. 2002). The approaches in crisp set theory allow only full or no membership (e.g., probability, evidence theory, spatial statistics, spatial analysis, data field), while the approaches in extended set theory allow partial membership to match the real world (e.g., fuzzy set, rough set, cloud model). SDM also simulates the networked thinking of the human brain and the optimal solutions of the evolution process (i.e., artificial neural network genetic algorithms). Rule induction, decision trees, and visualization techniques also benefit SDM.

5.1 Crisp Set Theory

Crisp set theory, which Cantor introduced in the nineteenth century, is the basis of modern mathematics. Probability and spatial statistics classically target the randomness in SDM. Evidence theory is an expansion of probability theory; spatial clustering and spatial analysis are extensions of spatial statistics. The data field concept will be introduced in Chap. 6.

5.1.1 Probability Theory

Probability theory (Arthurs 1965) is suitable for SDM with randomness on the basis of stochastic probabilities in the context of adequate samples and background information. Probability theory originated from research by European

© Springer-Verlag Berlin Heidelberg 2015
D. Li et al., *Spatial Data Mining*, DOI 10.1007/978-3-662-48538-5_5

mathematicians in the seventeenth century on problems related to gambling, which is the most emblematic random event. Later on, researchers found a large number of similar random phenomena in natural sciences, social sciences, engineering, industry, and agricultural production; they gradually applied probability theory to these areas. The Bernoulli Law of Large Numbers uses strict mathematical methods to demonstrate the stability of frequency, which can be used to associate probability with frequency. Central Limit Theorem was introduced by Laplace et al. Markov et al. then established general forms of both the Law of Large Numbers and the Central Limit Theorem to mathematically explain why many practical random phenomena follow a normal distribution.

By the twentieth century, Kolmogorov, a Russian mathematician, introduced the axiomatic system of probability, in which the basic concepts of probability were well defined and modern mathematical tools were introduced. Kolmogorov established probability theory as a branch of mathematics with strict theorization and removed its vagueness and non-strict forms in its early years. It also became the theoretical basis of random process. In the twentieth century, researchers began to study random process in order to solve a huge number of practical problems, especially in physics research. Brown, a biologist, discovered that pollen particles move in random directions in water. To explore the movement, Einstein and Wiener investigated the movement from the two aspects of physics and mathematics, respectively. Wiener brought forward the mathematical model for Brownian motion and carried out systematic research, thereby establishing him as the originator of random process. Erlang et al. studied the Poisson process in voice stream, which laid the foundation of queuing theory. Subsequently, the research for the birth-death process, the stationary process, the Markov process, and martingale theory enabled random process to become an important branch of applied probability theory.

Probability is different from likelihood. Given some situations, a big probability does not mean a high likelihood, nor does a small probability mean a low likelihood. However, if an event is impossible, the happening probability or likelihood is zero definitely. Generally, a small probability happens with a low likelihood, and a big probability coincides with a high likelihood. Here, probability can be treated as equal to likelihood, which is beneficial to deducing the probable distribution function of object attributes with various statistical methods. With the increasing attention to random phenomena, probability has attracted further research and applications.

Probability theory explores spatial datasets with randomness based on stochastic probability, which can enable understanding of the nature of random phenomena by mathematically modeling a large number of random events according to their features, properties, and regularities. The explored knowledge can be expressed as the conditional probability under given conditions when a certain assumption is positive (Arthurs 1965), which is often treated as decision-making background knowledge. When an error matrix is used to describe the uncertainties about the results of remote-sensing classification, the background knowledge is used to express the confidence levels. Lenarcik and Piasta combined probability theory with rough set in the system of probabilistic rough classifiers generation (ProbRough) by applying conditional attributes to deduce decision knowledge. Shi Wenzhong proposed a probability vector that different probabilities are used to show the distribution of spatial data uncertainty (Shi and Wang 2002). Using the decision-tree-based probability graphical model, the probability field denotes that the probability of categorical variables is the probability when the class observed at one location is positive (Zhang and Goodchild 2002). With the decision-tree-based probability graphical model, Frasconi et al. (1999) uncovered the database with graphic attributes and the discovered knowledge was used to supervise machine learning. Generally, the probability sample is used to carry out spatial analysis for the attributes of regional soil. For an area without probability samples, Brus (2000) designed a method using regressively interpolating probability samples to assess the general pattern of soil properties. According to the raster-structured land-cover classification model, evaluating how the uncertainty of input data impact the output results, Canters (1997) put forward the probability-based image membership vector that all the elements in one pixel element would affect and decide the category of the pixel together. Taking advantage of supervised machine learning, Sester (2000) supervised the interpretation of spatial data from a database with a given sample space automatically. In addition, Journel (1996) studied the uncertainty of random images.

5.1.2 Evidence Theory

Evidence theory (also known as Dempster-Shafer theory and significance theory) is an interval determined by the belief function and the plausibility function (Shafer 1976). As far as the existing evidence, the belief function measures a minimum to support the hypothesis, and the plausibility function measures the maximum degree at which the hypothesis cannot be denied. Under the condition that the unsupported interval from the evidence is null, evidence theory is identical to probability theory. Thus, evidence theory is the extension of probability. The probability is extended to an interval instead of an exact value in evidence theory. Given the same information for an uncertain parameter, the plausibility function and the belief function can be regarded as the upper probability and lower probability of possible values (Shafer 1976). Evidence theory was first suggested in the 1960s by Dempster, which Shafer further developed in the 1970s by formalizing a theoretical system dealing with uncertainty information and also made clear the borderline between uncertainty and the unknown. However, evidence theory cannot solve contradictory evidence or weak hypothetical support.

In SDM, evidence theory divides the entity into two parts: certainness to measure confidence and uncertainness to measure likelihood. Both measurements are non-additive. By using the orthogonal sum of the basic probability distribution function, evidence from all aspects can be merged into comprehensive evidence, such as the integration of multi-source information. Aided by the united rules of evidence theory, the spatial knowledge can be extracted from a dataset with more than one uncertain attribute. In addition, the comparative method also can be used in knowledge discovery with uncertain attributes (Chrisman 1997). The framework of evidence data mining (EDM) has two parts: mass functions and mass operators (Yang et al. 1994). The mass function represents the data and knowledge—that is, the data mass function and the rule mass function; several of the mass function operators are for knowledge discovery. EDM has been used to discover strong rules in relational databases, as well as to identify volcanoes from a large number of images of the Venus surface. The EDM framework makes the process of knowledge discovery equivalent to the operation of the mass function for data and rules. The methods based on evidence theory can easily deal with a null value or a missing attribute value and incorporate domain-knowledge in the course of knowledge discovery. At the same time, the essentials of the algorithms are parallel, which is advantageous when dealing with parallel, distributed, and heterogeneous databases. Evidence theory's advantages when dealing with uncertainty have potential applications in SDM.

5.1.3 Spatial Statistics

Spatial statistics identify spatial datasets with limited information by depicting disordered events with an ordered model (Cressie 1991), which is a commonly used way to effectively deal with numerical data along with spatial phenomenon (Zeitouni 2002). Spatial statistics primarily consist of geostatistics, spatial point patterns, and lattice data, along with the algorithms of spatial autocorrelation, spatial interpolation, spatial regression, spatial interaction, and simulation and modeling. Geostatistics is a spatial process indexed over a continuous space, spatial point patterns pertain to the location of "events" of interest, and lattice data are spatial data indexed over a lattice of points. When SDM is implemented in the context of randomness, spatial statistics may provide an auto-covariance structure, a variation function, or the similarity between a covariant and a local variant. If spatial entities are observed enough to calculate the maximum, minimum, mean, variance, mode, or histogram, then the prior probability of discovering the common geometric knowledge under the domain knowledge can be attained. With a solid theoretical foundation and a large number of sophisticated algorithms, spatial statistics can improve the performance of SDM in random process, estimating the range of decision-making uncertainty, analyzing the law of error propagation, analyzing the numeric data of spatial process, predicting the perspectives of development trends, and exploring the correlation among spatial entities. Moreover, spatial statistics excel at spatial multivariate analysis—that is, discriminant analysis, principal component analysis, factor analysis, correlation analysis, and multivariate regression analysis. When background knowledge is unavailable, clustering analysis is an alternative in SDM.

5.1.4 Spatial Clustering

Spatial clustering groups a set of data in a way that maximizes the similarity within clusters and minimizes the similarity between two different clusters (Han et al. 2000). It sorts through spatial raw data and groups them into clusters based on object characteristics. Without background knowledge, spatial clustering can directly find meaningful clusters in a spatial dataset on the basis of the object attribute. To discover spatial distribution laws and typical models for the entire dataset, spatial clustering, by using the measurement of a certain distance or similarity, is able to delineate clusters or dense areas in a large-scale multidimensional spatial dataset, which is a collection of data objects that are similar to one another within the same cluster and are dissimilar to the objects in other clusters (Kaufman and Rousseew 1990; Murray and Shyy 2000). As a branch of statistics, clustering is mainly based on geometric distance, such as Minkowski distance, Manhattan distance, Euclidean distance, Chebyshev distance, Cambera distance, and Mahalanobis distance. Different clustering distances may show some objects close to one another according to one distance but farther away according to the other distance (Ester et al. 2000). Derived from the matching matrix, some measurements therefore are given to compare various clustering results when different clustering algorithms perform on a set of data. Concept clustering is used when the similarity measurement among data objects is defined based on a concept description instead of a geometric distance (Grabmeier and Rudolph 2002).

Spatial clustering is different from classification in SDM. It discovers significant cluster structures from spatial databases without background knowledge. The clustering algorithms for pattern recognition groups the data points in a multidimensional feature space, and spatial clustering directly clusters graphs or images—the shape of which are complex as well as numerous. If multivariate statistical analysis is used at this time, the clustering will be inefficient and slow. Therefore, the clustering algorithms in SDM should be able to designate point, line, polygon, or arbitrary-shaped objects, along with self-adapted or user-defined parameters.

5.1.5 Spatial Analysis

Spatial analysis consists of the general methods that analyze spatial topology structure, overlaying objects, buffer zones, image features, distance measurements, etc. (Haining 2003). Exploratory data analysis uncovered the non-visual characteristics and exceptions by using dynamic statistical graphics and links to demonstrate data characteristics (Clark and Niblet 1987). Inductive learning, which is exploratory spatial analysis (Muggleton 1990) combined with attribute-oriented induction (AOI), explores spatial data to determine the preliminary characteristics for SDM. Image analysis can directly explore a large number of images or can perform as a preprocess phase of other knowledge discovery methods.

Reinartz (1999) addressed regional SDM, and Wang et al. (2000) introduced an approach to active SDM based on statistical information. Taking a spatial point as a basic unit, Ester et al. (2000) integrated SDM algorithms and a spatial database management system by processing the neighborhood relationships of many objects. Spatial patterns were discovered with neighborhood graphs, paths, and a small set of database primitives for their manipulation, along with neighborhood indices to speed up the primitive's processing. To reduce the search space for the SDM algorithms while searching for significant patterns, the filtering process was defined so that only certain classes of paths "leading away" from a starting object were relevant. In the diagnosis index of attribute exceptions from the possible causal relation, Mouzon et al. (2001) found that attribute uncertainty had influenced the exception diagnosis via the consistent algorithm and inductive algorithm.

5.2 Extended Set Theory

SDM is analogous to human thinking in that it is capable of mastering both certainty and uncertainty in the real world. The certainty leads to crisp set, and the uncertainty is crisp set extended. Crisp set has an accurate defined boundary, which is a binary logic of 0 or 1; therefore, a given element either completely belongs to a set or it does not. Crisp set allows only full membership or no membership at all, whereas there is partial membership in the real world. Most of the methods based on crisp set assign distinct attributes to spatial objects with their clear boundaries when, in reality, some spatial objects cannot be accurately described because they have indeterminate boundaries. Meanwhile, human thinking grasps the uncertainty by qualitatively inducing quantification, sufficiently transmitting the message with less cost, and efficiently making judgments and interpretations for complex objects in nature. Thus, crisp set theory has been extended in SDM to depict the uncertainty in the real world by using fuzzy set, rough set, and cloud model.

Crisp set is extended via membership and function mathematically. For the memberships, crisp set allows only full or no membership at all (i.e., $\{0, 1\}$), whereas fuzzy set allows partial membership with a close interval (i.e., $[0, 1]$), rough set allows partial membership with a set of one open interval and two distinct terminals (i.e. $\{0, (0, 1), 1\}$), and cloud model allows partial membership with random close intervals that are stochastically distributed in the space (i.e., random $[0, 1]$). The memberships are continuous in the close intervals but discrete in the set. As for the mathematical function, crisp set provides a crisp characteristic function, whereas fuzzy set provides a fuzzy membership function, rough set provides a pair of upper and lower approximations, and cloud model provides a model of mutual transformation between qualitative concepts and quantitative data by integrating randomness and fuzziness.

In the following section, fuzzy sets and rough sets are introduced. The cloud model will be introduced in Chap. 7.

5.2.1 Fuzzy Sets

A fuzzy set (Zadeh 1965) is used in SDM with fuzziness on the basis of a fuzzy membership function that depicts an uncertain probability (Li et al. 2006). Fuzziness is an objective existence. The more complicated the system is, the more difficult it is to describe it accurately, which creates more fuzziness. In SDM, a class and an object are treated as a fuzzy set and an element, respectively. Each object is assigned a membership in the class. If there are many objects with intermediate boundaries in the universe of interest, an object may be assigned a group of memberships. The closer the membership approximates 1, the more it is possible that the element belongs to the class. For instance, if there are soil, river, and vegetation in an image, a pixel will be assigned three memberships. The class uncertainty mostly derives from subjective supposition and objective vagueness. Without a precisely defined boundary, a fuzzy set is more suitable for fuzzy classification with spatial heterogeneous distribution of geographical uncertainty. To classify remote-sensing images, fuzzy classification can produce different intermediate results according to the classifier. For example, in statistical classifiers, there are likelihood values of a certain pixel belonging to the alternative classes; in neural network classifiers, there are class activation level values (Zhang and Goodchild 2002). Burrough and Frank (1996) presented a fuzzy Boolean logic model of uncertain data. Canters (1997) evaluated uncertain rules of estimating an area in fuzzy land cover classification. Wang and Wang (1997) proposed a fuzzy comprehensive method that combined fuzzy comprehensive evaluation with fuzzy cluster analysis for land price evaluation. Vazirgiannis and Halkidi (2000) used fuzzy logic to deal with the uncertainty in SDM.

Fuzzy comprehensive evaluation and fuzzy clustering analysis are the two essential techniques in fuzzy sets. When they are integrated into SDM, the knowledge is discovered well (Wang and Wang 1997) in the following steps:

1. A fuzzy set acquires the fuzzy evaluation matrix for each influential factor.
2. All of the fuzzy evaluation matrices multiply the corresponding weight matrices, the product matrix of which is the comprehensive matrix of all factors.
3. The comprehensive matrix is further used to create a fuzzy similar matrix, on the basis of which a fuzzy equivalent matrix is obtained.
4. Fuzzy clustering is implemented via the proposed maximum remainder algorithms.

However, by the time a fuzzy membership is determined, the subsequent calculation has abandoned the fuzziness that should be propagated to the final results continuously. Furthermore, because it focuses only on fuzzy uncertainty, fuzzy set may become invalid when there is more than one uncertainty in SDM (i.e., randomness and fuzziness).

5.2.2 Rough Sets

A rough set is used in SDM with incompleteness via lower and upper approximations (Pawlak 1991). It is an incompleteness-based reasoning method that creates a decision-making table by characterizing both the certainties and uncertainties. A rough set consist of an upper approximation set and a lower approximation set, and the standard for classifying objects is whether or not the information is sufficient in a given universe. Objects in the lower approximation set that have the necessary information definitely belong to the class, and objects within the universe but outside of the upper approximation set and without the necessary information definitely do not belong to the class. The difference set between the upper approximation set and the lower approximation set is the indeterminate boundary, in which the objects have insufficient information to determine whether or not they belong to the class. If two objects have exactly the same information, they are equivalent and cannot be distinguished from one another. Depending on whether statistical information is used, the existent rough set model may be algebraic and probabilistic (Yao et al. 1997). The basic unit of rough set is the equivalent class, as a grid in raster data, a point in vector data, or a pixel in an image. The more detailed the equivalent class is classified, the more accurately that a rough set can describe the objects, but at the expense of larger storage space and computing time (Ahlqvist et al. 2000). During its application in the field of approximate reasoning, machine learning, artificial intelligence, pattern recognition, and knowledge discovery, a rough set becomes more sophisticated and accomplished from the initial qualitative analysis (creating a minimum decision-making set) to the current focus on both qualitative analysis and quantitative calculation (computing rough probability, rough function and rough calculus). There is also a further interdisciplinary relationship, such as a rough fuzzy set, rough probabilistic set, and rough evidence theory.

Rough set-based SDM is a process of an intelligent decision-making analysis that can deal with inaccurate and incomplete information in spatial data. In SDM, a rough set may analyze attribute importance, attribute table consistency, and attribute reliability in spatial databases. The influences of attribute reliability on decision-making can be determined to simplify the spatial data, the attribute table, and the attribute dependence. By evaluating the absolute and relative uncertainty in the decision-making algorithm, the paradigm and causal relationship may be ascertained to generate the minimum decision-making and classification. To determine the patterns in the decision-making table, the soft computing of rough sets can simplify the attributes and regionalize the attribute values by deleting redundant rules (Pal and Skowron 1999). A rough set has been employed in applications such as distinguishing inaccurate spatial images and object-oriented software assessment; describing the uncertainty model; assessing land planning in a suburban area; selecting a bank location based on attribute simplification; roughly classifying remote-sensing images together with fuzzy membership functions, rough neighborhood, and rough precision; and generating the minimum decision-making knowledge of national agricultural data. These methods and applications in data mining have been summarized in the literatures and data mining systems based on rough set also have been developed (Polkowski and Skowron 1998a, b; Ahlqvist et al. 2000; Polkowski et al. 2000).

At present, the applicability of a rough set for the above uses has yet to be proven definitively. In addition, a rough set is not a specialized theory that originated from SDM. It provides the derivative topological relationships of the concept definition but does not address spatial relationships by undertaking actions such as superimposing multiple layers. As a result, it is necessary to further improve a rough set in the field of geospatial information sciences, such as georough space (Wang et al. 2002).

5.3   Bionic Method

Artificial neural networks (ANNs) and genetic algorithms are the typical bionic methods in SDM. The first simulates the neural network of a human brain to discover the networked patterns by using a self-adaptive nonlinear dynamic system, whereas the second imitates the process of biological evolution to determine the optimal solutions by using selection, crossover, and mutation as basic operators.

5.3.1   Artificial Neural Network

ANNs consist of a large number of neurons connected by abundant and sophisticated junctions. They are suitable for performing classification, clustering, and prediction for SDM in a complex environment and consist of a large number of neurons that are connected by abundant and sophisticated junctions. ANNs are highly nonlinear, ultra-large scale, continuous-temporal, and dynamic systems whose functions include distributed storage, associative memory, and massive parallel processing. ANNs also are capable of self-learning, self-organization, and self-adaption (Gallant 1993). The neural network is composed of an input layer, a middle layer, and an output layer, depending on the dynamic response of the network status to the external data input for information processing. A deep neural network is an ANN with multiple hidden layers of units between the input and output layers, which gives a method of deep learning. A large number of simple processing units (neurons) are networked to create the nonlinear functions of a complex system for determining patterns by learning from training samples. On the basis of the Hebb learning rules, the neural network can be divided into three types: forward feedback networks for prediction and pattern recognition (perceptron, back propagation model, function-like network, and fuzzy neural network), feedback networks for associative memory and optimization calculation (discrete model and continuous model of Hopfield), and self-organizing networks for clustering (ART model and Koholen model).

ANNs reason from the nonlinear systems in a blurry background (Miller et al. 1990). Rather than explaining a distinctive characteristic of a specific analysis at a specific condition, ANNs are capable of expressing that information, which is also called the connectionist method in artificial intelligence. Compared to traditional methods, ANNs afford high fault tolerance and robustness to the network by reducing the noise disturbance in pattern recognition. Moreover, the self-organizing and self-adapting capabilities of neural networks greatly relax the constraint and their neural network classification is more accurate than symbolic classification. Some commercial remote-sensing image processing software now offers neural network classification modules. Neural networks also can be used to classify, cluster, and forecast GIS data.

However, given a complex nonlinear system with many input variables, the convergence, stability, local minimum, and parameter adjustments of the network may encounter problems (Lu et al. 1996). For example, it is difficult to inter-infiltrate professional knowledge for defining network parameters (e.g., the number of neurons in the middle layer) and training parameters (e.g., learning rate and error threshold). Neural network classification requires scanning training data more often and hence takes more time. In addition, the discovered knowledge is hidden in the network structure instead of explicit rules. For intermediate results, it is unnecessary to convert the discovered knowledge into more refined rules, and the network structure itself can be treated as a kind of knowledge representation at this stage. However, at the final results stage of SDM, the discovered knowledge is not easily understood and explained for decision-making.

To overcome this weakness, the forward ANN algorithm can be improved to avoid slow training speeds, long learning times, and possible local minimums in the process of data mining by adding training data and hidden nodes at each step (Lee 2000). Lu et al. (1996) introduced NeuroRule to extract classification from neural networks via network training, network pruning, and rule extraction. Network training is similar to general BP network training. Network pruning aims at deleting redundant nodes and connections to obtain a concise network without increasing the error rate of classification. When extracting the rules, the activation values of the hidden units first are changed into a few discrete values without reducing classification accuracy. Then, the rules can be obtained according to the interdependences between the outputs and the hidden node values and between the activation values of the hidden units and the inputs. Taking full advantage of the capabilities of ANNs and using them in combination with other methods to overcome their shortcomings will facilitate their wider use in SDM.

5.3.2 Genetic Algorithms

A genetic algorithm (GA) is used to uncover the optimal solutions for classification, clustering, and prediction in SDM (Buckless and Petry 1994) by simulating the evolutionary process of "survival of the fittest" in natural selection by using a genetic mechanism as well as computer software. In the context of biological evolution and inheritance, a GA iteratively approximates the optimal solution generation by generation through the process of selection, crossover, and mutation. Selection creates a new population by selecting vigorous individuals from the old population; crossover creates a new individual by exchanging parts of two different individuals; and mutation changes a certain gene of some individuals. The iterating direction is guided by the adaptive function of deductive algorithms. When GA is executed, the problem to be resolved is first encoded to generate an initial solution. Then, the adaptive values are calculated to generate new solutions by selection, crossover, and mutation. The process is repeated until the best solution is found. Thus, GA has the advantages of robustness and domain independence for gradually searching for a global optimal solution. Robustness reduces the limitations of problems to the least condition and greatly improves the system capability of fault tolerance. The domain independence of GA makes its design especially suitable for the problems encountered with various properties in different fields.

GA treats an SDM task as a searching problem to iteratively conclude the optimal pattern that satisfies the adaptive value. It can fast search, compare, and deduce the optimal point in a vast volume of data with a learning mechanism. Jiang et al. (2001) introduced a tool to communicate information between different generations for interpreting the spatial structure of chromosomes. The combination of ANN and GA can optimize the connection strength and the network parameters.

5.4 Others

Given certain background knowledge, rule induction generalizes and synthesizes spatial data to obtain high-level patterns or characteristics in the form of a concept tree.

5.4.1 Rule Induction

Rule induction summarizes spatial data for uncovering the patterns among spatial datasets in order to acquire high-level patterns in the form of a concept tree, such as a GIS attribute tree and a spatial relationship tree, under a certain background. It focuses on searching generic rules from a large number of empirical data. The background knowledge can be provided by users or extracted automatically from datasets as one of the SDM tasks. When reasoning rules, an induction is different from a deduction or commonsense reasoning. Induction reasons on the basis of a large number of statistical facts and instances while deduction reasons on axioms and commonsense reasoning on acknowledged knowledge (Clark and Niblet 1987). As an extension of deductive reasoning and inductive reasoning, a decision rule is the relevance among the entire or partial data of a database. It emphasizes the optimal condition on the premise of induction, and the result therefore is the conclusion of induction. Decision rule reasoning under rough sets makes proper or similar decisions according to the condition. Decision rule includes association rules (if customers buy milk, then they also will buy bread 42 % of the time), sequential rules (a certain equipment item with one failure will also have the other failure 65 % of the time within one month), and serial rules (stock A and stock B will share the similar law of fluctuations in a season).

Association rules in SDM may be common or strong. Unlike common association rules, strong association rules are applied more widely and frequently, along with more profound significance. The algorithm of association rules mainly focuses on two aspects, such as increasing the efficiency and discovering more rules. For example, the SQL-like language that SDM uses to describe the mining process is in accordance with international standard query language, thereby making SDM somewhat standard and engineering-oriented. In the model of association rules, temporal information can be added to describe its time-effectiveness. The same kinds of records may be amalgamated according to the time interval among the data records and the item category of adjacent records. When association rules cannot describe the patterns in datasets, transfer rules may depict the systematic state transferred from the present moment to the next moment by a certain probability. The state of the next moment depends on the pre-state and transfer probability. AOI is propitious to data classification (Han et al. 1993). Learning from examples, also known as supervised learning, classifies the training samples beforehand in the context of the background knowledge. The selection of sample features has a direct impact on the effectiveness of its learning efficiency, results expression, and process intelligibility. On the other hand, unsupervised learning does not know the classification of the training samples in advance, and therefore learns from observation and discovery. The sequential rule is closely related to time; and the time interval constraint between the adjacent item sets can be used to extend the discovering of sequential patterns from a single-layer concept to a multi-layer concept, which goes forward one by one from top to down. When the database changes a little, the progressive discovery of sequential rules is able to use the previous results to accelerate the current mining process.

Koperski (1999) discovered strong association rules in geographical databases with a two-step optimization. Han et al. (2000) uncovered association rules with two parallel Apriori algorithms, such as intelligent data distribution and hybrid distribution. Eklund et al. (1998) discovered association rules for planning environments and monitoring the salinization of the secondary soil by integrating decision support systems with GIS data in the analysis of soil salinity. Aspinall and Pearson (2000) discovered association rules for protecting environments in Yellowstone National Park by comprehending the ecological landscape, GIS, and environmental models together. Clementini et al. (2000) explored association rules in spatial objects with broad boundaries at different levels. Levene and Vincent (2000) discovered the rules of functional independence and containing independence in relational databases.

5.4.2 Decision Trees

A decision tree generates a classification or decision set in the structure of a tree under certain spatial features. In SDM, the tree structure first creates a test function by using a training object set and then sets up the branches on the basis of different data. Within every branch set, lower nodes and sub-branches are iteratively set up. Leaf nodes are positive examples or negative examples (Quinlan 1993). Upon completion of this set up, the decision tree is trimmed to generate the rules to classify new objects, such as ID3 (Interactive Dichotomizer 3), C4.5, and See 5.0. The system classified the stars in the sky with a decision tree (Fayyad et al. 1996). Considering the non-spatial aggregation values of neighboring objects in a decision tree, Koperski (1999) presented a two-step decision classification by first extracting the spatial predicates with the relief algorithm of machine learning and then merged both the spatial and non-spatial predicates into the knowledge of classification decision. Marsala and Bigolin (1998) explored regional classification rules with a fuzzy decision tree in an object-oriented spatial database.

5.4.3 Visualization Techniques

Visualization explores and represents spatial data and knowledge visually in SDM. A visible picture is worth thousands of invisible words because SDM involves massive data, complex models, unintelligible programs, and professional results. By developing a computerized system, visualization converts the abstract patterns, relations, and tendencies of the spatial data to concrete visual representations that can be directly sensed by human vision (Slocum 1999). There are many techniques that make the process and results of SDM easily visible and understandable by users, such as charts, pictures, maps, animations, graphics, images, trees, histograms, tables, multi-media, and data cubes (Soukup and Davidson 2002). The visual expression and analysis of spatial data can supervise discovery operations, data location, result representation, and pattern evaluation and allows users to explore spatial data more clearly and reduces the complexity of modeling. If multi-dimensional visualization is processed in time according to the respective sequence, a comprehensive animation can be created to reflect the process and knowledge of SDM. Ankerst et al. (1999) produced a 3D-shaped histogram to express a similar search and classification in a spatial database. Maceachren et al. (1999) structuralized the knowledge from multi-source spatiotemporal datasets by combining geographical visualization with SDM.

5.5 Discussion

This chapter presented many methods and techniques that are usable in SDM, including crisp set theory, extended theory, bionic method, and others.

5.5.1 Comparisons

Crisp set theory, probability theory, and spatial statistics address SDM with randomness. Evidence theory expands probability theory, and spatial clustering and spatial analysis are extensions of spatial statistics. Based on the extended set theory, a fuzzy set focuses on fuzziness and rough set emphasizes on incompleteness. As far as its mathematical basis, probability theory and spatial statistics are random probability, a rough set is the approximation, a fuzzy set is the fuzzy membership, evidence theory is the evidence function, and the cloud model is the probability distribution of membership. Rules induction performs under certain background knowledge, and clustering algorithms perform without background knowledge. For classification, clustering, and prediction, ANN implements a self-adaptive nonlinear dynamic system, and GA determines the optimal solutions. A decision tree produces rules according to different features to express classification or a decision set in the form of a tree structure. Visualization extracts the patterns from spatial data in the form of visual representations. In addition, online SDM is built on multidimensional view, which is a verification of data mining, as well as an analysis tool based on a network that emphasizes operation efficiency and immediate response to user commands, for which its direct data source is spatial data warehouses.

In addition, SDM not only develops and completes its own methods and techniques, but also learns and absorbs the proven methods from other areas, such as data mining, database systems, artificial intelligence, machine learning, geographic information systems, remote sensing, etc.

5.5.2 Usability

Approaches generally are not isolated from one another, and SDM usually employs more than one approach in order to obtain more diverse, more accurate, and more reliable results for different requirements and the degree of sophistication needed is fully realized (Reinartz 1999). Increasing the complexity of the system are the skyrocketing amount of spatial data, continual refinements to spatial rules, and the higher accuracy demanded by users. SDM therefore requires constant refinement to keep the system relevant and effective. For example, SDM provides limited types of functions, such as spatial rules, algorithms, visualization modeling, and parallel computation. Therefore, the expansion of its functions would be a useful undertaking. SDM's efficiency when dealing with large amounts of data, its capability to resolve complicated problems, and the extensibility of the system can be improved.

Moreover, because SDM is an interdisciplinary subject, many factors need to be taken into consideration. In practical applications, the theory, method, and tools of SDM should be chosen according to the particular requirements of an application and should integrate the various operation techniques of algorithms. Combining various rules contributes to discovery of optimally useful rules, and providing various ways to create the same spatial rules is beneficial to optimizing the resolution process of complex problems. When evaluating spatial rules, adopting several checking methods may attain the maximal accuracy (Howard 2001).

References

Ahlqvist Q, Keukelaar J, Oukbir K (2000) Rough classification and accuracy assessment. Int J Geogr Inf Sci 14(5):475–496

Ankerst M et al (1999) 3D shape histograms for similarity search and classification in spatial databases. Lecture Notes in Computer Science, 1651, pp 207–225

Arthurs AM (1965) Probability theory. Dover Publications, London

Aspinall R, Pearson D (2000) Integrated geographical assessment of environmental condition in water catchments: linking landscape ecology, environmental modeling and GIS. J Environ Manage 59:299–319

Brus DJ (2000) Using nonprobability samples in design-based estimation of spatial means of soil properties. In: Proceedings of accuracy 2000, Amsterdam, pp 83–90

Buckless BP, Petry FE (1994) Genetic algorithms. IEEE Computer Press, Los Alamitos

Burrough PA, Frank AU (eds) (1996) Geographic objects with indeterminate boundaries. Taylor & Francis, Basingstoke

Canters F (1997) Evaluating the uncertainty of area estimates derived from fuzzy land-cover classification. Photogram Eng Remote Sens 63:403–414

Chrisman NC (1997) Exploring geographic information systems. Wiley, New York

Clark P, Niblet TT (1987) The CN2 induction algorithm. Mach Learn 3:261–283

Clementini E, Felice PD, Koperski K (2000) Mining multiple-level spatial association rules for objects with a broad boundary. Data Knowl Eng 34:251–270

Cressie N (1991) Statistics for spatial data. Wiley, New York

Eklund PW, Kirkby SD, Salim A (1998) Data mining and soil salinity analysis. Int J Geogr Inf Sci 12(3):247–268

Ester M et al (2000) Spatial data mining: databases primitives, algorithms and efficient DBMS support. Data Min Knowl Disc 4:193–216

Fayyad U, Piatetsky-Shapiro G, Smyth P, Uthurusamy R (eds) (1996) Advances in knowledge discovery and data mining. AAAI/MIT, Menlo Park, CA, pp 1–30

Frasconi P, Gori M, Soda G (1999) Data categorization using decision trellises. IEEE Trans Knowl Data Eng 11(5):697–712

Gallant SI (1993) Neural network learning and expert systems. MIT Press, Cambridge

Grabmeier J, Rudolph A (2002) Techniques of clustering algorithms in data mining. Data Min Knowl Disc 6:303–360

Haining R (2003) Spatial data analysis: theory and practice. Cambridge University Press, Cambridge

Han JW, Cai Y, Cercone N (1993) Data driven discovery of quantitative rules in relational databases. IEEE Trans Knowl Data Eng 5(1):29–40

Han EHS, Karypis G, Kumar V (2000) Scalable parallel data mining for association rules. IEEE Trans Knowl Data Eng 12(3):337–352

Howard CM (2001) Tools and techniques for knowledge discovery. PhD thesis. University of East Anglia, Norwich

Jiang W et al (2001) Bridging the information gap: computational tools for intermediate resolution structure interpretation. J Mol Biol 308:1033–1044

Journel AG (1996) Modelling uncertainty and spatial dependence: stochastic imaging. Int J Geogr Inf Sys 10(5):517–522

Kaufman L, Rousseew PJ (1990) Finding groups in data: an introduction to cluster analysis. Wiley, New York

Koperski K (1999) A progressive refinement approach to spatial data mining. PhD thesis, Simon Fraser University, British Columbia

Lee ES (2000) Neuro-fuzzy estimation in spatial statistics. J Math Anal Appl 249:221–231

Levene M, Vincent MW (2000) Justification for inclusion dependency normal form. IEEE Trans Knowl Data Eng 12(2):281–291

Li DR, Wang SL, Li DY (2006) Theory and application of spatial data mining, 1st edn. Science Press, Beijing

Li DR, Wang SL, Li DY, Wang XZ (2002) Theories and technologies of spatial data mining and knowledge discovery. Geomatics Inf Sci Wuhan Univ 27(3):221–233

Lu H, Setiono R, Liu H (1996) Effective data mining using neural networks. IEEE Trans Knowl Data Eng 8(6):957–961

Maceachren AM et al (1999) Constructing knowledge from multivariate spatiotemporal data: integrating geographical visualization with knowledge discovery in database methods. Int J Geogr Inf Sci 13(4):311–334

Marsala C, Bigolin NM (1998) Spatial data mining with fuzzy decision trees. In: Ebecken NFF (ed) Data mining. WIT Press/Computational Mechanics Publications, Ashurst Lodge, UK, pp 235–248

Miller WT et al (1990) Neural network for control. MIT Press, Cambridge

Mouzon OD, Dubois D, Prade H (2001) Using consistency and abduction based indices in possibilistic causal diagnosis. IEEE, pp 729–734

Muggleton S (1990) Inductive acquisition of expert knowledge. Turing Institute Press in association with Addison-Wesley, Wokingham

Murray AT, Shyy TK (2000) Integrating attribute and space characteristics in choropleth display and spatial data mining. Int J Geogr Inf Sci 14(7):649–667

Pal SK, Skowron A (eds) (1999) Rough fuzzy hybridization. Springer, Singapore

Pawlak Z (1991) Rough sets: theoretical aspects of reasoning about data. Kluwer Academic Publishers, London

Polkowski L, Skowron A (eds) (1998a) Rough sets in knowledge discovery 1: methodologies and applications. Studies in fuzziness and soft computing, vol 18. Physica-Verlag, Heidelberg

Polkowski L, Skowron A (eds) (1998b) Rough sets in knowledge discovery 2: applications, case studies and software systems. studies in fuzziness and soft computing, vol 19. Physica-Verlag, Heidelberg

Polkowski L, Tsumoto S, Lin TY (eds) (2000) Rough sets methods and applications: new developments in knowledge discovery in information systems. Physica-Verlag, Heidelberg

Quinlan JR (1993) C4.5: programs for machine learning. Morgan Kaufmann, San Mateo

Reinartz T (1999) Focusing solutions for data ming: analytical studies and experimental results in real-world domains. Springer, Berlin

Sester M (2000) Knowledge acquisition for the automatic interpretation of spatial data. Int J Geogr Inf Sci 14(1):1–24

Shafer G (1976) A mathematical theory of evidence. Princeton University Press, Princeton

Shi WZ, Wang SL (2002) GIS attribute uncertainty and its development. J Remote Sens 6(5):393–400

Slocum TA (1999) Thematic cartography and visualization. Prentice Hall, Prentice

Soukup T, Davidson I (2002) Visual data mining: techniques and tools for data visualization and mining. Wiley, New York

Vazirgiannis M, Halkidi M (2000) Uncertainty handling in the data mining process with fuzzy logic. In: IEEE, pp 393–398

Wang SL, Li DR, Shi WZ, Wang XZ (2002) Theory and application of Geo-rough space. Geomatics Inf Sci Wuhan Univ 27(3):274–282

Wang XZ, Wang SL (1997) Fuzzy comprehensive method and its application in land grading. Geomatics Inf Sci Wuhan Univ 22(1):42–46

Wang J, Yang J, Muntz R (2000) An approach to active spatial data mining based on statistical information. IEEE Trans Knowl Data Eng 12(5):715–728

Yang JB, Madan G, Singh L (1994) An evidential reasoning approach for multiple-attribute decision making with uncertainty. IEEE Trans Syst Man Cybern 24(1):1–18

Yao YY, Wong SKM, Lin TY (1997) A review of rough set models. In: Lin Y, Cercone N (eds) Rough sets and data mining analysis for imprecise data. Kluwer Academic Publishers, London, pp 47–75

Zadeh LA (1965) Fuzzy sets. Inf Control 8(3):338–353

Zeitouni K (2002) A survey of spatial data mining methods databases and statistics point of views. In: Becker S (ed) Data warehousing and web engineering. IRM Press, London, pp 229–242

Zhang JX, Goodchild MF (2002) Uncertainty in geographical information. Taylor & Francis, London

Chapter 6
Data Field

A field exists commonly in physical space. The concept of a data field, which simulates the methodology of a physical field, is introduced in this chapter for depicting the interaction between the objects associated with each data point of the whole space. The field function mathematically models how the data contributions for a given task are diffused from the universe of a sample to the universe of a population when interacting between objects. In the universe of discourse, all objects with sampled data not only radiate their data contributions but also receive data contributions from other objects. Depending on the given task and the physical nature of the objects in the data distribution, the field function may be derived from the physical field. All of the equipotential lines depict the interesting topological relationships among the objects, which visually indicate the interacted characteristics of objects. Thus, a data field bridges the gap between a mining model and a data model.

6.1 From a Physical Field to a Data Field

Anaxagoras, a Greek philosopher, thought physical nature was "a portion of everything in everything." Visible or invisible particles mutually interact; sometimes, the interaction is untouched between physical particles. Michael Faraday first put forward the field concept to describe this interaction. He believed that the untouched interactions between physical particles had to be performed via a transforming media that he suggested was the field. To visually portray an electronic field configuration, he further introduced the concept of an electronic field line—that is, drawing a vector of attribute length and direction at every point of space. Each field line has a starting point and an ending point, which originates at the source and terminates at the sink. In a charge-free region, every field line

© Springer-Verlag Berlin Heidelberg 2015
D. Li et al., *Spatial Data Mining*, DOI 10.1007/978-3-662-48538-5_6

is continuous that originates on a positive charge and terminates on a negative charge, and they do not intersect. In space, a particle makes a field, and a field acts on another particle (Giachetta et al. 2009). Because the interaction of particles in physics can be described with the help of a field of force, we wonder whether it is possible to formalize the self-cognition of humans in terms of the idea of field. If so, we perhaps can establish a virtual cognitive field to model the interactions of data objects.

6.1.1 Field in Physical Space

In physical space, interacted particles lead to various fields. Due to the interacting particles, the field can be either a classical field or a quantum field. Depending on the interacting range, the field may be long-range or short-range. As an interacting range increases, the strength of the long-range field slowly diminishes to the point of being undetectable (e.g., gravitational fields and electromagnetic fields), while the strength of the short-range field rapidly attenuates to zero (e.g., nuclear field). Based on the interacting transformation, the field is often classified as a scalar field, vector field, tensor field, or spinor field. At each point, a scalar field's values are given by a single variable on the quantitative difference, a vector field is specified by attaching a vector with magnitude and direction, a tensor field is specified by a tensor, and a spinor field is for quantum field theory. A vector field may be with or without sources. A vector field with constant zero divergence is a field without sources (e.g., a magnetic field); otherwise, it is a field with sources (e.g., an electrostatic field). In basic physics, the vector field is often studied as a field with sources, such as Newton's law of universal gravitation and Coulomb's law. When it is put in the field with sources, an arbitrary particle feels a force with strength (Giachetta et al. 2009). Moreover, according to the values of a field variable, whether or not it changes over time at each point, the field with sources can be further grouped into time-varying fields and stable active fields. The stable field with sources is also known as a potential field.

Field strength is often represented by using the potential. In a field, the potential difference between an arbitrary point and another referenced point is a determinate value on a unit particle. The physical potential is the work performed by a field force when a unit particle is moved from an arbitrary point to another referenced point in the field. It is a scalar variable of the amount of energy transferred by the force acting at a distance. The energy is the capacity to do the work, and the potential energy refers to the ability of a system to do work by virtue of its position or internal structure. The distribution of the potential field corresponds to the distribution of the potential energy determined by the relative position between interacted particles. Specifying the potential value at a point in space requires parameters such as object mass or electrical charge, and distance between a point and its field source, while it is independent of the direction of the distance vector. The strength of the field is visualized via the equipotential, which refers to a region in space where every point in it is at the same potential.

In physics, it is a fact that a vacuum is free of matter but not free of field. Modern physics even believes that field is one of the basic forms of particle existence, such as a mechanical field, nuclear field, gravitational field, thermal field, electromagnetic field, and crystal field. As field theory developed, field was abstracted as a mathematical concept to depict the distribution rule of a physical variable or mathematical function in space (Landau and Lifshitz 2013). The potential function is often seen simply as a monotonic function of spatial location, having nothing to do with the existence of the particle. That is, a field of the physical variable or mathematical function in a space exists if every point in the space has an exact value from the physical variable or mathematical function.

6.1.2 Field in Data Space

Data space is an exact branch of physical space. In a data space, all objects are mutually interacting via data. From a generic physical space to an exact data space, the physical nature of the relationship between the objects is identical to that of the interaction between the particles. If the object described with data is taken as a particle with mass, then a data field can be derived from a physical field.

Inspired by the field in physical space, the field was introduced to illuminate the interaction between objects in data space. In the data space, all objects are mutually interacting via data. Given a logic database with N records and M attributes, the process of knowledge discovery can begin with the underlying distribution of the N data points in the M-dimensional universal space. If each data object is viewed as a point charge or mass point, it will exert a force on any other objects in its vicinity, and the interactions of all the data objects will form a field. Thus, an object described with data is treated as a particle with mass. The data refer to the attributes of the object determined from observation, while the mass is the physical variable of matter determined from its weight or from Newton's second law of motion. Surrounding the object, there is a virtual field simulating the physical field around particles. Each object has its own field. Given a task, the field gives a virtual media to show the mutual interaction between objects without touching each other. An object transforms its field to other objects, and it also receives all their fields. Depending on their contributions to a given task, the field strength of each object is different, which then uncovers a different interaction. By using this interaction, objects may be self-organized under the given data and task. Some patterns that are previously unknown, potentially useful, and ultimately understood may be further extracted from datasets.

6.2 Fundamental Definitions of Data Fields

Definition 6.1 In data space $\Omega \subseteq R^P$, let dataset $D = \{x_1, x_2, ..., x_n\}$ denote a P-dimensional independent random sample, where $x_i = (x_{i1}, x_{i2}, ..., x_{iP})^T$ with $i = 1, 2, ..., n$. Each data object x_i is taken as a particle with mass m_i. x_i radiates its data energy and is also influenced by others simultaneously. Surrounding x_i, a virtual field is derived from the corresponding physical field. Thus, the virtual field is called a data field.

6.2.1 Necessary Conditions

In the context of the definition, a data field exists only if the following necessary conditions are characterized (i.e., short-range with source and temporal behavior):

1. The data field is a short-range field. For an arbitrary point in the data space Ω, all the fields from different objects are overlapped to achieve a superposed field. If there are two points—x_1 with enough data samples and x_2 with less or no data samples—the summarized strength of the data field at point x_1 must be stronger than that of x_2. That is, the range of the data field is so short that its magnitude decreases rapidly to 0 as the distance increases. The rapid attenuation may reduce the effect of noisy data or outlier data. The effect of the superposed field may further highlight the clustering characteristics of the objects in close proximity within data-intensive areas. The potential value of an arbitrary point in space does not rely on the direction from the object, so the data field is isotropic and spherical in symmetry. The interaction between two faraway objects can be ignored.
2. The data field is a field with sources. The divergence is to measure the magnitude of a vector field at a given point in terms of a signed scalar. If it is positive, it is called a source. If it is negative, it is called a sink. If it is zero in a domain, there is no source or sink, and it is called solenoidal (Landau and Lifshitz 2013). The data field is smooth everywhere away from the original source. The outward flux of a vector field through a closed surface is equal to the volume integral of the divergence on the region inside the surface.
3. The data field has temporal behavior. The data on objects are static independent of time. For a stable field with sources independent of time in space, there exists a scalar potential function $\varphi(x)$ corresponding to the vector field function $F(x)$ that describes the intensity function. Both functions can be interconnected with a differential operator (Giachetta et al. 2009), $F(x) = \nabla\varphi(x)$.

6.2.2 Mathematical Model

The distribution law of a data field can be mathematically described with a potential function. In a data space $\Omega \subseteq R^P$, a data object x_i brings about a virtual field with the mass m_i. If an arbitrary point $x = (x_1, x_2, \ldots, x_P)^T$ exists in Ω, the scalar potential function of the data field on x_i is defined as Eq. (6.1):

$$\varphi(x_i) = m_i \times K\left(\frac{\|x - x_i\|}{\sigma}\right) \tag{6.1}$$

where m_i is the mass of x_i. $K(x)$ is the unit potential function. $\|x - x_i\|$ is the distance between x_i and x in the field. σ is an impact factor.

Each data object x_i in D has its own data field in Ω. All of the data fields are superposed on point x. In other words, any data object is affected by all the other objects in the data space. Therefore, the potential value at x in the data fields on D in Ω is defined as follows.

Definition 6.2 Given a data field created by a set of data objects $D = (x_1, x_2, \ldots, x_n)$ in space $\Omega \subseteq R^P$, the potential at any point $x \in \Omega$ can be calculated as

$$\varphi(x) = \varphi_D(x) = \sum_{i=1}^{n} \varphi_i(x) = \sum_{i=1}^{n} \left(m_i \times K\left(\frac{\|x - x_i\|}{\sigma}\right) \right) \tag{6.2}$$

For the fact that the vector intensity always can be constructed out of scalar potential by a gradient operator, the vector intensity $F(x)$ at x can be given as

$$F(x) = \nabla\varphi(x) = \frac{2}{\sigma^2} \sum_{i=1}^{n} \left((x_i - x) \cdot m_i \cdot K\left(\frac{\|x - x_i\|}{\sigma}\right) \right)$$

Because the term $\frac{2}{\sigma^2}$ is a constant, the above formula can be simply written as

$$F(x) = \sum_{i=1}^{n} \left((x_i - x) \cdot m_i \cdot K\left(\frac{\|x - x_i\|}{\sigma}\right) \right) \tag{6.3}$$

6.2.3 Mass

Mass m_i is the mass of object x_i ($i = 1, 2, \ldots, n$). It represents the strength of the data field from x_i; and it meets $m_i \geq 0$ and normalization condition $\sum_{i=1}^{n} m_i = 1$. If each data object is supposed to be equal in mass, indicating the same influence over space, then a simplified potential function can be derived from Eq. (6.4):

$$\varphi(x) = \frac{1}{n} \sum_{i=1}^{n} K\left(\frac{\|x - x_i\|}{\sigma}\right) \tag{6.4}$$

6.2.4 Unit Potential Function

The unit potential function $K(x)$ ($\int K(x)\mathrm{d}x = 1$, $\int xK(x)\mathrm{d}x = 0$) expresses the law
that x_i always radiates its data energy in the same way in its data field. In fact,
the potential function reflects the density of the data distribution. According to
the properties of probability density function, it can be proven that the difference
between the potential function and the probability density function is only a nor-
malization constant if $K(x)$ has finite integral in Ω; that is, $\int_{\Omega} K(x)\mathrm{d}x = M < +\infty$
(Giachetta et al. 2009; Wang et al. 2011).

   $K(x)$ is a basis potential. As the data field must be short-range with a source and
temporal behavior, the selection of $K(x)$ may need to consider such conditions as
the nature inside the data characteristics, the feature of data radiation, the standard
to determine the field-strength, the application domain of the data field, etc.
(Li et al. 2013). Because the scalar operation is simpler and more intuitive than the
vector operation, the following criteria are recommended for selecting the potential function of the data field so that the potential $\varphi(x)$ of an arbitrary data object x_i
at an arbitrary position x satisfies three conditions:

1. $K(x)$ is a continuous, smooth, and finite function that is defined in Ω,
2. $K(x)$ is isotropic, and
3. $K(x)$ is a monotonically decreasing function on the distance between x_i and x.
   $K(x)$ is maximum when the distance is 0, while $K(x)$ approaches 0 when the
   distance tends to infinity.

Generally speaking, a function that matches the above three criteria can define the
potential function of a data field.

   In this book, nuclear-like potential function is taken as $K(x)$ in the form of
Gaussian potential. It is known that a nucleon generates a centralized force field in
an atomic nucleus. In space, the nuclear force acts radially toward or away from the
source of the force field, and the potential of the nuclear field decreases rapidly to 0.
Thus, a potential function of a data field is studied to simulate a nuclear field; that is,
a nuclear-like field. There are three methods to compute the potential value of a point
in a nuclear field: Gaussian potential, square-well potential, and exponential potential (Li et al. 2013). Because Gaussian distribution is ubiquitous and Gaussian function matches the physical nature of a data field, adopting the Gaussian function—that
is, a nuclear-like potential function—is preferred for modeling the distribution of
data fields. In terms of nuclear-like potentials, at the arbitrary point $x \in \Omega$, the potential of the data field from data object x_i is shown in Eq. (6.5); the potential superposition of n basis potentials from all the data objects in dataset D is shown in Eq. (6.6).

$$\varphi(x_i) = m_i \times e^{-\left(\frac{\|x-x_i\|}{\sigma}\right)^2} \tag{6.5}$$

$$\varphi(x) = \sum_{i=1}^{n}\left(m_i \times e^{-\left(\frac{\|x-x_i\|}{\sigma}\right)^2}\right) \tag{6.6}$$

6.2.5 Impact Factor

Once the form of potential function $K(x)$ is fixed, the distribution of the associated data field is primarily determined by the impact factor σ under the given data set D in space Ω.

The impact factor $\sigma \in (0, +\infty)$ controls the interacted distance between data objects. Its value has a great impact on the distribution of the data field. The effectiveness of the impact factor is from various sources, such as radiation brightness, radiation gene, data amount, distance between the neighbor equipotential lines, and grid density of Descartes coordinate. They all make their contributions to the data field. As a result, the distribution of a potential value is determined by the impact factor, and the different potential function has a much smaller influence on the estimation.

In nature, the optimal choice of σ is a minimization problem of a univariate, nonlinear function, which can be solved by standard optimization algorithms, such as a simple one-dimensional searching method, stochastic searching method, or simulated annealing method. Li and Du (2007) introduced the optimization algorithm of impact factor σ.

Take an example in two-dimensional space. Figure 6.1 shows the equipotential distributions of 5 data field from the same five data objects with a variance of σ. When σ is very small (Fig. 6.1a), the interaction distance between two data objects is very short. $\varphi(x)$ is equivalent to the superposition of n peak functions, each center of which is the data object. The potential value around every data object is very small. The extreme case is that there is no interaction between two objects, and the potential value is $1/n$ at the position of data object. Otherwise, when σ is very large (Fig. 6.1c), the data objects are strongly interacting. $\varphi(x)$ is equivalent to the superposition of n basic functions, which change slowly when the width is large. The potential value around every data object is very large. The extreme case is that the potential value is approximately 1 at the position of the data object. When the integral of unit potential function is finite, there is a normalized constant between the potential function and the probability intensity function at most (Giachetta et al. 2009). As can be seen from the abovementioned two extreme

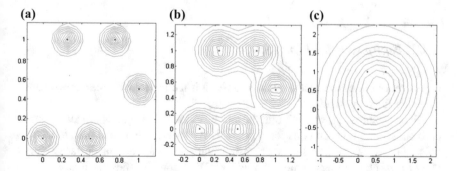

Fig. 6.1 Equipotential distributions of data field on 5 objects with different σ **a** $\sigma = 0.05$ **b** $\sigma = 0.2$ **c** $\sigma = 0.8$

conditions, the potential distribution of a data field may not uncover the required rule on inherent data characteristics. Thus, the selection of impact fact σ does portray the inherent distribution of data objects (Fig. 6.1b).

6.3 Depiction of Data Field

In physics, a field can be depicted with the help of field force lines for a vector field and isolines (or iso-surfaces) for a scalar field. The field lines, equipotential lines (or surfaces), are also adopted to represent the overall distribution of the data fields.

6.3.1 Field Lines

A field line is a line with an arrowhead and length. The line arrowhead indicates the direction, and the line length indicates the field strength. It is a locus that is defined by a vector field and a starting location within the field. A series of field lines visually portray the field of a data object. The density of the field lines further shows the field strength. All the data field lines centralize the data object and the field of the source is the strongest. When approaching the source, the length of the field lines becomes longer and the density thicker. Otherwise, given the distance $\|x - x_i\|$ leaving the field source, the larger the distance grows, the weaker the field becomes. The same is true with the field line—that is, leaving the data object, the length of the field lines becomes shorter and the density scarcer. Their distribution shows that the data field is a distribution of spherical symmetry. Figure 6.2 is a plot of force lines for a data force field produced by a single object, which indicates that the radial field lines always are in the direction of the spherical coordinate towards the source. The force lines are uniformly distributed and spherically symmetric, decreasing as the radial distance $\|x - x_i\|$ further increases, which indicates a strong attractive force on the sphere centered on the object.

6.3.2 Equipotential Line (Surface)

Equipotential lines (surfaces) come into being if the points with the same potential value are lined up together.

Given potential ψ, an equipotential can be extracted as one where all the points on it satisfy the implicit equation $\varphi(x) = \psi$. By specifying a set of discrete potentials $\{\psi_1, \psi_2, \ldots, \psi_n\}$, a series of nested equipotential lines or surfaces can be drawn for better understanding of the associated potential field as a whole form. Let $\psi_1, \psi_2, \ldots, \psi_n$ respectively be the potentials at the positions of the objects $x_1, x_2, \ldots, x_n$. The potential entropy can be defined as (Li and Du 2007)

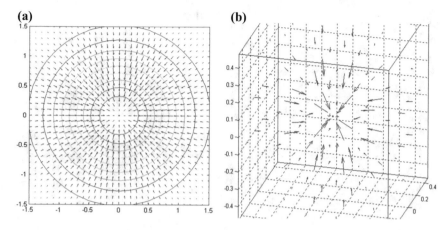

Fig. 6.2 The plot of force lines for a data force field created by a single object ($m = 1$, $\sigma = 1$) **a** 2-dimensional plot, **b** 3-dimensional plot

$$H = -\sum_{i=1}^{n} \frac{\psi_i}{Z} \log\left(\frac{\psi_i}{Z}\right) \tag{6.7}$$

where $Z = \sum_{i=1}^{n} \psi_i$ is a normalization factor. For any $\sigma \in [0, +\infty]$, potential entropy H satisfies $0 \leq H \leq \log(n)$, and $H = \log(n)$ if and only if $\psi_1 = \psi_2 = \cdots = \psi_n$.

The equipotential lines are mathematical entities that describe the spherical circles at a distance $\|x - x_i\|$ on which the field has a constant value, such as contour lines on map tracing lines of equal altitude. They always cross the field lines at right angles to each other, in no specific direction at all. The potential difference compares the potential from one equipotential line to the next. In the context of equal data masses, the area distributed with intensive data will have a greater potential value when the potential functions are superposed. At the position with the largest potential value, the data objects are the most intensive. Figure 6.3a is a map of the equipotential lines for a potential field produced by a single data object in a two-dimensional space, which consists of a family of nested, concentric circles or spheres centered on the data object. As can be seen from Fig. 6.3a, the equipotentials closer to the object have higher potentials and are farther apart, which indicates strong field strength near and around the object.

In a multi-dimensional space, the equipotential lines are extended to the equipotential surfaces because the field becomes a set of nested spheres with the center of the coordinate of the data object, and the radius is $\|x - x_i\|$. Figure 6.3b is another map of the equipotential surfaces for another 3D space. Obviously, the equipotential surfaces corresponding to different potentials are quite different in topology.

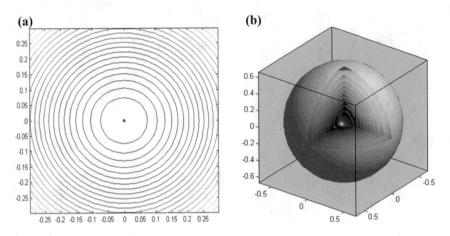

Fig. 6.3 The equipotential map for a potential field produced by a single data object ($\sigma = 1$) **a** equipotential lines, **b** equipotential surfaces

6.3.3 Topological Cluster

Every topology of equipotentials containing data objects actually corresponds to a cluster with relatively dense data distribution. The nested structure composed of different equipotentials may be treated as the clustering partitions at different density levels.

Figure 6.4 shows the distribution of field lines and equipotential lines on data objects in two-dimensional space. As can be seen from Fig. 6.4, the equipotential lines are always perpendicular to the field lines associated with the data field, and all the data objects in the field are centralized, surrounding five positions (i.e., A, B, C, D, and E) under the forces from the data objects. The field lines always point to the five positions with local maximum potential in the data field (Fig. 6.4a), and the self-organized process discovers the hierarchical clustering-characteristics of data objects with the field level by level (Fig. 6.4b). Under a set of suitable potential values, the corresponding set of equipotential lines show the clustering feature of data objects by considering different data-intensive areas as the centers. When the potential is 0.0678 at level 1, all the data objects are self-grouped into the five clusters A, B, C, D, and E. As the potential decreases, the clusters that are in close proximity are merged. Cluster A and cluster B form the same lineage AB at level 2. Cluster D and cluster E are amalgamated as cluster DE at level 3. When the potential is 0.0085 at level 5, all the clusters join the generalized cluster ABCDE (Fig. 6.4a). During the merging process from the five clusters A, B, C, D, and E at the bottom level 1 to the generalized cluster ABCDE at the top level 5, a clustering spectrum automatically comes into being (Fig. 6.4b). In other words, the center in the data-intensive area has the potential maximum. From the local maximums to the final global maximum, five positions are further self-organized cluster by cluster.

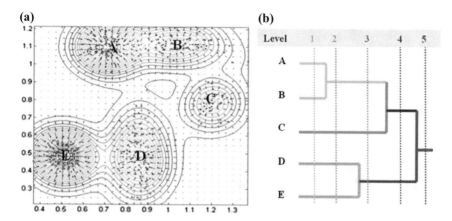

Fig. 6.4 Self-organized characteristics of 390 data objects in the field ($\sigma = 0.091$) **a** self-organized level by level, **b** characterized level

References

Giachetta G, Mangiarotti L, Sardanashvily G (2009) Advanced classical field theory. World Scientific, Singapore

Landau LD, Lifshitz EM (2013) Statistical physics, 3rd edn, part 1: volume 5 (course of theoretical physics, volume 5). Pergamon Press, Oxford

Li DY, Du Y (2007) Artificial intelligence with uncertainty. Chapman & Hall/CRC, London

Li DR, Wang SL, Li DY (2013) Spatial data mining theories and applications, 2nd edn. Science Press, Beijing, China

Wang SL, Gan W, Li DY, Li DR (2011) Data field for hierarchical clustering. Int J Data Warehouse Min 7(3):235–246

Chapter 7
Cloud Model

In SDM, mutually transforming between a qualitative concept and quantitative data is a bottleneck. In this chapter, the cloud model acts as a transforming model between a qualitative concept and its quantitative data (Li and Du 2007). It has three numerical characters: *Ex*, *En*, and *He*. Among various cloud models, a normal cloud model is ubiquitous. Cloud generators are introduced as a forward cloud generator, backward cloud generator, and precondition cloud generator. Uncertain reasoning focuses on one-rule reasoning and multi-rule reasoning.

7.1 Definition and Property

The cloud model is described by linguistic values to represent the uncertain relationship between a specific qualitative concept and its quantitative expression. It is introduced to reflect the uncertainties of the concepts in natural languages. The cloud, as a reflection of the connection of randomness and fuzziness, constructs the mapping between quality and quantity (Li et al. 2009).

7.1.1 Cloud and Cloud Drops

Definition 7.1 Let U be a universal set described by precise numbers and C be the qualitative concept related to U. If there is a number $x \in U$, which randomly realizes the concept C, and the certainty degree of x for C, i.e., $\mu(x) \in [0, 1]$, is a random value with stabilization tendency:

$$\mu: U \to [0, 1] \quad \forall x \in U \quad x \to \mu(x)$$

© Springer-Verlag Berlin Heidelberg 2015
D. Li et al., *Spatial Data Mining*, DOI 10.1007/978-3-662-48538-5_7

then the distribution of x on U is defined as a cloud, and every x is defined as a cloud drop.

For a better understanding of the cloud, we use the united distribution of (x, μ) to express the concept C; that is, $C(x, \mu)$.

7.1.2 Properties

The cloud has the following properties:

1. The universal set U can be either one-dimensional or multi-dimensional.
2. The random realization in the definition is the realization in terms of probability. The certainty degree is the membership in the fuzzy set, and it also has the distribution of probability. All of these properties show the consecutiveness of fuzziness and randomness.
3. For any $x \in U$, the mapping from x to [0, 1] is multiple. The certainty degree of x on C is a probability distribution rather than a fixed number.
4. The cloud is composed of cloud drops, which are not necessarily in order. One cloud drop is one realization of the qualitative concept. The more cloud drops there are, the better the overall feature of this concept is represented.
5. The more probable the cloud drop appears, the higher the certainty degree is— and hence, the more the cloud drop is contributing to the concept.

The cloud model leads the research on intelligence at the basic linguistic value, focusing on an approach to quantitate the qualitative concept to make it more intuitive and universal. In this way, the concept is transformed into a number of quantitative values—in other words, transformed into the points in the space of a universal set. This process, as a discrete transition process, is random. The selection of every single point is a random event and can be described by the probability distribution function. The certainty degree of the cloud drop is fuzzy and can be described by the probability distribution function as well. This cloud model is flexible and boundless, and its global shape rather than its local region can be observed. These attributes are similar to the cloud in natural phenomena, so we call this mathematic transition between data the "cloud."

7.1.3 Integrating Randomness and Fuzziness

The cloud model integrates randomness and fuzziness. The uncertainty of the concept can be represented by multiple numerical characters. We can say that the mathematical expectation, variance, and high-order moment in probability theory reflect several numerical characters of the randomness but not the fuzziness. Membership is a precise description of the approach to the fuzziness, which does not take the randomness into account. A rough set measures the uncertainty

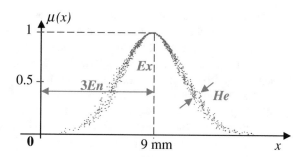

Fig. 7.1 A cloud model of "displacement is 9 mm around". Reprinted from Wang and Shi (2012), with kind permission from Springer Science + Business Media

by the research of two precise sets that are based on the background of precise knowledge rather than the background of uncertain knowledge. In the cloud model approach, we can express the uncertainty of the concept by higher-ordered entropy apart from the expectation, entropy, and hyperentropy in order to conduct research in more depth.

Figure 7.1 shows an interpretation of the concept of "displacement is 9 mm around" with the cloud model, where x is the displacement and $\mu(x)$ is the certainty to the concept. In Fig. 7.1, a piece of the cloud is composed of many drops represented by data, any one of which is a stochastic mapping in the universe of discourse from a qualitative fuzzy concept, along with the membership of the data belonging to the concept. It is observable as a whole shape but fuzzy in detail, similar to a natural cloud in the sky. When the cloud drop of the piece of cloud is created, the mapping of each cloud drop is random and its membership is fuzzy, which is the process that integrates randomness and fuzziness; "displacement is 9 mm around" is a piece of the cloud. It is composed of many cloud drops represented by data $\{\ldots, 8 \text{ mm}, 9 \text{ mm}, 10 \text{ mm}, \ldots\}$, any one of which is a stochastic mapping in the universe of "displacement" from a qualitative fuzzy concept of "displacement is 9 mm around," along with the membership $\{\ldots, 0.9, 1, 0.9, \ldots\}$ of the data $\{\ldots, 8 \text{ mm}, 9 \text{ mm}, 10 \text{ mm}, \ldots\}$ belonging to the concept. Also, "displacement is 9 mm around" is a concept, and $\{\ldots, 8 \text{ mm}, 9 \text{ mm}, 10 \text{ mm}, \ldots\}$ are the data of the concept. The generated data may be a stochastic process of $\{\ldots\ldots, 8 \text{ mm}, 9 \text{ mm}, 10 \text{ mm}, \ldots\}$, and their memberships $\{\ldots, 0.9, 1, 0.9, \ldots\}$ on the concept are fuzzy.

7.2 The Numerical Characteristics of a Cloud

The overall property of a concept can be represented by the numerical characters, which are the overall quantitative property of the qualitative concept. They are of great significance to understanding the connotation and the extension of the concept.

In the cloud model, we employ the expected value Ex, the entropy En, and the hyperentropy He to represent the concept as a whole.

1. The expected value *Ex*: The mathematical expectation of the cloud drop distributed in the universal set. In other words, it is the point that is the most representative of the qualitative concept or the most classical sample while quantifying the concept.
2. The entropy *En*: The uncertainty measurement of the qualitative concept. It is determined by both the randomness and the fuzziness of the concept. In one aspect, as the measurement of randomness, *En* reflects the dispersing extent of the drops; in the other aspect, it is also the measurement of "this and that," representing the value region in which the drop is acceptable by the concept. As a result, the connection of randomness and fuzziness is reflected by using the same numerical character.
3. The hyperentropy *He*: This is the uncertainty measurement of the entropy—that is, the entropy of the entropy—which is determined by both the randomness and fuzziness of the entropy.

To express the qualitative concept by the quantitative method, we generate cloud drops under the numerical characters of the cloud; the reverse (i.e., from the quantitative expression to the qualitative concept) extracts the numerical characters from the group of cloud drops. Figure 7.1 shows the three numerical characteristics of the linguistic term "displacement is 9 mm around." In the universe of discourse, *Ex* is the position corresponding to the center of the cloud gravity, the elements of which are fully compatible with the spatial linguistic concept; *En* is a measure of the concept coverage (i.e., a measure of the spatial fuzziness), which indicates how many elements could be accepted to the spatial linguistic concept; and *He* is a measure of the dispersion on the cloud drops, which also can be considered as the entropy of *En*.

7.3 The Types of Cloud Models

There are various implementation approaches for the cloud model, resulting in different kinds of clouds, such as the normal cloud model, the symmetric cloud model, the half cloud model, and the combined cloud model.

Normal distribution is one of the most important distributions in probability theory. It is usually represented by two numerical characters—the mean and the variance. The bell-shaped membership function is the most frequently used function in the fuzzy set, and it is generally expressed as $\mu(x) = e^{-\frac{(x-a)^2}{2b^2}}$. The normal cloud is a brand-new model developed based on the normal distribution and the bell-shaped membership function.

There are various mathematical characters representing uncertainties, and these variations also exist in the real world. For example, the symmetric cloud model represents the qualitative concept with symmetric features (Fig. 7.2a). The half cloud model depicts the concept with uncertainty on one side (Fig. 7.3b). As a result, we can construct various kinds of cloud models.

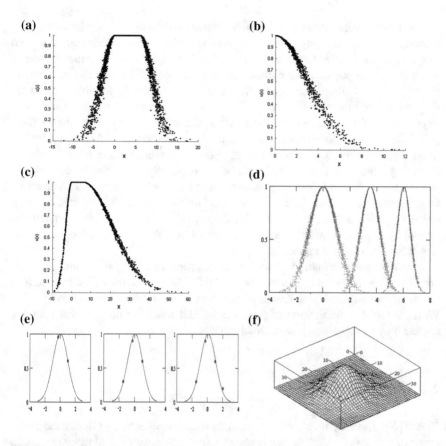

Fig. 7.2 Various cloud models: **a** Symmetric cloud model. **b** Half cloud model. **c** Combined cloud model. **d** Floating cloud. **e** Geometry cloud. **f** Multi-dimensional cloud model

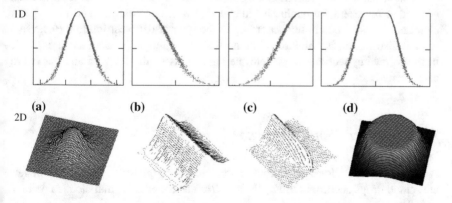

Fig. 7.3 Spatial objects represented by a cloud model: **a** point, **b** line, **c** directed line, **d** polygon

The clouds constructed by other given clouds are called virtual clouds, such as a combined cloud, floating cloud, and geometric cloud. A combined cloud is used to synthesize linguistic terms into a generalized term. If we use the mechanism of combined cloud construction recursively from low concept levels to high concept levels, we can obtain the concept hierarchies for the linguistic variables, which are very important in data mining (Fig. 7.2c). The floating cloud mechanism is used to generate default clouds in the blank areas of the universe by other given clouds (Fig. 7.2d). If we consider a universe as a linguistic variable and we want to represent linguistic terms by clouds, the only essential work is to specify the clouds at the key positions. Other clouds can be automatically generated by the floating cloud construction method. A geometric cloud is a whole cloud constructed by least square fitting on the basis of some cloud drops—that is, the expected curve of geometry cloud is generated by the least square (Fig. 7.2e). The cloud model can be multidimensional as well. Figure 7.2f shows the joint relationship of the multidimensional cloud and its certainty degree.

In geospatial information science, there are three basic objects: point, line, and area. The line may be non-directed or directed. If they are represented by a combination of cloud models, the spatial objects may be illustrated as shown in Fig. 7.3. With the further combination of points, lines, and areas, various types of complex objects may be represented with cloud models.

7.4 Cloud Generator

Given {*Ex, En, He*} in a concept, the data may be generated to depict the cloud drops of a piece of cloud, which is called a forward cloud generator; given certain datasets, {*Ex, En, He*} also may be generated to represent a concept called a backward cloud generator. The backward cloud generator can be used to extract the conceptual knowledge from numeric datasets, and the forward cloud generator can be used to represent the qualitative knowledge with quantitative data. The type of cloud model must be chosen according to the type of data distribution. Because a normal distribution is ubiquitous, the normal cloud model is taken as an example in the following sections to present the algorithms of the forward and backward cloud generators.

7.4.1 Forward Cloud Generator

A forward normal cloud generator is a mapping from quality to quantity. It generates cloud drops according to *Ex*, *En*, and *He*. The forward normal cloud is defined as follows:

Definition 7.2 Let U be a quantitative universal set described by precise numbers and C the qualitative concept related to U. Also, there is a number $x \in U$, which is a random realization of the concept C. If x satisfies $x \sim N(Ex, En'^2)$, where $En' \sim N(En, He^2)$, and the certainty degree of x on C is

$$\mu = e^{-\frac{(x-Ex)^2}{2(En')^2}}$$

then the distribution of x on U is a normal cloud.

Algorithm 7.1 Forward normal cloud generator CG (Ex, En, He, n)
Input: (Ex, En, He), and the number of cloud drops n;
Output: n of cloud drops x and their certainty degree μ, i.e., $Drop(x_i, \mu_i)$, $i = 1, 2, ..., n$;
Steps:

(1) Generate a normally distributed random number En'_i with expectation En and variance He^2; i.e., $En'_i = NORM(En, He^2)$.
(2) Generate a normally distributed random number x_i with expectation Ex and variance En'_i; i.e., $x_i = NORM(Ex, En'^2_i)$.
(3) Calculate $\mu_i = e^{-\frac{(x_i-Ex)^2}{2(En'_i)^2}}$.
(4) x_i with certainty degree of μ_i is a cloud drop in the domain.
(5) Repeat steps (1)–(4) until n cloud drops are generated.

The key in this algorithm is the combined relationship—that is, within the two random numbers, one number is the input of the other in generation. The variance should not be 0 while generating normal random numbers. That is why En and He are required to be positive in the CG algorithm. If $He = 0$, step (1) will always provide an invariable number En; as a result, x will become a normal distribution. If $He = 0$, $En = 0$, the generated x will be a constant Ex and $\mu \equiv 1$. By this means, we can say that certainty is the special case of the uncertainty.

The foundation of the algorithm lies in the way random numbers with normal distribution are generated. In a majority of the programming languages, there are functions to generate random numbers with uniform distribution in $[0, 1]$. However, to generate normally distributed random numbers by uniformly distributed numbers, the seed of the uniformly random function must be investigated as it determines whether the generated random numbers are the same each time. There are many statistical computations in the literature for the problem of generating normally distributed random numbers or other distributed random numbers by uniformly distributed ones.

This algorithm is applicable to the dimensional space, as well as a two-dimensional (2D) or higher dimensional space. For example, for a 2D space with the given numerical characters of a 2D normal cloud—that is, expectations (Ex, Ey), entropies (Enx, Eny), and hyperentropies (Hex, Hey)—cloud drops can be generated by a 2D forward normal cloud generator.

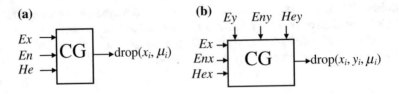

Fig. 7.4 Forward cloud generator: **a** 1D, **b** 2D

Algorithm 7.2 2D forward normal cloud generator

Input: (*Ex, Ey, Enx, Eny, Hex, Hey, n*)
Output: *Drop*(x_i, y_i, μ_i), $i = 1,\dots,n$;
Steps:

    (1)   Generate a 2D normally distributed random vector (Enx'_i, Eny'^2_i) with expectation (*Enx, Eny*) and variance (*Hex*^{2}, *Hey*^{2}).

    (2)   Generate a 2D normally distributed random vector (x_i, y_i) with expectation (*Ex, Ey*) and variance (Enx'^2_i, Eny'^2_i).

    (3)   Calculate $\mu_i = e^{-\left[\frac{(x_i - Ex)^2}{2Enx'^2_i} + \frac{(y_i - Ey)^2}{2Eny'^2_i}\right]}$.

    (4)   *Let* (x_i, y_i, μ_i) be a cloud drop, and one quantitative implementation of the linguistic value is represented by this cloud. In this drop, (x_i, y_i) is the value in the universal domain corresponding to this qualitative concept and μ_i is the degree measure of the extent to which (x_i, y_i) belongs to this concept.

    (5)   Repeat steps (1)–(4) until *n* cloud drops are generated.

Figure 7.4a shows the algorithm of the forward normal cloud generator. The corresponding 2D forward normal cloud generator is shown in Fig. 7.4b.

7.4.2 Backward Cloud Generator

Backward cloud generator (CG^{-1}) is the model for transition from quantitative value to qualitative concept. It maps a quantity of precise data back into the qualitative concept expressed by *Ex, En, He*, as shown in Fig. 7.5.

There are two kinds of basic algorithms of the CG^{-1} based on statistics, which are classified by whether or not the certainty degree is utilized.

Algorithm 7.3 Backward normal cloud generator with the certainty degree

Input: x_i and the certainty degree μ_i, $i = 1, 2, \dots, n$
Output: (*Ex, En, He*) representative of the qualitative concept

Fig. 7.5 Backward cloud generator

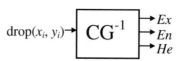

$$\text{drop}(x_i, y_i) \rightarrow \boxed{\text{CG}^{-1}} \begin{matrix} \rightarrow Ex \\ \rightarrow En \\ \rightarrow He \end{matrix}$$

Steps:

(1) Calculate the mean of x_i for Ex, i.e., $Ex = MEAN(x_i)$.
(2) Calculate the standard variance of x_i for En, i.e., $En = STDEV(x_i)$.
(3) For each couple of (x_i, μ_i), calculate

$$En_i' = \sqrt{\frac{-(x_i - Ex)^2}{2 \ln \mu_i}}.$$

(4) Calculate the standard variance of En'_i for He, i.e., $He = STDEV(En'_i)$.

In the algorithm, $MEAN$ and $STDEV$ are the functions for the mean and standard variance of the samples.

There are drawbacks to this one-dimensional (1D) backward cloud generator algorithm, which are as follows:

1. The certainty degree μ is required for recovering En and He; however, in practical applications, only a group of data representative of the concept usually can be obtained, but not the certainty degree μ;
2. It is difficult to extend this algorithm to higher dimensional situations, and the error in the higher dimensional backward cloud is larger than that of the 1D backward cloud.

To overcome the drawbacks of the algorithm requiring the certainty degree, the following backward cloud algorithm utilizes only the value of x_i for backward cloud generation based on the statistical characters of the cloud.

Algorithm 7.4 Backward normal cloud generator without certainty degree

Input: Samples x_i, $i = 1, 2, ..., n$
Output: (Ex, En, He) representative of the qualitative concept
Steps:

(1) $\bar{X} = \frac{1}{n} \sum_{i=1}^{n} x_i, S^2 = \frac{1}{n-1} \sum_{i=1}^{n} (x_i - \bar{X})^2$ // the mean and variance of x_i.
(2) $Ex = \bar{X}$.
(3) $En = \sqrt{\frac{\pi}{2}} \times \frac{1}{n} \sum_{i=1}^{n} |x_i - Ex|$.
(4) $He = \sqrt{S^2 - En^2}$.

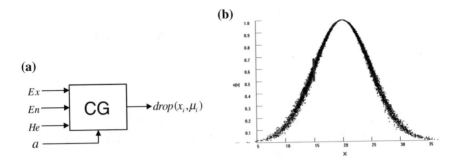

Fig. 7.6 Precondition cloud generator (**a**) and joint membership distribution (**b**)

7.4.3 Precondition Cloud Generator

Consider a universe U, in which the normal cloud model C can be expressed as C (*Ex, En, He*). If it is possible to generate the membership distribution of a specific point a, which is in universe U through the cloud generator, then the cloud model generator is called a precondition cloud generator. This is illustrated in Fig. 7.6a.

Algorithm 7.5 Precondition cloud generator

Input: The numeral characteristics of quality concept (*Ex, En, He*), the specific value x and the number of output cloud drops n.
Output: The cloud drops $(x_0, \mu_0), (x_1, \mu_1), \ldots (x_n, \mu_n)$
Steps:
    Begin
    For $(i = 0; i < n; i++)$

        $En_{n_i} = Norm(En, He)$

$$\mu_i = e^{\frac{-(x-Ex)^2}{2(En_{n_i})^2}}$$

        Return $drop(x, \mu_i)$
    End For
    End
    The joint distribution of membership, which is generated by specific value x and the precondition cloud generator, is illustrated in Fig. 7.6. All of the cloud drops are located in the straight line of $X = x$.

7.5 Uncertainty Reasoning

Uncertainty reasoning may include one-rule reasoning and multi-rule reasoning.

7.5.1 One-Rule Reasoning

If there is only one factor in the rule antecedent, we call the rule a one-factor rule—that is, "If A, then B." CG_A is the X-conditional cloud generator for linguistic term A, and CG_B is the Y conditional cloud generator for linguistic term B. Given a certain input x, CG_A generates random values μ_i. These values are considered as the activation degree of the rule and input to CG_B. The final outputs are cloud drops, which form a new cloud.

Combining the algorithm of the X and Y conditional cloud generators (Li et al. 2013), the following algorithm is introduced for one-factor one-rule reasoning.

Algorithm 7.6 One-factor one-rule reasoning

Input: x, Ex_A, En_A, He_A
Output: drop(x_i, μ_i)
Steps:

   (1) $En'_A = G(En_A, He_A)$ // Create random values that satisfy the normal distribution probability of mean En_A and standard deviation He_A.

   (2)   Calculate $\mu = e^{\frac{-(x-Ex_A)^2}{2(En'_A)^2}}$.

   (3) $En'_B = G(En_B, He_B)$ // Create random values that satisfy the normal distribution probability of mean En_B and standard deviation He_B.

   (4)   Calculate $y = Ex_B \pm \sqrt{-2\ln(\mu)}En'_B$, let (y, μ) be cloud drops. If $x \le Ex_A$, "−" is adopted, while if $x > Ex_A$, "+" is adopted.

   (5)   Repeat steps (1)–(4), generating as many cloud drops as needed.

Figure 7.7 is the output cloud of a one-factor one-rule generator with one input. It can be seen that the cloud model-based reasoning generated uncertain results. The uncertainty of the linguistic terms in the rule is propagated during the reasoning process. Because the rule output is a cloud, the final result can be obtained in several forms: (1) one random value; (2) several random values as sample results; (3) the expected value, which is the mean of many sample results; and (4) the linguistic term, which is represented by a cloud model. The parameters of the model are obtained by an inverse cloud generator method.

If we input a number of values to the one-factor rule and draw the inputs and outputs in a scatter plot, we can obtain the input-output response graph of the one-factor

Fig. 7.7 Output cloud of one-rule reasoning

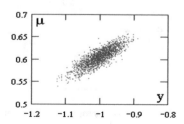

Fig. 7.8 Input-output
response of one-rule
reasoning

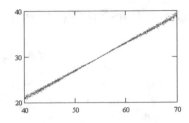

one-rule reasoning (Fig. 7.8). The graph looks like a cloud band, not a line. The band is more focused closer to the expected values; farther away from the expected value, the band is more dispersed. This is consistent with human intuition. The above two figures and discussion show that the cloud model based on uncertain reasoning is more flexible and powerful than the conventional fuzzy reasoning method.

If the rule antecedent has two or more factors, such as "If A_1, A_2,..., A_n, then B," the rule is called a multi-factor rule. In this case, a multi-dimensional cloud model represents the rule antecedent. Figure 7.9 is a two-factor one-rule generator, which combines a 2D X-conditional cloud generator and a 1D Y-conditional cloud generator. It is not difficult to apply the reasoning algorithm on the basis of the cloud generator; consequently, multi-factor one-rule reasoning is conducted in a similar way.

7.5.2 Multi-rule Reasoning

Usually, there are many rules in a real knowledge base. Multi-rule reasoning is frequently used in intelligent GIS or spatial decision-making support systems. Figure 7.10 is a one-factor multi-rule generator, and the algorithm is as follows.

Algorithm 7.7 One-factor multi-rule generator reasoning

Input: x, Ex_{Ai}, En_{Ai}, He_{Ai}
Output: y

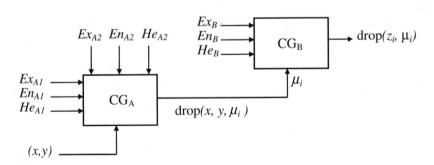

Fig. 7.9 A two-factor one-rule generator

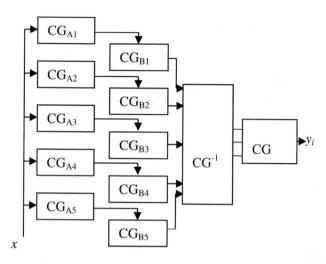

Fig. 7.10 One-factor five-rule generator

Steps

(1) Given input x, determine how many rules are activated. If Ex_{Ai} $-3En_{Ai} < x < Ex_{Ai} +3En_{Ai}$, then rule i is activated by x.

(2) If only one rule is activated, output is a random y by the one-factor one-rule reasoning algorithm. Go to step (4).

(3) If two or more rules are activated, each rule first outputs a random value by the one-factor one-rule reasoning algorithm, and a virtual cloud is constructed by the geometric cloud generation method. A cloud generator algorithm is run to output a final result y with the three numerical characters of the geometric cloud. Because the expected value of the geometric cloud is also a random value, the expected value can be taken as the final result for simplicity. Go to step (4).

(4) Repeat step (1)–(3) to generate as many outputs as required.

The main concept of the multi-rule reasoning algorithm is that when several rules are activated simultaneously, a virtual cloud is created by the geometric

Fig. 7.11 Two activated rules

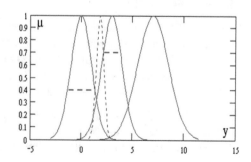

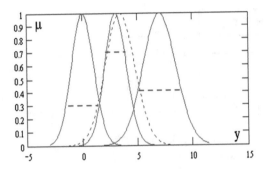

Fig. 7.12   Three activated rules

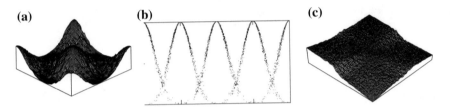

Fig. 7.13   An example of multi-factor multi-rule reasoning: **a** Input. **b** Output. **c** Input-output response

cloud method. Because of the least square fitting property, however, the final output is more likely to be close to the rule of high activated degree. This is consistent with human intuition. Figure 7.11 is a situation of two rules activated, and Fig. 7.12 has three rules activated. Only the mathematical expected curves are drawn for clearness, and the dash curves are the virtual cloud. The one-factor multi-rule reasoning can be easily extended to multi-factor multi-rule reasoning on the basis of multi-dimensional cloud models.

The following is an illustrative example of multi-factor multi-rule reasoning. Suppose there are five rules to describe the terrain features qualitatively, and rule input is the linguistic term of location that is represented by a 2D cloud (Fig. 7.13a). Rule output is the linguistic term of elevation that is represented by a 1D cloud (Fig. 7.13b).

- Rule 1: If location is southeast, then elevation is low.
- Rule 2: If location is northeast, then elevation is low to medium.
- Rule 3: If location is central, then elevation is medium.
- Rule 4: If location is southwest, then elevation is medium to high.
- Rule 5: If location is northwest, then elevation is high.

Figure 7.13c is the input-output response surface of the rules. The surface is an uncertain surface and the roughness is uneven. Closer to the center of the rules (the expected values of the clouds), the surface is smoother, showing the small uncertainty; at the overlapped areas of the rules, the surface is rougher, showing large uncertainty. This proves that multi-factor multi-rule reasoning also represents and propagates the uncertainty of the rules as does the one-factor one-rule reasoning does.

References

Li DR, Wang SL, Li DY (2013) Spatial data mining theories and applications, 2nd edn. Science Press, Beijing

Li DY, Du Y (2007) Artificial intelligence with uncertainty. Chapman & Hall/CRC, London

Li DY, Liu CY, Gan WY (2009) A new cognitive model: cloud model. Int J Intell Syst 24(3):357–375

Wang SL, Shi WZ (2012) Data mining and knowledge discovery. In: Kresse W, Danko David (eds) Handbook of geographic information. Springer, Berlin

Chapter 8
GIS Data Mining

SDM can extend finite data in GIS to infinite knowledge. In this chapter, the Apriori algorithm, concept lattice, and cloud model are discussed and utilized in case studies to discover spatial association rules and attribute generalization. First bank operational income analysis and location selection assessment are used to demonstrate how inductive learning uncovers the spatial distribution rules; urban air temperature and agricultural data are used to demonstrate how a rough set discovers decision-making knowledge; and the monitoring process of the Baota landslide is used to show the practicality and creditability of applying the cloud model and data fields under the SDM views to support decision-making. Fuzzy comprehensive clustering, which perfectly integrates fuzzy comprehensive evaluation and fuzzy clustering analysis, and mathematical morphology clustering, which can handle clusters of arbitrary shapes and can provide knowledge of outliers and holes, are also discussed. The results show that SDM is practical and credible as decision-making support.

8.1 Spatial Association Rule Mining

Spatial association rule mining extracts association rules whose support and confidence are not less than the user-defined *min_sup* and *min_conf*, respectively. The problem to be solved by data mining can be decomposed into two sub-problems: (1) how to locate all the frequent itemsets whose support is at or more than the *min_sup*, and (2) how to generate all the association rules whose confidence is at or more than the *min_conf* for the frequent itemsets. The solution to the second sub-problem is straightforward, so the focus here is to develop a new efficient algorithm to solve the first sub-problem.

© Springer-Verlag Berlin Heidelberg 2015
D. Li et al., *Spatial Data Mining*, DOI 10.1007/978-3-662-48538-5_8

8.1.1 The Mining Process of Association Rule

Suppose $I = \{i_1, i_2, ..., i_n\}$ is a set of items—that is, an itemset. An itemset is a collection of items: k itemset is an itemset containing k items. If $A \subseteq I$, $B \subseteq I$, and $A \cap B = \emptyset$, then the association rule is an implicative form like $A \Rightarrow B$. D is the set of transactional data relevant to the mining task. Each transaction T with an identifier TID is the itemset enabling $T \subseteq I$. A frequent itemset is one where the occurrence frequency is no smaller than the product of min_sup (the minimum support threshold) and the total number of transactions in the transaction set D. Another threshold is min_conf (the minimum confidence threshold).

The relationship is mapped among a collection of objects or attributes. The composite operation of the two mappings is defined as a closed operation $\gamma(l) = \alpha(l) \circ \beta(l) = \beta(\alpha(l))$ for the itemset l, whose closed operation $\gamma(l)$ is the largest collection of all objects that contain l.

Definition 8.1 For frequent itemset l, $\gamma(l) = l$ means that the set is a closed itemset. When the condition also satisfies min_supp, it is called a frequent closed itemset (Zaki and Ogihara 2002).

According to the definition, it is intuitive for an itemset l to generate the frequent closed itemset with the following method. First, perform the operation α to identify the set of all the objects that contain l. If the percentage containing such objects versus all the collections is no smaller than the degree of the support threshold, then it is evident that the collection is frequent; Second, perform the operation β and then calculate the joint itemset of all the objects in the collection. However, the efficiency of such an implementation is very low.

In the process of association rule mining, after fully understanding the data, the user must define the objective, prepare the data, select the appropriate min_sup and min_conf, and understand these association rules. Thus, the mining process of association rules includes the following steps:

(1) Prepare the dataset.
(2) Set the min_sup and min_conf.
(3) Find all the frequent itemsets whose support is no smaller than the min_sup with the mining algorithm.
(4) Generate all the strong rules whose confidence is no smaller than the min_conf according to the frequent itemsets.
(5) Adjust the support threshold and confidence threshold for regenerating the strong rules if the amount of generated rules is too many or too few.
(6) Refine the association rules of interest by using professional knowledge to understanding the generated rules.

Within these steps, the most complex and time-consuming job is step (3) of generating the frequent itemsets. Based on the frequent itemsets, step (4) to generate association rules is relatively simple, but how to avoid the generation of too many redundant and excessive rules also needs to be carefully considered. Other steps may be thought of as related and auxiliary.

The non-redundant rule cannot be reasoned from other rules. Among the generalized association rules, a large number of rules may be generated; if a too-small threshold is set, the generated rules may be beyond the processing capability. In these rules, there are still a number of redundant rules. Because the redundant rules can be derived by other rules, it is of great importance to remove them and create non-redundant rules, which have identical support and confidence, and few antecedent conditions but many subsequent conclusions, which will help uncover as much as possible of the most useful information.

Definition 8.2 For rule r: $l_1 \Rightarrow l_2$, if $\neg \exists\, r': l_1' \Rightarrow l_2'$, $support(r) \leq support(r')$, $confidence(r) \leq confidence(r')$, and $l_1' \subseteq l_1$, $l_2' \subseteq l_2$, then the rule r is called a non-redundant rule (Zaki and Ogihara 2002).

8.1.2 Association Rule Mining with Apriori Algorithm

The *Apriori* algorithm utilizes prior information in frequent itemsets to generate association rules (Agrawal and Srikant 1994). The basic idea is to browse the transactional data many times after first counting the individual projects and determining the frequent itemsets (a collection of items). In each browsing thereafter, the first priority is to generate the candidate frequent itemsets according to the last-browsed frequent itemsets; then, this frequent dataset is browsed and the actual support of the candidate itemsets is calculated. This process is iterated until new frequent itemsets are discovered. Priori information is used to generate the first candidate frequent itemsets, and the repeated browsing of data need only be implemented on the local data thereafter.

Suppose C_k denotes the candidate k-itemset and L_k represents the k-itemset whose occurrence frequency is no smaller than the product of the min_sup multiplied by the total number of transaction in C_k (namely, k-frequent itemset; Han et al. 2012). In the *Apriori* algorithm, a non-frequent $(k - 1)$-itemset cannot be a subset of the frequent k-itemset. Because the connotation of the concept consists of transactions containing k-itemset is more than the connotation of the concept that consists of transactions containing $(k - 1)$-itemset, its extension will inevitably decrease and the number of transactions is likely to decrease as well. Thus, according to the property, the non-frequent itemsets in $(k - 1)$-itemset can be deleted before the k-itemset is generated. By deleting the non-frequent itemsets in $(k - 1)$-itemset, the $(k - 1)$-frequent itemset can be generated. The basic process of this algorithm is as follows:

1. Create C_1 (1-candidate itemset) from items in the transactional database D.
2. Create L_1 (1-frequent itemset) by deleting the non-frequent subsets in C_1 when browsing D.
3. Perform the recursive process by joining L_{k-1} $((k - 1)$-frequent itemset) with itself to generate C_k (k-candidate itemset), then browse D to generate L_k (k-frequent itemset) until no frequent itemset is generated.

4. Create all the non-void subsets s ($s \subseteq l$) for each frequent itemsets l.
5. Extract the association rules "$s \Rightarrow l - s$" if freq($s \cup (l - s)$)/freq(s) $\geq$ *min_conf*, where freq is the support frequency, *min_conf* is the *conf_conf*.

In the process of joining them, the items in the itemset are arranged alphabetically. When joining L_k with L_k, if the preceding $k - 1$ items in some two elements are identical, they are joinable; otherwise, they are unjoinable and are left alone. For instance, join L_1 with L_1 to generate C_2 (candidate 2-itemset), then browse D to delete the non-frequent subsets in C_2 and generate L_2 (2-frequent itemset). The *Apriori* algorithm can easily explore the association rules from the dataset. For example, in Fig. 8.1, assume that six customers {T_1, T_2, T_3, T_4, T_5, T_6,} bought a total of six kinds of products {a, b, c, d, e, f} in a retail outlet, which created transactional data D. Let freq count the support frequency. If the support threshold is 0.3 and the confidence threshold is 0.8, then the process of generating frequent itemsets will proceed. Table 8.1 shows the extracted 20 association rules.

However, the *Apriori* algorithm has its weaknesses. First, in the process of generating k-frequent itemsets from k-candidate itemsets, the database D is re-browsed once more, which results in browsing the database too many times. If the database is excessively massive, the algorithm will be very time-consuming. Second, when association rules are extracted, it is important to calculate all the subsets of the frequent itemsets, which is time-consuming as well. Third, many redundant rules are generated. We offer the concept lattice as a solution to these problems.

TID	item	scan D and count each candidate itemset	Itemset	freq	compare with support threshold	Itemset	freq	create C2 from L1 scan D and count each candidate itemset	Itemset	freq	Itemset	freq
T1	a, b, d		(a)	5		(a)	5		{a, b}	4	{b, f}	2
T2	a, b, c, d		(b)	5		(b)	5		{a, c}	2	{c, d}	2
T3	a, b, d, e		(c)	2		(c)	2		{a, d}	5	{c, e}	1
T4	b, e, f		(d)	5		(d)	5		{a, e}	2	{c, f}	0
T5	a, b, d, f		(e)	3		(e)	3		{a, f}	1	{d, e}	2
T6	a, c, d, e		(f)	2		(f)	3		{b, c}	1	{d, f}	1
									{b, d}	4	{e, f}	1
									{b, e}	2		

	Create 2-frequent itemset L2				Create 3-candidate itemset C3				Create 3-frequent itemset L3	
compare	Itemset	freq	Itemset	freq	create C3 from L2, scan D and count each candidate itemset	Itemset	freq	Itemset	freq
with	{a, b}	4	{b, e}	2		{a, b, c}	1	{a, d, e}	2	compare with
support	{a, c}	2	{b, f}	2		{a, b, d}	4	{a, d, f}	1	support
threshold	{a, d}	5	{c, d}	2		{a, b, e}	1	{a, e, f}	0	threshold
	{a, e}	2	{d, e}	2		{a, b, f}	1	{b, d, e}	1	
	{b, d}	4				{a, c, d}	2	{b, d, f}	1	
						{a, c, e}	1	{b, e, f}	1	
						{a, c, f}	0	{c, d, e}	1	

Create 3-frequent itemset L3

compare	Itemset	freq
with	{a, b, d}	4
support	{a, c, d}	2
threshold	{a, d, e}	2

Fig. 8.1 Stepwise creation of a candidate set and frequent itemset on dataset D

Table 8.1 Association rules extracted with the *Apriori* algorithm

Rule	Support	Confidence	Rule	Support	Confidence	Rule	Support	Confidence
$a \Rightarrow b$	0.67	0.8	$b \Rightarrow a$	0.67	0.8	$c \Rightarrow a$	0.33	1.0
$a \Rightarrow d$	0.83	1.0	$d \Rightarrow a$	0.83	1.0	$a \Rightarrow bd$	0.67	0.8
$ab \Rightarrow d$	0.67	1.0	$ac \Rightarrow d$	0.33	1.0	$b \Rightarrow d$	0.67	0.8
$d \Rightarrow b$	0.67	0.8	$b \Rightarrow ad$	0.67	0.8	$ad \Rightarrow b$	0.67	0.8
$c \Rightarrow ad$	0.33	1.0	$ae \Rightarrow d$	0.33	1.0	$f \Rightarrow b$	0.33	1.0
$c \Rightarrow d$	0.33	1.0	$d \Rightarrow ab$	0.67	0.8	$bd \Rightarrow a$	0.67	1.0
$cd \Rightarrow a$	0.33	1.0	$de \Rightarrow a$	0.33	1.0			

8.1.3 Association Rule Mining with Concept Lattice

The concept lattice (also called formal concept analysis) was introduced by Rudolf Wille in 1982. It is a set model that uses a mathematical formula to directly express human understanding of the concept's geometric structure. The concept lattice provides a formal tool to take the intension and extension of a concept as the concept unit. The extension is the instances covered by the concept, and the intension is the description of the concept as well as the common characteristics of the instances covered by the concept. Based on a binary relationship, the concept lattice unifies the intension and extension of a concept and reflects the generalization/characterization connection between an object and an attribute. Each vertex of the concept lattice is a formal concept composed of an extension and an intension. A Lattice Hasse diagram simply visualizes the generalization/characterization relationship between the intension and extension of a concept. Creating a Hasse diagram is similar to the process of concept classification and clustering. Discovered from such features as spectrum, texture, shape, and spatial distribution, the patterns are described by intension sets and the inclusion of extension sets. Therefore, a concept lattice is suitable as the basic data structure of rules mining.

SDM may be treated as the process of concept formation from a database. A concept lattice first analyzes the formal context triple (G, M, I), where G is the set of (formal) objects, M is the set of (formal) attributes, and I is the relationship between the object G and the attribute M; that is, $I \subseteq G \times M$. The formal concept of formal context (G, M, I) defines a pair (A, B), where, $A \subseteq G$ and $B \subseteq M$. Set A, B as the intension and extension of formal concept (A, B), respectively. The subconcept-superconcept relationship among these concepts is formally denoted as: $H_1 = (A_1, B_1) \leq H_2 = (A_2, B_2)$: $\Leftrightarrow A_1 \subseteq A_2$ ($\Leftrightarrow B_1 \supseteq B_2$). The set of all the concepts of formal context (G, M, I) constitutes a complete concept lattice; that is, $\underline{B}$ (G, M, I). A Lattice Hasse diagram is created under partial relationship. If $H_1 < H_2$, and no other element H_3 is in the lattice such that $H_1 < H_3 < H_2$, there is an edge from H_1 to H_2.

The algorithm to construct the concept lattice may automatically generate frequent closed itemsets. In the constructed concept lattice, a frequent concept lattice node is denoted as (O, A), where O is a collection of objects with the maximum common attributes in connotation collection *A*, where *A* is a collection of objects with the maximum common attributes in object collection *O*. Obviously, the connotation of each frequent concept lattice node is a frequent closed itemset. These frequent closed itemsets are generated automatically during the process of constructing the concept lattice, and the frequent concept lattice node may further result in association rules. Compared with the *Apriori* algorithm, the concept lattice reduces the number of association rules by decreasing the quantity of the frequent itemsets and helps generate non-redundant rules by using a productive subset of the frequent closed itemsets.

8.1.3.1 Productive Subset of a Frequent Closed Itemset

Definition 8.3 For the closed itemset *l* and itemset *g*, if $g \subseteq I$ meets $\gamma(g) = l$, and $\neg \exists\, g' \subseteq I, g' \subseteq g$, meets $\gamma(g') = l$, then *g* is called a productive subset of *l*.

The above is the formal definition of a productive subset (Zaki and Ogihara 2002). The productive subset is the sub-itemset designated to construct the frequent closed itemset—that is, the reduced connotation set (Xie 2001). To get the productive subset, all the nonblank subsets are first uncovered from the connotation sets of each frequent closed node. Sequentially, a subset is deleted if it is a child node to a parent node or another subset is its proper subset. The remaining subsets are the productive subsets of the frequent closed nodes, which give birth to non-redundant association rules. Figure 8.2 illustrates the nodes of the frequent concept lattice and their productive subsets G_f by creating the concept lattice from Fig. 8.1 and Table 8.1.

8.1.3.2 Generation of Non-redundant Rules

According to the range of confidence values, there are two non-redundant association rules: one with confidence of 100 %, the other with confidence less than 100 %.

1. Rules with confidence of 100 %: $\{r: g \Rightarrow (f/g)| f \in FC \wedge g \in G_f \wedge g \neq f\}$ where *f* is a frequent closed itemset, G_f is its productive subset, *f/g* is the itemset

Fig. 8.2 Nodes of frequent closed concept lattice and their productive subsets

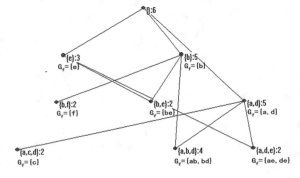

Table 8.2 Association rules with confidence of 100 %

Rule	Support	Confidence	Rule	Support	Confidence	Rule	Support	Confidence
{f} ⇒ {b}	0.33	1.00	{a} ⇒ {d}	0.83	1.00	{d} ⇒ {a}	0.83	1.00
{a, b} ⇒ {d}	0.67	1.00	{b, d} ⇒ {a}	0.67	1.00	{a, e} ⇒ {d}	0.33	1.00
{d, e} ⇒ {a}	0.33	1.00	{c} ⇒ {a, d}	0.33				

Table 8.3 Association rules with confidence less than 100 %

Rule	Support	Confidence	Rule	Support	Confidence	Rule	Support	Confidence
{b} ⇒ {a, d}	0.67	0.80	{a} ⇒ {b, d}	0.67	0.80	{d} ⇒ {a, b}	0.67	0.80

created by removing itemset g out of f, and FC is the frequent closed itemset. In the rule, the occurrence frequency of the antecedent is equal to the occurrence frequency of the antecedent together with the subsequent frequency (Zaki and Ogihara 2002). Under the definition of the productive subset, the occurrence frequency of a productive subset is equal to the frequency of the connotation set of the concept nodes. Therefore, the rules can be directly generated from the productive subset. If $support\_count(G_f) = support\_count(f)$, then the confidence $(r: g \Rightarrow (f/g)) = support\_count(f)/support\_count(G_f) = 100$ %. Table 8.2 presents eight association rules with confidence of 100 %, which are extracted from Fig. 8.2 with a support threshold of 0.3.

2. Rules with confidence less than 100 %. These rules are generated by analyzing their inclusion relationship from top to bottom. For two adjacent frequent closed nodes in a concept lattice, if they are a parent–child relationship and their frequent closed itemsets and productive subset are the parent node: (f_1, G_{f1}), child nodes: (f_2, G_{f2}), then the rule is: $\{r: g \Rightarrow (f_2 - g)|g \in G_{f1}\}$(Zaki and Ogihara 2002). Table 8.3 shows three association rules with confidence less than 100 %, which are extracted from Fig. 8.2 with the support threshold of 0.3.

To sum Tables 8.2 and 8.3, the 11 rules are extracted with the concept lattice. Comparatively, 20 rules are extracted in Table 8.1 with the *Apriori* algorithm. The reason for this difference in the number of rules is that the concept lattice can automatically delete the redundant rules.

8.1.3.3 Direct Extraction Method for Rules

Using the concept lattice, the association rules to meet *min_supp* and *min_conf* can be automatically extracted from the formal context. It is unnecessary sometimes to generate all the rules because only a specific rule is of interest. Then, the direct extraction method may deduce these rules by analyzing the relationship between some designated concept lattice nodes.

Assume that in the *Hasse* diagram, there are two concept lattice nodes of interest: $C_1 = (m_1, n_1)$ and $C_2 = (m_2, n_2)$, respectively, corresponding to the extensive cardinality $|m_1|$ and $|m_2|$. Also, $fr(n)$ represents the function of the occurrence frequency of the objects containing connotation n in the data set. The relationship between the concept nodes is used to directly compute the confidence and support of the association rules (Clementini et al. 2000).

1. The method to compute support is as follows:

   (a) If there is a direct connection between two nodes (i.e., there is a line connection between the two nodes in the *Hasse* diagram), then: $s(n_1 R n_2) = \text{fr}(n_1 \cup n_2)/\text{fr}(\emptyset)$, where $\text{fr}(n_1 \cup n_2)$ denotes the contemporary occurrence frequency of n_1 and n_2 in the dataset D, and $\text{fr}(\emptyset)$ means the frequency of the "void set" connotation attribute in the dataset D; in fact, it is equal to the cardinality of the dataset D.

   (b) If there is no direct connection between the two nodes, then it is necessary to find a common child node n_3 between the two, then: $s(n_1 R n_2) = \text{fr}(n_3)/\text{fr}(\emptyset)$.

2. The method to compute confidence is as follows:

   (a) If there is a direct connection between two nodes (i.e., there is a line connection between the two nodes in the *Hasse* diagram), then the confidence between n_1 and n_2 is $C(n_1 R n_2) = \text{fr}(n_1 \cup n_2)/\text{fr}(n_1) = |m_2|/|m_1|$, where $\text{fr}(n_1)$ ($\text{fr}(n_1) \neq 0$) is the occurrence frequency of n_1 in the dataset D, and $\text{fr}(n_1 \cup n_2)$ is the contemporary occurrence frequency of n_1 and n_2 in the dataset D. If $n_2 \subseteq n_1$, then $C(n_1 R n_2) = 1$.

   (b) If there is no direct connection between the two nodes, then it is necessary to find a common child node n_3 between the two, then $C(n_1 R n_2) = C(n_1 R n_3) \times C(n_2 R n_3)$.

The confidence and support are set according to the specific circumstances. When both of them meet the *min_supp* and *min_conf*, the association rules are of interest.

8.1.3.4 Performance of Concept Lattice

The time complexity is an aspect of the concept lattice. When the concept lattice is applied in data mining, the formal context of mining objects first are transformed into the form of a single-valued attribute table; then, the corresponding concept lattice nodes are created from the intersection and union of the contextual concepts. For a formal context with n attributes, the concept lattice can create a maximum with 2^n concept nodes. The concept lattice does not seem practical in data mining. In fact, even if the number of features of the formal context is huge, the actual object often has only a smaller fraction of the features. The number of actual lattice nodes also is often much less than the upper limit because the actual data often contain some unlikely concurrent feature combinations. Although it takes a little time to manipulate the intersection and union on the concept nodes, the concept lattice runs much faster than the *Apriori* algorithm when creating frequent itemsets for association rules.

The complexity of $O(2^n)$ is for a general concept lattice. A specific concept lattice may be simplified under the following aspects: (1) reduce the number of attributes by selecting the focus attributes of interest (i.e., connotation reduction of concept); (2) reduce the number of objects by selecting the representative samples from the vast object collections; and (3) reduce the space complexity by selecting efficient coding techniques for the data distribution.

For large-scale applications, creating the concept lattice may produce a lot of nodes to occupy many processing and storage resources. The pruning strategy may reduce the number of lattice nodes. For example, only if the occurrence frequency is greater than the support threshold, the frequent concept lattice node is generated, which is called a semi-lattice or partial-lattice. When the data are excessively complex, the plotted *Hasse* diagram may be too complex to observe. The nested *Hasse* diagram may reduce the visual complexity by treating a closely related concept sub-lattice as a whole when constructing the *Hasse* diagram plotting.

8.1.4 Association Rule Mining with a Cloud Model

In general, the frequent itemsets of spatial association rules exist at a high conceptual level, and it is difficult to uncover them at a low conceptual level. In particular, when the attribute is numeric and the mining is on the original conceptual level, strong association rules will not be generated if *min_sup* and the *min_conf* are larger, while many uninteresting rules will be produced if the thresholds are relatively smaller. In this case, the attribute needs to be elevated to a higher conceptual level via attribute generalization, and then the association rules are extracted from the generalized data. The cloud model flexibly partitions the attribute space by simulating human language. Every attribute is treated as a linguistic variable that sometimes is represented with two or more attributes, which is regarded as a multidimensional linguistic variable. For each linguistic variable, several language values are defined, and overlapping is allowed among the neighboring language values. In the overlapping area, an identical attribute value may be assigned to different clouds under different probabilities. The cloud to represent a linguistic variable can be assigned by the user interactively or acquired by the cloud transform automatically.

Cloud model-based attribute generalization can be a pre-processing step for mining association rules. The Apriori algorithm is implemented after the attributes first are generalized. This pretreatment is highly efficient for the linear relationship between the time spent and the size of the database. During the process of attribute generalization, the database only needs to be browsed once in accordance with the maximum membership of each attribute value assigned to the corresponding linguistic value. After attribute generalization, different tuples from the original data may be merged into one tuple if they become identical at the high concept level, which reduces the size of the database significantly. Also, a new attribute "count" is added to record how many original tuples are combined in a merged tuple. It is not a real attribute but rather is for the purpose of computing the support of the itemset in the mining process. Due to the characteristics of the cloud and the overlapping between adjacent clouds, the new attribute "count" will be different at different times in the attribute generalization relationship table when the final mining results are stable.

A geographical and economic database is used below as an experimental case to verify the feasibility and efficacy of cloud models in mining association rules. The interesting item in this case is that the relationship between the geospatial information and the economic conditions, due to a small amount of actual data, is set as the basis; and a large number of simulation data are produced by generating random numbers (10,000 total records). The six attributes are x, y, z (elevation), road network density, distance to the sea, and per capita income. These attributes are all numeric, the positions x and y are in the Cartesian coordinate, and the road density is denoted in road length per km^2. If the original digital data are directly used, it is difficult to discover the rules. So, the cloud model first is used to generalize the attributes. Specifically, attributes x and y are treated as a language variable called "location," and its definition of eight two-dimensional language values are "southwest," "northeast," "north-by-east," "southeast," "northwest," "north," "south," and "central"—most of which are swing cloud. The remaining attributes are regarded as one-dimensional language variables and three language values are defined, such as "low," "medium," and "high" for expressing terrain, road network density, and per capita annual income; and "near," "median," and "far" are used to show the distance to the ocean. The language values "low" and "near" are semi-fall cloud, and "high" and "far" are semi-rise cloud. There is some overlapping between adjacent cloud models. Due to the irregularity of Chinese territory in shape, the eight numerical character values of the cloud model for expressing "location" are given manually, and the hyper-entropy value is taken as 1/10,000 of the entropy. The remaining one-dimensional clouds are generated by the above-mentioned method automatically. After attribute generalization, the amount of data is considerably reduced. The generalized attributes are shown in Table 8.4. Depending on the characters of the cloud model, the value of "count" in the generalized attribute table is slightly different at different times, and the digits in the table represent average results after a plurality of generalization is taken.

With the *min_sup* of 6 % and *min_conf* of 75 % in the generalized database, the result is eight four-itemsets. Eight association rules are discovered, among which the subsequent rule is the per capita annual income, and the antecedent is the sum of the other three attributes. The association rules are expressed with a productive method, the contents of which are as follows:

- Rule 1: If the location is "southeast," the road network density is "high," and the distance to the ocean is "near," then the per capita annual income is "high."
- Rule 2: If the location is "north-by-east," the road network density is "high," and the distance to the ocean is "near," then the per capita annual income is "high."

Table 8.4 The generalized attributes

Location	Terrain	Road network density	Distance to the ocean	Per capita annual income	Count (%)
Southeast	Medium	High	Near	High	4
Southeast	Medium	High	Near	Medium	2
Southeast	Low	High	Near	High	8
Southwest	High	Low	Far	Low	12
Southwest	High	Low	Far	Medium	3
South	Medium	High	Medium	Medium	4
South	Low	High	Medium	High	2
South	Low	High	Medium	Medium	6
Northwest	Medium	Low	Far	Low	8
Northwest	Medium	Low	Far	Medium	2
Northwest	High	Low	Far	Low	5
Central	Medium	High	Medium	Medium	5
Central	Low	High	Medium	Medium	6
Northeast	Low	High	Medium	Medium	9
Northeast	Low	High	Near	High	1
Northeast	Medium	Low	Far	Low	3
North-by-east	Low	High	Near	High	8
North-by-east	Low	High	Near	Medium	2
North	Medium	Medium	Medium	Medium	7
North	Medium	Medium	Far	Medium	3

- Rule 3: If the location is "northeast," the road network density is "high," and the distance to the ocean is "medium," then the per capita annual income is "medium."
- Rule 4: If the location is "north," the road network density is "medium," and the distance to the ocean is "medium," then the per capita annual income is "medium."
- Rule 5: If the location is "northwest," the road network density is "low," and the distance to the ocean is "far," then the per capita annual income is "low."
- Rule 6: If the location is "central," the road network density is "high," and the distance to the ocean is "medium," then the per capita annual income is "medium."
- Rule 7: If the location is "southwest," the road network density is "low," and the distance to the ocean is "far," then the per capita annual income is "low."
- Rule 8: If the location is "south," the road network density is "high," and the distance to the ocean is "medium," then the per capita annual income is "medium."

These association rules are visualized in Fig. 8.3. Different rules are plotted with the color gradient ellipse gradually decreasing from the oval center outward with

Fig. 8.3 Geographic-
economic association rules

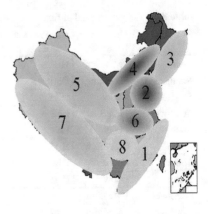

Fig. 8.4 Geographic-
economic association rules
generalization

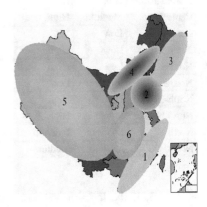

a decreasing membership degree, and the marked digits in the ovals are the rule numbers. In order to find multi-level association rules, further generalization of the location attribute by a virtual cloud is implemented. The combined cloud "west" is from the "northwest" and the "southwest," and the "central-south" is from the "south" and "central." Therefore, the eight rules in Fig. 8.3 are reduced to six rules in Fig. 8.4, where rules 5 and 7 are merged as new rule 5, rules 6 and 8 are merged as new rule 6, and the remaining rules 1, 2, 3, and 4 are unchanged.

- Rule 5: If the location is "west," the road network density is "low," and the distance to the ocean is "far," then the per capita annual income is "low."
- Rule 6: If the location is "central-south," the road network density is "high," and the distance to the ocean is "medium," then the per capita annual income is "medium."

The results indicate the efficacy of the cloud model in the association rule mining pretreatment process. The combination of the cloud model and the *Apriori* algorithm also demonstrate that the cloud model is capable of discovering association rules from numeric data at multiple conceptual levels.

8.2 Spatial Distribution Rule Mining with Inductive Learning

To implement inductive learning in the spatial database, first it is necessary to determine the learning granularity. The learning granularity may be an object that is a point, line, or area in a graphical database or a pixel that refers to the imagery element. The granularity depends on the learning purpose and the structure of the spatial database and therefore the learning attributes need to be determined. The learning samples of the spatial database can be taken from all the data of the region of interest. When there is a very large amount of data, in order to improve the learning speed, the learning samples can be randomly selected by a sampling method. In accordance with the identified granularity, the selected attributes, and the learning samples, the learning data are reorganized into a similar form of relational data table, which is input for the inductive learning. The output is in the form of equivalent productive rules that are easy for reading and application (Fig. 8.5).

Spatial distribution rule mining employs inductive learning in the following example to extract knowledge from bank income data and other relevant layers and to uncover the relationship between the banking income and the geographical location, transportation, population, income status, age, and race distribution. The result may be used to prepare a comparative evaluation of bank income, estimate the operational and management status, and further support the selection of a new bank location. The flow chart is shown in Fig. 8.5.

Atlanta city is chosen as the test region, and involves the spatial data of bank distribution maps, the road network, and the census plot, using the map projection of Albers homolosine projection in kilometer units. The attributes of the bank layer include the deposits of 117 banks in 1993 and 1994, and the times when they were established. The attributes of the census plot layer include the plot area (square miles), the population number in 1993, the population growth rate, the

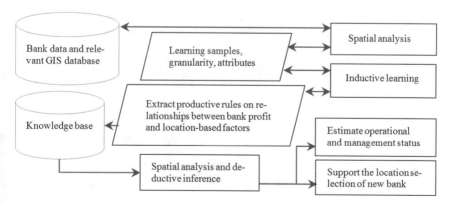

Fig. 8.5 Spatial distribution rule mining with inductive learning bank example

race percentage, the mean income, the median income, the gender percentage, the mean age, and the median age. The learning granularity is a spatial point object (i.e., each bank). First, the spatial and non-spatial information of the multiple layers are agglomerated into the bank attribute table by spatial analysis, thereby forming the data for learning. Then, by inductive learning the hidden relationship between the bank's operational income and a variety of geographic factors is uncovered; and the spatial analysis and deductive inference are combined to speculate the bank's management status and to make an assessment for the selection of a new bank locations. The GIS software utilized is ArcView along with a secondary development, and the inductive learning software is C5.0.

In order to facilitate the inductive learning, based on the deposits in 1993 and 1994, the bank operational income is graded as "good," "average," or "poor." If the "deposits in 1994" > "deposits in 1993" and "deposits in 1994" > "median deposits in 1994," then the operational income is "good"; if "deposits in 1994" < "deposits in 1993" and "deposits in 1994" < "median deposits in 1994," then the operational income is "bad"; and the remaining operational income is "average." The results includes 49 "average," and there are 34 "good" and "poor," which are shown by different colors in Fig. 8.6.

As the bank profit has a pronounced relationship with how many other banks are in the area and how close by they are, the distance to the nearest bank for each bank is first calculated and the number of banks within a certain distance (1 km), which are added as two indicators into the bank attributes. Then, the closest distance to the road network as well as the attribute data of the location where the bank is located are obtained via spatial analysis and are added to the bank attribute table. In addition, the coordinates of the bank also are added to the bank attribute table. In this way, data mining is carried out in the new agglomerated bank attribute table. There are a total of 23 attributes in the inductive learning; with the

Fig. 8.6 Bank, road network, and census lot

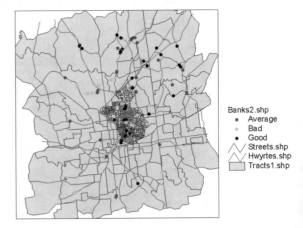

exception of the bank operational income attribute "class" being a discrete value, the values of the other attributes are continuous.

The bank operational income is treated as the decision attribute (classifying attribute), and the other attributes are the condition attributes. Using inductive learning to uncover the bank attribute table, a total of 22 rules are obtained (Table 8.5).

Table 8.5 The association rules between bank operational income and geographical factors

Rule No.	Rule content
Rule 1	(Cover 19) PCT ASIAN > 3.06, AVG INC > 36483.52, DIST CLOSEST BANK > 0.663234 ⇒ class Good (0.857)
Rule 2	(Cover 4) SQ MILES ≤ 0.312, POP GROWTH > -6.62 ⇒ class Good (0.833)
Rule 3	(Cover 2) NO CLOSE BANK > 18, X COORD > 1065.441 ⇒ class Good (0.750)
Rule 4	(Cover 18) YEAR ESTABLISHED ≤ 1962, POP GROWTH > −6.62, PCT ASIAN > 0.88, X COORD > 1064.672 ⇒ class Good (0.700)
Rule 5	(Cover 17) YEAR ESTABLISHED > 1924, PCT BLACK ≤ 4.09 ⇒ class Good (0.526)
Rule 6	(Cover 8) PCT BLACK > 4.09, MID AGE > 35.43 X COORD ≤ 1064.672 ⇒ class Good (0.900)
Rule 7	(Cover 5) POP GROWTH ≤ −6.62, X COORD ≤ 1065.441 ⇒ class Average (0.857)
Rule 8	(Cover 4) YEAR ESTABLISHBD ≤ 1965, PCT BLACK ≤ 4.09, PCT ASIAN ≤ 3.06 ⇒ class Average (0.833)
Rule 9	(Cover 4) PCT OTHER > 1. 32, DIST CLOSEST BANK ≤ 0.376229 ⇒ class Average (0.833)
Rule 10	(Cover 4) YEAR ESTABLISHED ≤ 1924, POP GROWTH > -6.62, DIST CLOSEST BANK ≤ 0.179002 ⇒ class Average (0.833)
Rule 11	(Cover 4) POP GROWTH ≤ -6.62, NO CLOSE BANK ≤ 18 ⇒ class Average (0.833)
Rule 12	(Cover 9) 1951 < YEAR ESTABLISHED ≤ 1962, PCT ASIAN ≤ 0.88, AVG AGE > 31. 34 ⇒ class Average (0.800)
Rule 13	(Cover 8) YEAR ESTABLISHED > 1951, MIN DIST Road > 0.093013, PCT BLACK > 4.09, X COORD > 1064.672 ⇒ class Average (0.800)
Rule 14	(Cover 7) YEAR ESTABLISHED > 1962, PCT BLACK > 4.09, DIST CLOSEST BANK ≤ 0.050138 ⇒ class Average (0.778)
Rule 15	(Cover 2) PCT BLACK ≤ 4.09, PCT MALE ≤ 42.71 ⇒ class Average (0.750)
Rule 16	(Cover 2) PCT ASIAN > 3.06, AVG INC ≤ 36483.52 ⇒ class Average (0.750)

(continued)

Table 8.5 (continued)

Rule No.	Rule content
Rule 17	(Cover 5) PCT ASIAN $\leq$ 3.06, MED AGE $\leq$ 35.43, DIST CLOSEST BANK > 0.376229, X COORD $\leq$ 1064.672 $\Rightarrow$ class Bad (0.857)
Rule 18	(Cover 4) YEAR ESTABLISHED > 1960, PCT OTHER $\leq$ 1.32, MED AGE $\leq$ 35.43, X COORD $\leq$ 1064. 672 $\Rightarrow$ class Bad (0.833)
Rule 19	(Cover 3) PCT ASIAN $\leq$ 3.06, AVG AGE $\leq$ 31. 34 $\Rightarrow$ class Bad (0.800)
Rule 20	(cover 3) 1924 < YEAR ESTABLISHED $\leq$ 1951, SQ MILES > 0.312 AVG AGE $\leq$ 36.22 $\Rightarrow$ class Bad (0.800)
Rule 21	(Cover 20) YEAR ESTABLISHED > 1962, MIN DIST Road $\leq$ 0.093013 SO MILES > 0.312, PCT BLACK > 4.09, PCT ASIAN $\leq$ 3.06, AVG AGE > 34.1, DIST CLOSEST BANK > 0.050138, X COORD > 1064.672 $\Rightarrow$ class Bad (0.773)
Rule 21	(Cover 2) YEAR ESTABLISHED > 1981, PCT BLACK $\leq$ 4.09, PCT ASIAN > 0.82, PCT MALE > 42.71 $\Rightarrow$ class Bad (0.750)

Evaluation of learning result

Rules		(a)	(b)	(c)	Classified as
No.	Errors	33	1		(a): class Good
22	12(10.3 %)	3	42	4	(b): class Average
		3	1	30	(c): class Bad

The learning error rate is 10.3 %, for a learning accuracy rate of 89.7 %. These rules reveal the relationship between the operational income and the geographical factors of the Bank of Atlanta locations (Fig. 8.6; Table 8.5). These relationships are relatively complex: the closely located banks both have "good" operational income and "average" or "bad" operational income, and the banks established at a similar time both have "good" operational income and "average" or "bad" operational income. Precise rules are formed by combining a number of factors. In rule 1, if the bank location has a proportion of Asian population greater than 3.06 %, the average annual income is greater than 36,483.52, and the nearest bank distance from another is greater than 0.663234 km, then its operational income is "good." The rule produces five bank cases at a confidence of 0.857. For rule 2, if the area where the bank location is located is less than or equal to 0.312 square miles and the population growth rate is greater than −6.62 %, then the bank operational income is "good." Rule 4 reveals that if the bank was established earlier than or equal to 1962, has a population growth rate greater than −6.62 %, an Asian population proportion greater than 0.88 %, and the x-coordinate is greater than 1064.672 (located in the eastern part), then the bank operational income is "good." For rule 19, if the location where the bank is located has a proportion of the Asian population less than or equal to 3.06 % and the average age of the plot is less than or equal to 31.34 years, then the bank

operational income is "poor." By comparing rules 1, 17, and 19 in plots, it is concluded that the proportion of the Asian population has a major impact on the operational income of the bank. Further analysis of the bank's operational income can
be implemented in accordance with the results of the inductive learning. For the
correctly classified banks, if the rules they obey cover more bank cases (e.g., rule
4, rule 21), then their operational income and geographical factors have a relatively
stronger correlation. From the results of the inductive learning, there are 12 misclas-
sified banks, i.e., the operational income categories from the rule reasoning are different from the actual categories, which therefore are exceptions to these rules. Because
the inductive learning only utilizes the geographical information (as well as the bank
founding time), it is reasonable to think that the reason for these exceptions is caused
by the internal operations of the bank. If the bank's operational income is "good"
or "average" while its actual operational income is "poor," then the bank should be
considered as having poor operation and management and ought to improve its management or staff mobility. If the results of a bank's operational income obtained by
rule reasoning is "bad" while its actual operational income is "average" or "good,"
the bank's management methods are worthy of further analysis. Figure 8.7 shows the
operational "good" and "bad" banks speculated by the exception of the rule.

The knowledge discovered by inductive learning is used not only to infer the
operation and management of a bank, but also to inform location assessments for
new bank locations as well. When a new bank location is evaluated, the attributes
of the candidate locations are entered into the rules, generating the operational
income by reasoning, which can be used as an important reference for location
selection. Figure 8.9 simulates this process. In the buffering zone within 0.5 km of
the road network, uniformly spaced new bank locations are generated by the same
method as the inductive learning. The attributes of the bank distance to the closest
road and the attribute data of the location where the bank is located are added to
the new bank's attributes. When the distance to the nearest bank for each new bank
is calculated and the number of banks within a certain distance of the new bank

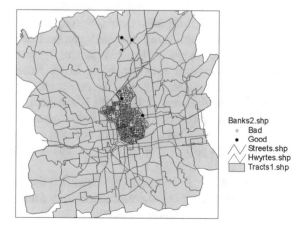

Fig. 8.7 "Good" and "bad"
operational banks

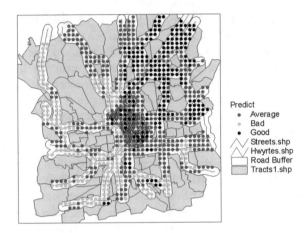

Fig. 8.8 The new bank location evaluation map

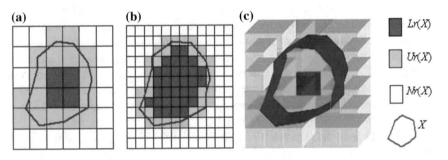

Fig. 8.9 Rough sets illustrations. Reprinted from Ref. Wang et al. (2004), with kind permission from Springer Science+Business Media. **a** $p \times p$, **b** $2p \times 2p$, **c** 3D

is determined, which are two of the indicators, only existing banks are taken into consideration as it is assumed there are no other new banks.

The attribute data of the new bank can be reasoned with the rules by inductive learning to obtain predictive values of "good," "average," or "bad" for the operational income of each new bank. In the reasoning process, if several rules are activated simultaneously, then the output category of high confidence (sum) is regarded as the final output. If no rule is activated, then the default output value is "average." An overlay of a new bank layer and other relevant layers is carried out, with different operational incomes in different colors to produce a forecast map of the new bank's projected operational income or as the new bank location assessment map as shown in Fig. 8.8.

In Fig. 8.8, it can be clearly observed that the northeast and the central urban areas are the preferable geographical locations for the new bank location, but there are only a few good plots in the south.

It should be noted that the above speculation of the bank management and location assessment results are only considered as having relative accuracy, which means they should be treated as a decision-making guide and reference only. Various instances of uncertainty exist within several processes in the knowledge acquisition and application process. The first source of uncertainty is that the test data involved in the learning is not rich enough. For example, if only the total deposits are known, they can be used to represent the operational income. However, since there is no information about the number of bank staff, it is not possible to depict the operational efficiency. Furthermore, if the bank depositor distribution data are not available, only the household statistics of the bank locations can be used in inductive learning in accordance with the proximity principle, which cannot consider the situation where the place of work and place of residence is not identical, etc.

The second source of uncertainty is based on the bank deposits in two consecutive years to define the operational income of "good," "average," "poor," which means different definitions will produce different learning results. In addition, the first uncertainty source has a certain relationship with the second uncertainty source. Therefore, the more abundant the bank operational data are, the more reasonable the definition of the operational income will be; however, a great deal of operational data is not available for commercial and research purposes.

The third source of uncertainty is that the knowledge discovered from inductive learning is often subject to different rules and therefore different degrees of coverage (the number of covered cases) and ultimately different degrees of confidence in applying the knowledge. On the other hand, the knowledge acquired by inductive learning may not be complete as well because appropriate rules are not in place during location investigations for reasoning the new bank information of a small number of candidate locations. The only output possible then becomes the default value of "average."

The aforementioned example shows that inductive learning can uncover the hidden relationships between the bank operational income and a variety of geographical factors from the bank data and related layer data and that further speculation about operational status and new location selection assessment can be made with this knowledge. Despite the uncertainties that exist, these analyses and evaluation results are significant information for bank management and decision-making. GIS spatial analysis also plays an important role in this example (e.g., buffer analysis, distance analysis, and inclusion analysis), which are applied for extracting the spatial information needed by the inductive learning from multiple layers. Also, inductive learning techniques provide knowledge for subsequent spatial analysis and decision-making, which discovers the required knowledge so that spatial analysis is no longer dependent on expert input. Inductive learning and spatial analysis promote each other and improve the intelligence level of spatial analysis and decision support. Although the example here is based on the case of the bank operational analysis and location selection assessment here, combining inductive learning and spatial analysis is very appropriate for a wide variety of evaluation and spatial decision support applications in the resources and facilities area. The quantitative description of uncertainty and its visualization in the combined process of GIS/inductive learning for SDM are topics worthy of further study.

8.3 Rough Set-Based Decision and Knowledge Discovery

Rough set provides a means for GIS attribute analysis and knowledge discovery with incompleteness via lower and upper approximations. Under rough sets, SDM is executed in the form of a decision-making table (Pawlak 1991).

Given a universe of discourse U and U is a finite and non-empty set. Suppose an arbitrary set $X \subseteq U$. X^c is the complement set of X, and $X \cup X^c = U$. There is equivalence relation $R \subseteq U \times U$ on U. $U/R(x)$ is the equivalence class set composed of the disjointed subsets of U partitioned by $R(x)$, $R(x)$ is the equivalence class of R including element x, and (U, R) formalizes an approximate space. The definitions for lower and upper approximation are as follows.

- Lower approximation (interior set) of X on U: $Lr(X) = \{x \in U | [x]_R \subseteq X\}$,
- Upper approximation (closure set) of X on U: $Ur(X) = \{x \in U | [x]_R \cap X \neq \Phi\}$.

If the approximate space is defined in the context of region, then

- Positive region of X on U: $Pos(X) = Lr(X)$,
- Negative region of X on U: $Neg(X) = U - Ur(X)$
- Boundary region of X on U: $Bnd(X) = Ur(X) - Lr(X)$

The lower approximation $Lr(X)$ is the set of spatial elements that definitely belong to the spatial entity X, while the upper approximation $Ur(X)$ is the set of spatial elements that possibly belongs to X. The difference between the upper approximation and the lower approximation is the uncertain boundary $Bnd(X) = Ur(X) - Lr(X)$. It is impossible to decide whether an element in $Bnd(X)$ belongs to the spatial entity due to the incompleteness of the set. $Lr(X)$ is certainly "Yes," $Neg(X)$ is surely "No," while both $Ur(X)$ and $Bnd(X)$ are uncertainly "Yes or No." With respect to an element $x \in U$, it is certain that $x \in Pos(X)$ belongs to X in terms of its features, but $x \in Neg(X)$ does not belong to X while $x \in Bnd(X)$ cannot be ensured by means of available information whether it belongs to X or not. Therefore, it can be seen that $Lr(X) \subseteq X \subseteq Ur(X) \subseteq U$, $U = Pos(X) \cup Bnd(X) \cup Neg(X)$, and $Ur(X) = Pos(X) \cup Bnd(X)$ (Fig. 8.9).

X is defined if $Lr(X) = Ur(X)$, while X is rough with respect to $Bnd(X)$ if $Lr(X) \neq Ur(X)$. A subset $X \in U$ defined with the lower approximation and upper approximation is called a rough set. The rough degree is

$$\delta(X) = \frac{R_{\text{card}}(Ur(X) - Lr(X))}{R_{\text{card}}(X)} = \frac{R_{\text{card}}(Bnd(X))}{R_{\text{card}}(X)}$$

where $R_{\text{card}}(X)$ denotes the cardinality of set X. X is crisp when $\delta(X) = 0$.

An object is described by a set of attributes $A = (a_1, a_2, ..., a_m)$. In the context of SDM, the attributes may be divided into condition attributes $C = (a_1, a_2, ..., a_c)$ and decision attributes $D = (a_{c+1}, a_{c+2}, ..., a_m)$, $A = C \cup D$, $C \Rightarrow D$.

Suppose there is a subset S of C, $S = \{a_p | 1 \leq p \leq c\}$, $S \subseteq C$. If $C - S \Rightarrow D$, then S is superfluous; and $a_p(1 \leq p \leq c)$ in S is called the superfluous attribute on the decision attributes D, under which a_p can be reduced. The attribute reduction may decrease the number of dimensions.

Condition attributes C and decision attributes D create a decision-making table. Knowledge discovery from a decision-making table is undertaken in the following steps: (1) observed data arrangement; (2) data conversion; (3) decision table management—attribute value discretization; (4) decision table simplification—attribute reduction and value reduction; (5) decision algorithm minimization; and (6) decision control.

8.3.1 Attribute Importance

Table 8.6 shows the relationships among the attributes *location, terrain*, and *road network density*, which was processed by induction within mainland China and is used to analyze the dependence and importance of an attribute (Li et al. 2006, 2013).

In Table 8.6, U consists of 13 objects {1, 2, ..., 13}, and the attribute set is $A = \{location, terrain, road network density\}$. Objects 3 and 4 are undiscriminating for attribute *location*, and Objects 9 and 11 are undiscriminating for the attribute *terrain*, ... The partitioning generated by the attributes is as follows:

$U/R(location) = \{\{1\},\{2\},\{3, 4\},\{5, 6\},\{7, 8\},\{9, 10\},\{11\},\{12, 13\}\}$
$U/R(terrain) = \{\{1, 4, 6, 8, 10, 13\}, \{2, 3, 5, 7, 12\}, \{9, 11\}\}$
$U/R(road\ network\ density) = \{\{1\}, \{2, 3, 5, 6, 7, 8, 12, 13\}, \{4, 9, 10, 11\}\}$
$U/R(location, terrain) = \{\{1\}, \{2\}, \{3\}, \{4\}, \{5\}, \{6\}, \{7\}, \{8\}, \{9\}, \{10\}, \{11\}, \{12\}, \{13\}\}$

Table 8.6 An example of attribute class

U	1	2	3	4	5	6	7
Location	North	North-by-east	Northeast	Northeast	Southeast	Southeast	South
Terrain	Medium	Low	Low	Medium	Low	Medium	Low
Road network density	Medium	High	High	Low	High	High	High
U	8	9	10	11	12	13	
Location	South	Northwest	Northwest	Southeast	Central	Central	
Terrain	Medium	High	Medium	High	Low	Medium	
Road network density	High	Low	Low	Low	High	High	

Accordingly, there are:

$\text{Pos}_{location}$(*road network density*) = {1, 2, 5, 6, 7, 8, 9, 10, 11, 12, 13}, $\delta_{location}$(*road network density*) = 11/13;

$\text{Pos}_{terrain}$(*road network density*) = {2, 3, 5, 7, 9, 11, 12}, $\delta_{terrain}$(*road network density*) = 7/13;

$\text{Pos}_{\{location,\ terrain\}}$ (*road network density*) = {1, 2, 3, 4, 5, 6, 7, 8, 9, 10, 11, 12, 13}, $\delta_{\{location,\ terrain\}}$(*road network density*) = 13/13 = 1.

Because the dependency of *road network density* to {*location, terrain*} is 1, then the dependency relation is {*location, terrain*} $\Rightarrow$ {*road network density*}.

$\delta_{\{location,\ terrain\}}$(*road network density*) $-$ $\delta_{\{location,\ terrain\}-\{location\}}$(*road network density*) = $\delta_{\{location,\ terrain\}}$(*road network density*) $-$ $\delta_{\{terrain\}}$(*road network density*) = 6/13

$\delta_{\{location,\ terrain\}}$(*road network density*) $-$ $\delta_{\{location,\ terrain\}-\{terrain\}}$(*road network density*) = $\delta_{\{location,\ terrain\}}$(*road network density*) $-$ $\delta_{\{location\ \}}$(*road network density*) = 2/13

As 6/13 > 2/13, the attribute *location* is more important than the attribute *terrain*.

8.3.2 Urban Temperature Data Mining

To study the temperature patterns in various regions, the characteristic rules are summarized from the inductive results by further analyzing the numeric temperature data of major cities in China (Di 2001). The experimental data, taken from The *Road Atlas of China Major Cities*, are the monthly mean temperatures of a total of 37 cities. As the dispersion degree of the data is relatively large and difficult to use for the purpose of extracting general rules, a histogram of temperature data was obtained from statistics. For the sake of simplicity, the histogram equalization method was used to group the temperature data into three clusters of low (-19.4 to 9.8), medium (9.8 ~ 20.3), and high (20.3 to 29.6). The 12 months are grouped into four quarters, and the temperature of each quarter is the mean of the three months. The city names are summarized based on their positions as north, northeast, south, northwest, southwest, etc. Table 8.7 shows the results. Because there was a large number of tuples in the data table, the data were further summarized (i.e., medium or low into medium-low and medium or high into medium-high; Table 8.8). In Table 8.8, the northwest and southwest groups are identical, which are further summarized as west. The fully summarized results are shown in Table 8.9, from which the characteristic rules were uncovered; the results are consistent with subjective feelings about the distribution of temperature patterns in China.

- (location, north) $\Rightarrow$ (spring, medium-low) $\wedge$ (summer, high) $\wedge$ (autumn, medium-low) $\wedge$ (winter, low) $\wedge$ (annual mean, medium-low)
- (position, northeast) $\Rightarrow$ (spring, low) $\wedge$ (summer, height) $\wedge$ (autumn, low) $\wedge$ (winter, low) $\wedge$ (annual mean, low)

Table 8.7 A generalized result of temperature data

Location	Spring	Summer	Autumn	Winter	Annual average	Count
North	Low	High	Low	Low	Low	1
North	Medium	High	Low	Low	Low	1
North	Medium	High	Medium	Low	Medium	6
North	Low	High	Medium	Low	Medium	1
Northeast	Low	High	Low	Low	Low	3
South	Medium	High	Medium	Low	Medium	9
South	Medium	High	High	Medium	Medium	1
South	Medium	High	High	Medium	High	1
South	High	High	High	Medium	High	5
Northwest	Low	High	Low	Low	Low	1
Northwest	Medium	High	Low	Low	Low	2
Northwest	Low	Medium	Low	Low	Low	1
Northwest	Medium	High	Medium	Low	Medium	1
Southeast	Low	Medium	Low	Low	Low	1
Southeast	Medium	High	Medium	Low	Medium	3

Table 8.8 A further generalized result of temperature data

Location	Spring	Summer	Autumn	Winter	Annual average	Count
North	Medium-low	High	Medium-low	Low	Medium-low	9
Northeast	Low	High	Low	Low	Low	3
South	Medium-high	High	Medium-high	Medium-low	Medium-high	16
Northwest	Medium-low	Medium-high	Medium-low	Low	Medium-low	5
Southeast	Medium-low	Medium-high	Medium-low	Low	Medium-low	4

Table 8.9 The temperature summarization

Location	Spring	Summer	Autumn	Winter	Annual average
North	Medium-low	High	Medium-low	Low	Medium-low
Northeast	Low	High	Low	Low	Low
South	Medium-high	High	Medium-high	Medium-low	Medium-high
West	Medium-low	Medium-high	Medium-low	Low	Medium-low

- (position, south) ⇒ (spring, medium-high) ∧ (summer, high) ∧ (autumn, medium-high) ∧ (winter, medium-low) ∧ (annual mean, medium-high)
- (location, west) ⇒ (spring, medium-low) ∧ (summer, medium-high) ∧ (autumn, medium-low) ∧ (winter, low) ∧ (annual mean, medium-low)

Table 8.9 can be seen as a decision table where the spring, summer, autumn, winter, and annual mean temperatures are the condition attributes, and the location (a decision attribute) is determined according to the temperature characteristics.

Table 8.10 The simplified attribute table

Location	Spring	Summer	Winter
North	Medium-low	High	Low
Northeast	Low	High	Low
South	Medium-high	High	Medium-low
West	Medium-low	Medium-high	Low

Table 8.11 The final result

Location	Spring	Summer
North	Medium-low	High
Northeast	Low	–
South	Medium-high	–
West	–	Medium-high

Then, each condition attribute is removed to determine whether the decision table remains consistent and also to determine whether the condition attribute is omissible relative to the decision attribute. First, all the record values of attributes "spring," "autumn," and "annual mean" are identical, which can be preserved only once. Then, the decision table can be simplified as Table 8.10, where the attribute "winter" is omissible because the decision table is still consistent if the attribute "winter" is removed; however, "spring" or "summer" are not omissible because the decision table will be inconsistent, i.e., identical conditions versus different decisions if the attributes "spring" or "summer" are removed. Then, the values of various condition attributes can be further simplified, removing the omissible attribute values, which are represented as "–," and the final simplified results are shown in Table 8.11, from which the minimum decision rules (distinguishing rules) are finally extracted.

- (spring, medium-low) ∧ (summer, high) ⇒ (location, north);
- (spring, low) ⇒ (location, northeast);
- (spring, medium-high) ⇒ (location, south);
- (summer, medium-high) ⇒ (location, west).

8.3.2.1 Agricultural Statistical Data Mining

The data for this experiment are agricultural data from the statistical yearbook of China's 30 main provinces (cities and districts) from 1982 to 1990 (Di 2001; Li et al. 2013).

With the scatter plot of exploratory data analysis, every province (city, district) first is intuitively observed with some preliminary regular patterns and its data are selected for further processing. Exploratory data analysis can analyze the relationship between one attribute and the other attributes; for example, when the agricultural population increases, the general trend of total agricultural output also increases. However, it is difficult to discover more rules using the scatter plot only. It is also difficult to describe more rules with relevant statistical methods because the proximal population can vary greatly in its output, and the proximal output can vary greatly for a certain population number. In addition, the regularity of the scatter plot of arable land area and total agricultural output is poor, as well as the agricultural investment and the agricultural output. Thus, using the relationship between a variety of agricultural factors and the agricultural total output, the data for 1990 were selected for creating a decision table. The condition attributes include arable land area, agricultural population, and agricultural investment; and the decision attribute is the total agricultural output (Table 8.12).

Table 8.12   Attribute table with agricultural value

Name	Arable land area	Agricultural population	Agricultural investment	Agricultural output
Heilongjiang	8826.53	20084.6	343	24,540
Liaoning	3470.40	22724.4	327	27,380
Jilin	3935.53	14883.2	195	18,910
Shandong	6867.87	68459.6	216	64,750
Hebei	6560.47	52315.5	279	35,760
Beijing	414.47	3952.96	132	7020
Tianjin	432.27	3830.47	120	5490
Henan	6944.40	74522.5	479	50,200
Shanxi	3701.80	22177.7	167	12,480
Shaanxi	3541.07	26659.6	123	17,000
Gansu	3477.13	18682.5	286	10,310
Inner Mongolia	498.53	14877.7	156	15,690
Ningxi	795.00	3542.94	94	2470
Xinjiang	3072.93	10005.9	325	14,470
Qinghai	572.00	3101.09	72	2450
Zhejiang	1731.13	35426.6	173	33,680
Jiansu	4562.33	52707.6	192	58,050
Shanghai	324.00	4188.87	176	6820
Anhui	4373.00	48175.6	333	37,090
Hubei	3486.60	41728.7	395	40,220
Hunan	3318.60	51824	210	39,740
Jiangxi	2355.47	30530.3	113	25520
Guangdong	2524.67	47690.1	468	60,070
Guangxi	2578.47	36766.4	137	25,220
Fujian	1238.47	24994.3	117	22,870
Hainan	433.53	5146.28	212	6870
Sichuan	6307.20	92187.2	340	63,710
Guizhou	1854.00	28402.1	87	14,550
Yunnan	2822.80	32399.8	223	21,170
Xizang	223.00	1882.24	30	1720

As shown in Table 8.12, a cloud model and rough set were combined to discover the implicit classification rules. First, a cloud model was used to deduce the attributes. Then, the attribute values of a province and city name were generalized into the region in which they are located: northeast, north, northwest, east, central, south, and southwest. Other numeric attributes were generalized into three linguistic values with the maximum variance method (Di 2001): "small," "medium," and "big" for the arable land area and agricultural investment; "many," "medium," and "few" for the agricultural population; and "high," "medium," and "low" for the agricultural output. These numeric attributes were discretized and represented with the cloud model. Table 8.13 shows the results of the generalized attributes, in which a new attribute of "count" is added to record the number of merged records.

Table 8.13 Generalized attribute table on agricultural information

No.	Region	Arable land area	Agricultural population	Agricultural investment	Agricultural output	Count
1	Northeast	Big	Medium	Big	Medium	1
2	Northeast	Medium	Medium	Big	Medium	1
3	Northeast	Medium	Few	Medium	Low	1
4	North	Big	Many	Medium	High	1
5	North	Big	Many	Medium	Medium	1
6	North	Small	Few	Small	Low	1
7	North	Small	Few	Small	Low	1
8	North	Big	Many	Big	High	1
9	North	Medium	Medium	Medium	Low	1
10	Northwest	Medium	Medium	Small	Low	1
11	Northwest	Medium	Medium	Medium	Low	1
12	Northwest	Medium	Few	Medium	Low	1
13	Northwest	Small	Few	Small	Low	1
14	Northwest	Medium	Few	Big	Low	1
15	Northwest	Small	Few	Small	Low	1
16	East	Small	Medium	Medium	Medium	1
17	Eat	Medium	Many	Medium	High	1
18	East	Small	Few	Medium	Low	1
19	East	Medium	Many	Big	Medium	1
20	Central	Medium	Medium	Big	Medium	1
21	Central	Medium	Many	Medium	Medium	1
22	Central	Medium	Medium	Small	Medium	1
23	South	Medium	Many	Big	High	1
24	South	Medium	Medium	Small	Medium	1
25	South	Small	Medium	Small	Medium	1
26	South	Small	Few	Medium	Low	1
27	Southeast	Big	Many	Big	High	1
28	Southeast	Small	Medium	Small	Low	1
29	Southeast	Medium	Medium	Medium	Medium	1
30	Southeast	Small	Few	Small	Low	1

Based on Table 8.13, the initial decision rules were generated by a rough set.

$U/R(region)$ = {{1, 2, 3}, {4, 5, 6, 7, 8, 9}, {10, 11, 12, 13, 14, 15}, {16, 17, 15, 19}, {20, 21}, {23, 24, 25, 26}, {27, 28, 29, 30}}

$U/R(arable\ land\ area)$ = {{1, 4,5,8,27}, {2, 3, 9, 10, 11, 12, 14, 17, 19, 20, 21, 22, 23, 24, 29}, {6, 7, 13, 15, 16, 18, 25, 26, 28, 30}}

$U/R(agricultural\ population)$ = {{1, 2, 9, 10, 11, 16, 20, 22, 24, 25, 25, 29}, {3, 6, 7, 12, 13, 14, 15, 18, 26, 30} {4, 5, 8, 17, 19, 21, 23, 27}} is

$U/R(agricultural\ investment)$ = {1, 2, 8, 14, 19, 20, 23, 27}, {3, 4, 5, 9, 11, 12, 16, 17, 18, 21, 26, 29}, {6, 7, 10, 13, 15, 22, 24, 25, 28, 30}}

$U/R(agricultural\ output)$ = {{1, 2, 5, 16, 19, 20, 21, 22, 24, 25, 29}, {3, 6, 7, 9, 11, 12, 13, 14, 15, 18, 26, 28, 30}, {4, 8, 17, 23, 27}}

The calculation of positive region is as follows:

$Pos_{region}(agricultural\ output)$ = {10, 11, 12, 13, 14, 15, 20, 21, 22}

$Pos_{arable\ land\ area}(agricultural\ output)$ = Φ

$Pos_{agricultural\ population}(agricultural\ output)$ = {3, 6, 7, 12, 13, 14, 15, 15, 26, 30}

$Pos_{agricultural\ investment}(agricultural\ output)$ = Φ

$Pos_{\{region,\ agricultural\ population,\ agricultural\ investment\}}(agricultural\ output)$ = {1, 2, 3, 6, 7, 8, 9, 10, 11, 12, 13, 14, 15, 16, 17, 18, 19, 20, 21, 22, 23, 24, 25, 26, 27, 28, 29, 30}

$Pos_{\{region,\ arable\ land\ area,\ agricultural\ population\}}(agricultural\ output)$ = $Pos_{\{region,\ agricultural\ population,\ agricultural\ investment\}}(agricultural\ output)$

$Pos_{\{region,\ arable\ land\ area,\ agricultural\ population,\ agricultural\ investment\}}(agricultural\ output)$ = $Pos_{\{region,\ agricultural\ population,\ agricultural\ investment\}}(agricultural\ output)$

As seen from the above results, the cardinal number of the positive region is always smaller than the cardinal number of the universe of discourse 30. There is no dependence on the γ value of 1—that is, the decision table is inconsistent. The reason for the inconsistency is that record 4 and record 5 share identical condition attributes but entirely different decision attributes. Also, the attribute of agricultural population is important in the condition attributes, but the two attributes of arable land area and agricultural investment are redundant for decision-making; therefore, one of them can be removed, which then reduced the attribute of arable land area. Merging the same records resulted in Table 8.14. For convenience, the tuples were renumbered.

To remove the redundant attribute values of the condition attributes in Table 8.14, each condition attribute value within each record was carefully evaluated to determine whether the removal of the value would change the decision-making results; and if there was no change, then the value was redundant and could be removed. For example, for record 8 and record 12, the removal of the two attributes of agricultural population and agricultural investment did not affect the agricultural output; therefore, all the attribute values of the two attributes were removed. By removing all redundant attribute values and merging identical records in Table 8.14, the final simplified decision table was obtained (Table 8.15).

Table 8.14 Decision table without redundant attributes

No.	Region	Agricultural population	Agricultural investment	Agricultural output	Count
1	Northeast	Medium	Big	Medium	2
2	Northeast	Few	Medium	Low	1
3	North	Many	Medium	High	1
4	North	Many	Medium	Medium	1
5	North	Few	Small	Low	2
6	North	Many	Big	High	1
7	North	Medium	Medium	Low	1
8	Northwest	Medium	Small	Low	1
9	Northwest	Medium	Medium	Low	1
10	Northwest	Few	Medium	Low	1
11	Northwest	Few	Small	Low	2
12	Northwest	Few	Big	Low	1
13	East	Medium	Medium	Medium	1
14	East	Many	Medium	High	1
15	East	Few	Medium	Low	1
16	East	Many	Big	Medium	1
17	Central	Medium	Big	Medium	1
18	Central	Many	Medium	Medium	1
19	Central	Medium	Small	Medium	1
20	South	Many	Big	High	1
21	South	Medium	Small	Medium	2
22	South	Few	Medium	Low	1
23	Southeast	Many	Big	High	1
24	Southeast	Medium	Small	Low	1
25	Southeast	Medium	Medium	Medium	1
26	Southeast	Few	Small	Low	1

In Table 8.15, every record corresponds to a decision rule, and the recorded count value can be regarded as the support of the rules, in which rule 3 and rule 4 are inconsistent rules and the other 14 rules are consistency rules. For example, rule 2 can be rephrased as "if the agricultural population is few, then the agricultural output is low"; rule 7 can be rephrased as "if the region lies in the northwest, then the agricultural output is low"; and rule 3 and rule 4 can be merged and rephrased as "if the region lies in the north, the agricultural population is large, and the agricultural investment is medium, then the agricultural output is high or medium." In fact, the decision table by itself is an expression form of the decision knowledge. These rules reveal the macro-laws between regional location, agricultural population, and agricultural investment with total agricultural output, and it is therefore important for agricultural development decision-making.

Table 8.15 Decision table with final simplified attributes

No.	Region	Agricultural population	Agricultural investment	Agricultural output	Count
1	–	Medium	Big	Medium	2
2	–	Few	–	Low	6
3	North	Many	Medium	High	1
4	North	Many	Medium	Medium	1
5	North	Many	Big	High	1
6	North	Medium	Medium	Low	1
7	Northwest	–	–	Low	6
8	East	Medium	–	Medium	1
9	East	Many	Medium	High	1
10	East	Many	Big	Medium	1
11	Central	–	–	Medium	3
12	South	Many	–	High	1
13	South	Medium	–	Medium	2
14	Southeast	Many	–	High	1
15	Southeast	Medium	Small	Low	1
16	Southeast	Medium	Medium	Medium	1

To sum up, rough set is able to refine the decision table greatly without changing the results. Key elements are preserved but the omissible attributes are reduced, which may significantly accelerate the speed of decision-making. If richer data are collected, the discovered rules will be more abundant and rewarding.

8.4 Spatial Clustering

Spatial clustering assigns a set of objects into clusters by virtue of their observations so that the objects that are similar to one another within the same cluster and dissimilar to the objects in other clusters, or it groups a set of data in a way that maximizes the similarity within clusters and also minimizes the similarity between two different clusters. It is an unsupervised technique without the knowledge of what causes the grouping and how many groups exist (Grabmeier and Rudolph 2002; Xu and Wunsch 2005). Spatial clustering helps in the discovery of densely populated regions by exploring spatial datasets. For big data, an effective clustering algorithm can handle noise properly, detect clusters of arbitrary shapes, and perform stably and independently of expert experience and the order of the data input (Fränti and Virmajoki 2006). Clustering algorithms may be implemented on partition, hierarchy, distance, density, grid, model, etc. (Parsons et al. 2004; Zhang et al. 2008; Horng et al. 2011; Silla and Freitas 2011).

1. *Partition-based algorithms*

   The algorithms are used by the distance iterative clustering model, which relies on the distance from the target to the clustering center for clustering convex and far distances between cluster with slight disparity in the diameter difference (Ester et al. 2000), such as Partitioning Around Medoids (PAM). Ng and Han (2002) proposed an improved k-medoid random search algorithm, the Clustering Large Applications Based on RANdomized Search (CLARANS) algorithm, which is a specialized SDM algorithm. Concept clustering is an extension of segmentation, which partitions data into different cluster by a set of object-describing conceptual values, rather than the geometric distance-based approach, to realize the similarity measurement between data objects (Pitt and Reinke 1988). The common numeric concept hierarchical algorithm discretizes the domain interval of numerical attributes, forming multiple sub-intervals as the leaf nodes at the conceptual level. It carries out conceptual segmenting for all the data in the numerical field, and each segment is represented by a concept value; then, all of the data in the original fields are substituted by the corresponding concept value of each segment, which creates a concept table. The supervised conceptualization methods include ChiMerge (Kerber 1992), Chi2 (Liu and Rudy 1997), minimum entropy, and minimum error; the unsupervised conceptualization methods includes uni-interval, uni-frequency, and k-means.

2. *Hierarchy-based algorithms*

   The algorithms decompose datasets iteratively into a subset of a tree diagram until every subset contains only one target, and the construction process is suitable by the splitting or merging methods. It uncovers a nested sequence of clusters with a single, all-inclusive cluster at the top and single-point clusters at the bottom. In the sequence, each cluster is nested into the next cluster. Hierarchical clustering is either agglomerative or divisive. Agglomerative algorithms start with each element as a disjoint set of clusters and merge them into successively larger clusters (Sembiring et al. 2010). Divisive algorithms begin with the whole set and proceed to divide it into successively smaller clusters (Malik et al. 2010). Examples can be found in Balanced Iterative Reducing and Clustering using Hierarchies (BIRCH) (Zhang et al. 1996), Clustering Using Representatives CURE) (Guha et al. 1998), and Clustering In Quest (CLIQUE) (Agrawal and Srikant 1994), all of which utilize the hierarchical algorithm and the correlation between the dataset to achieve clustering. Accounting for both interconnectivity and closeness in identifying the most similar pair of clusters, CHAMELEON yields accurate results for highly variable clusters using dynamic modeling (George et al. 1999); however, it is not suitable for grouping large volume data (Li et al. 2008). Some clustering algorithms were further hybridized, such as clustering feature tree (CBCFT) hybridizing BIRCH with CHAMELEON (Li et al. 2008), Bayesian hierarchical clustering for evaluating marginal likelihoods of a probabilistic model (Heller and Ghahramani 2005), and the support vector machine (Horng et al. 2011). Lu et al. (1996) proposed the automatic generation algorithm AGHC and the AGPC for numeric conceptual hierarchy based on the hierarchy and partitioning clustering algorithm.

3. *Density-based algorithms*

   The algorithms detect clusters by calculating the density of data objects nearby. Because the density of data objects makes no assumptions on the shape of clusters, density-based algorithms are able to determine the clusters of arbitrary shapes. Density-based Spatial Clustering of Applications with Noise (DBSCAN) detects clusters by counting the number of data objects inside an area of a given distance (Ester et al. 1996). The Statistical Information Grid-based Method (STING) and STING+ are continuous improvements of the DBSCAN (Wang et al. 2000). DENsity-based CLUstEring (DENCLUE) modeled the overall density as the sum of the influential functions of data objects and then used a hill-climbing algorithm to detect clusters inside the dataset (Hinneburg and Keim 2003). DENCLUE has high time complexity; however, more efficient algorithms are sensitive to noise or rely too much on parameters.

4. *Grid-based algorithms*

   The algorithms quantize the feature space into a finite amount of grids and then implement the calculations on the quantized grids. The effect of grid-based algorithms is independent from the number of data objects and depends only on the amount of quantized grids in each dimension. For example, WaveCluster (Sheikholeslami et al. 1998) applied wavelet transform to discover the dense regions in the quantized feature space. The operation of quantizing grids makes grid-based algorithms very efficient when handling large datasets. However, at the same time, they bring a bottleneck on accuracy. In addition, there is mathematical morphology clustering for raster data (Li et al. 2006).

5. *Model-based algorithms*

   The algorithms focus on fitting the distribution of given data objects with a specific mathematic model. As an improvement of k-means, the EM algorithm (Dempster et al. 1977) assigns each data object to a cluster according to the mean of the feature values in that cluster. Its steps of calculation and assignment repeat iteratively until the objective function obtains the required precision. Knorr and Ng (1996) discovered clustering proximity and common characteristics. Ester et al. (1996) employed clustering methods to investigate the category interpretation knowledge of mining in large spatial databases. Tung et al. (2001) proposed an algorithm dealing with the obstruction of a river and highway barrier in spatial clustering of SDM. Murray and Shyy (2000) proposed an interactive exploratory clustering technique.

8.4.1 Hierarchical Clustering with Data Fields

A data field is a novel method for unsupervised clustering under hierarchical demands. Data objects are grouped into clusters by simulating their mutual interactions and opposite movements in data fields despite the clusters having different geometric shapes, proximity, varying densities, and noise. In the data field, the self-organized process of equipotential lines on many data objects discovers their hierarchical clustering-characteristics with approximate linear performance. To optimize the impact factor, the interacted distance between data objects is controlled, and mass estimation selects core data objects with nonzero masses for the initial clusters. The clustering results are achieved by iteratively merging the core data objects, hierarchy by hierarchy (Wang et al. 2011, 2014; Wang and Chen 2014).

To illustrate the concept of data fields and their application in hierarchical clustering, its performance was compared against that of k-means (MacQueen 1967), BIRCH (Zhang et al. 1996), CURE (Guha et al. 1998), and CHAMELEON (George et al. 1999). The data set (Guha et al. 1998) contained 100,000 points in two dimensions that form five clusters with different geometric shapes, proximity, and varying densities, along with significant noise and special artifacts. A total of 80,000 data points were stochastically sampled from the 100,000 points, and the remaining 2,000 were left for testing. All the clustering concepts were implemented with Matlab 7 software on a personal computer (Intel Core 4G CPU, 1G memory). To denote different clusters, the points in the different clusters are represented by different colors in Fig. 8.10. As a result, the points that belong to the same cluster are the same color as well.

As can be seen from Fig. 8.10, data fields exhibited better clustering quality and diminished the contribution of noisy data more effectively without a user-specified threshold (Fig. 8.10b). In the data field that describes the interactions among the objects, the data objects are grouped into clusters by simulating

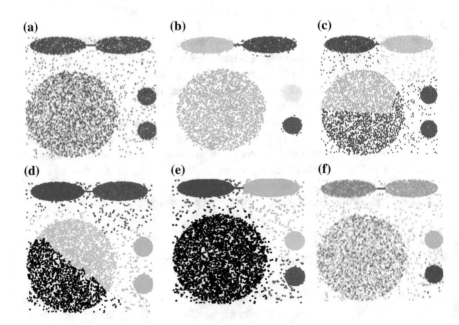

Fig. 8.10 Comparison of results from different clustering algorithms. **a** Original data set, **b** data field (5 clusters), **c** k-means (5 clusters), **d** BIRCH (5 clusters), **e** CURE (5 clusters), **f** CHAMELEON (5 clusters)

their mutual interactions and opposite movements. Every topology of equipotential lines containing data objects actually corresponds to a cluster with relatively dense data distribution. The nested structures, composed of different equipotential lines, are treated as the clustering partitions at different density levels. Although the clusters had different geometric shapes, proximity, and varying densities, along with significant noise and special artifacts, data field correctly identified the genuine five clusters, a big green circle, a small green circle, a small pink circle, a small red ellipse plus a line, and a small light-blue ellipse. The clusters correspond to the earliest iteration at which data field identified the genuine clusters and placed each in a single cluster. Moreover, the outliers were removed by using 3σ range of neighboring clusters, and only non-noise data objects were amalgamated to the corresponding core objects separately. Furthermore, the parameters were extracted from the data set, and the user was no longer forced to specify them. Therefore, the extracted parameters not only depended on the data set objectively but also avoided the subjective differences specified by different users, which sensibly improved the accuracy of the clustering result.

In comparison, the k-means method assumes that clusters are hyper-ellipsoidal and of similar sizes; it attempts to break a dataset into clusters such that the partition optimizes a given criterion. In this experiment, the k-means algorithms failed to find clusters that varied in size as well as concave shapes (Fig. 8.10c). They were unable to find the right clusters for the data set (Fig. 8.10a) and the worst case was that the data points in the big circle cluster are incorrectly grouped into three partitions; two small circle clusters are merged in the same cluster. The noisy data were amalgamated into their neighboring clusters separately. BIRCH was the only algorithm that found clusters with spherical shapes because the clustering feature is defined through the radius, as shown in Fig. 8.10d). While its results (Fig. 8.10d) were better than k-means (Fig. 8.10c), BIRCH's merging scheme still made multiple mistakes for the data set (Fig. 8.10a). The data points in the big circle are wrongly partitioned into two clusters, two small circle clusters were placed together in a cluster, and the noisy data were merged in different clusters based on their distance to the corresponding clusters. CURE measures the similarity between two clusters by the similarity of the closest pair of points belonging to different clusters. From 10 representative points and $\alpha = 0.3$, it was able to correctly find the clustering structure of the data distribution (Fig. 8.10e), and designated five clusters as did data field (Fig. 8.10b). The data points in the big circle cluster are not incorrectly partitioned, and the two small circle clusters are different from each other. Compared to BIRCH, CURE became much better for the data set (Fig. 8.10a), but it did not effectively handle noisy data as it allocated all of it to the nearest cluster.

8.4.2 Fuzzy Comprehensive Clustering

Fuzzy comprehensive clustering integrates both fuzzy comprehensive evaluation and fuzzy clustering analysis in a unified way. First, the participating factors are selected and their weights are determined by Delphi hierarchical processing (Wang and Wang 1997); then, the membership is evaluated and the fuzzy comprehensive evaluation matrix X is obtained. Second, the comprehensive evaluation matrix B of all the factors obtained is determined based on the equation $Y = A - X$, and the results of the grading evaluation of the assessed spatial entities are integrated into a comprehensive grading evaluation matrix. A is the factor weight matrix. Third, the elements in the comprehensive grading evaluation matrix are used to obtain fuzzy similar matrix R and fuzzy equivalent matrix (transitive closed matrix) $t(R)$. Finally, according to $t(R)$, the maximum residual method or mean absolute distance method, which are based on fuzzy confidence, are used in clustering to obtain the final clustering knowledge.

Suppose that the set of influential factors is $U = \{u_1, u_2, \ldots, u_m\}$, with the matrix of weights $A = (a_1, a_2, \ldots, a_m)^{\mathrm{T}}$, and the set of sub-factors $u_i = \{u_{i1}, u_{i2}, \ldots, u_{iki}\}$ with the matrix of weights $A_i = (a_{i1}, a_{i2}, \ldots, a_{iki})^{\mathrm{T}}$ $(i = 1, 2, \ldots, m)$ simultaneously (Fig. 8.11). The set of grades is $V = \{v_1, v_2, v_3, \ldots, v_n\}$, including n grades.

In the context of the given grades, the fuzzy membership matrix of the sub-factors of factor U_i may be described as

Fig. 8.11 Factors and their weights

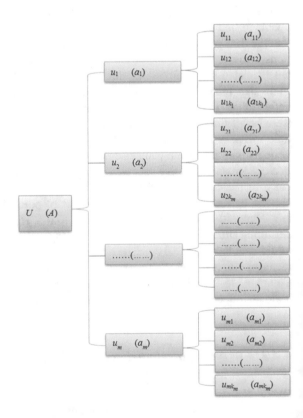

$$X_i = \begin{pmatrix} x_{11} & x_{12} & \cdots & x_{1n} \\ x_{21} & x_{22} & \cdots & x_{2n} \\ \cdots & \cdots & \cdots & \cdots \\ x_{k_i 1} & x_{k_i 2} & \cdots & x_{k_i n} \end{pmatrix}$$

The comprehensive evaluation matrix $Y_i = A_i \cdot X_i$.

The fuzzy evaluation matrix of U (i.e., all factors) on the spatial entity p is

$$Y^{(p)} = A \cdot Y = A \cdot (Y_1, Y_2, \ldots, Y_i, \ldots, Y_m)^{\mathrm{T}}$$

When p $(p \geq 1)$ spatial entities are evaluated at the same time, obtain a total matrix of fuzzy comprehensive evaluation. Let the number of spatial entity be l $(p = 1, 2, \ldots, l)$, then $Y_{l \times n} = (Y^{(1)}, Y^{(2)}, \ldots, Y^{(i)}, \ldots, Y^{(l)})^{\mathrm{T}}$

$$Y_{l \times n} = \begin{pmatrix} y_{11} & y_{12} & \cdots & y_{1n} \\ y_{21} & y_{21} & \cdots & y_{2n} \\ \cdots & \cdots & \cdots & \cdots \\ y_{l1} & y_{l1} & \cdots & y_{ln} \end{pmatrix}.$$

Take the element y_{ij} $(i = 1, 2, \ldots, l; j = 1, 2, \ldots, n)$ of matrix $Y_{l \times n}$ as the original data, and the fuzzy similar matrix can be created—that is, $R = (r_{ij})_{l \times l}$—which indicates the fuzzy similar relationships among the entities.

$$r_{ij} = \frac{\sum_{k=1}^{n} (y_{ik} \times y_{jk})}{\left(\sum_{k=1}^{n} y_{ik}^2 \right)^{\frac{1}{2}} \times \left(\sum_{k=1}^{n} y_{jk}^2 \right)^{\frac{1}{2}}}$$

The fuzzy similar matrix $R = (r_{ij})_{l \times l}$ will need to be changed into the fuzzy equivalent matrix $t(R)$ when clustering. The fuzzy matrix for clustering is a fuzzy equivalent matrix $t(R)$, and it shows a fuzzy equivalent relationship among the entities instead of the fuzzy similar relationships. The fuzzy equivalent matrix has three characteristic conditions of self-reverse, symmetry, and transfer. However, the fuzzy similar matrix R frequently satisfies two characteristic conditions of self-reverse and symmetry but not transfer. The fuzzy similar matrix R may be changed into the fuzzy equivalent matrix $t(R)$ via the self-squared method, for there must be a minimum natural number $k(k = 1, 2, \ldots, l,$ and $k \leq l)$ that makes equivalent matrix $t(R) = R^{2^k}$ if and only if $R^{2^k} = R^{2^{k+1}} = R^{2^{k+2}} = \cdots = R^{2^l}$ (Wang and Klir 1992).

$$t(R)_{l \times l} = \begin{pmatrix} t_{11} & t_{12} & \cdots & t_{1l} \\ t_{21} & y_{21} & \cdots & t_{2l} \\ \cdots & \cdots & \cdots & \cdots \\ t_{l1} & t_{l1} & \cdots & t_{ll} \end{pmatrix}$$

The fuzzy confidential level α is a fuzzy probability that two or more than two entities belong to the same cluster. Under the umbrella of the fuzzy confidential level α, the maximum remainder algorithms is described as follows:

1. Select fuzzy confidential level α.
2. Summarize the elements, column by column, in the fuzzy equivalent matrix $t(\underline{R})$, excluding the diagonal elements.

$$T_j = \sum_{i=1}^{l} t_{ij}, (i \neq j, \quad i,j = 1, 2, \ldots, l)$$

3. Compute the maximum and the ratio

$$T_{\max}^{(1)} = \max(T_1, T_2, \ldots, T_l), \quad K_j^{(1)} = \frac{T_j}{T_{\max}^{(1)}}$$

4. Put the $K_j^{(1)} \geq \alpha$ of entity j into the first cluster.
5. Repeat the above steps (3) and (4) in the remained T_j until the end.

The resulting cluster only gives a group of entities with similar or identical characteristics. In the context of fuzzy sets, an algorithm of maximum characteristics is proposed to decide the grades. In all the l spatial entities, suppose that there are h ($h \leq l$) spatial entities that are grouped into ClusterZ.

$$ClusterZ = \{ Entity\_1, Entity\_2, \ldots, Entity\_h \}$$

Then, the fuzzy evaluation values y_{ij} in $Y_{l \times n}$ of each cluster are added respectively, column by column, (i.e., grade by grade) in the set $V = \{v_1, v_2, v_3, \ldots, v_n\}$. The grade of ClusterZ is the grade with the maximum sum of every column—that is, all the entities in ClusterZ belonging to the same grade.

$$Grade\_ClusterZ = \max(Sum\_v_1, Sum\_v_2, \ldots, Sum\_v_n)$$

$$= \max \left(\sum_{i=1}^{h} y_{i1}, \sum_{i=1}^{h} y_{i2}, \ldots, \sum_{i=1}^{h} y_{in} \right)$$

The resulting grade is furthermore the grade of all the entities in ClusterZ.

The case study of fuzzy comprehensive clustering then was applied to land evaluation in Nanning City for 20 land parcels with different characteristics stochastically distributed (Table 8.16; Fig. 8.12). The set of grades is $V = \{I, II, III, IV\}$.

Before the land evaluation process began, the influential factors were selected and weighted by using Delphi hierarchical processing in the context of the local reality and the experience of experts. The trapezoidal fuzzy membership function was chosen to compute the membership that each influential factor belongs to every grade. Then, the matrix of fuzzy comprehensive evaluation $Y_{20 \times 4}$ could be obtained (Table 8.16).

Based on the matrix of $Y_{20 \times 4}$, fuzzy comprehensive evaluation traditionally determines the land grades on maximum fuzzy membership (Table 8.17).

Comparatively, fuzzy comprehensive clustering further takes the elements y_{ij} ($i = 1, 2, \ldots, 20; j = 1, 2, 3, 4$) of $Y_{20 \times 4}$ as the original data to create the fuzzy similar matrix $R_{20 \times 20}$. In the context of $R_{20 \times 20}$, fuzzy equivalent matrix $t(R_{20 \times 20})$

Table 8.16 Land parcels and their evaluation matrix

No.	Name	Grade I	Grade II	Grade III	Grade IV
1	Nanning railway station	0.671	0.001	0.328	0.005
2	Nanhua building	0.885	0.099	0.016	0.021
3	Train department	0.176	0.162	0.662	0.300
4	Yong river theatre	0.787	0.057	0.155	0.041
5	Medical school of women and children health	0.141	0.654	0.204	0.834
6	Nanning shop building	0.508	0.002	0.491	0.001
7	Yongxin government division	0.009	0.594	0.404	0.006
8	Minsheng shop	0.723	0.135	0.142	0.213
9	Provincial procuratorate	0.454	0.423	0.121	0.210
10	No. 1 Middle School of Nanning railway station	0.180	0.193	0.626	0.232
11	Xinyang road primary school	0.265	0.215	0.519	0.002
12	No.25 middle school	0.125	0.419	0.455	0.013
13	Municipal engineering corporation	0.335	0.558	0.092	0.034
14	Dragon palace hotel	0.058	0.117	0.704	0.902
15	Provincial school of broadcasting and TV	0.160	0.041	0.798	0.030
16	Provincial architecture institute	0.210	0.123	0.666	0.041
17	Training school of water supply corporation	0.023	0.335	0.641	0.102
18	No.2 Tingzi Middle School	0.065	0.441	0.523	0.210
19	Jiangnan government division	0.016	0.038	0.945	0.060
20	Baisha paper mill	0.031	0.622	0.424	0.604

The header of the table reads: Land Parcel | Fuzzy evaluation matrix $Y_{20 \times 4}$

is computed. With the clustering threshold of 0.99, the clustering process is shown in Table 3 by using the maximum remainder of algorithms-based clustering. In the end, the grades of each land parcel are obtained via maximum characteristics algorithms (Table 8.18), and they are also mapped in Fig. 8.12.

A comparison of the results of the proposed comprehensive clustering (Table 8.19; Fig. 8.12) and the single fuzzy comprehensive evaluation (Table 8.17) is presented below.

The results of the traditional fuzzy comprehensive evaluation did not match the reality of Nanning city for some of the land parcels; for example, the following errors were made for land parcels 14, 9, and 20:

- Land parcel 14, which is grade I, was misidentified as grade II. Land parcel 14 is the Dragon Palace Hotel and is located in the city center with shops, infrastructure, and large population density.
- Land parcel 9, which is grade II, was misidentified as grade I. Land parcel 9 is the Provincial Procuratorate, which is obstructed from the city center by the Yong River.

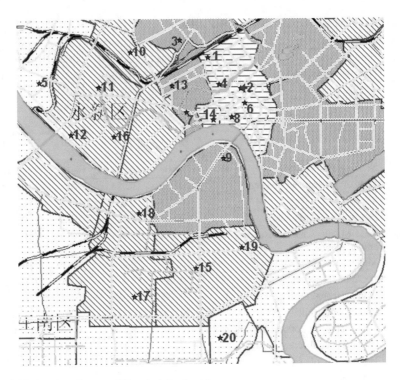

Fig. 8.12 Land grade map of Nanning city (parts)

Table 8.17 The results of traditional fuzzy comprehensive evaluation

Grade	I	II	III	IV
Land parcel	1, 2, 4, 6, 8, 9	7, 13, 20	3, 10, 11, 12, 15, 16, 17, 18, 19	5, 14

– Land parcel 20, which is grade IV, was misidentified as grade II. Land parcel 20
  is the Baisha Paper Mill, which is in a suburban area with bad infrastructure.

These errors occurred because traditional fuzzy comprehension evaluation cannot
differentiate close membership values, especially when the difference between the
first maximum membership and the second maximum membership is small. When
the grade with the maximum membership value is chosen as the land grade, some
important land information, which is hidden in the land's influential factors, may
be lost, such as obstructers of a river, lake, road, railway (Tung et al. 2001) or
the spread of land position advantage. The subordinate grade will be cut off once
the maximum membership value is chosen even though the maximum membership may represent the essential grade of a land parcel. For example, as can be
seen from the matrix $Y_{20 \times 4}$ (Table 8.17), land parcel 20, the Baisha Paper Mill, has
membership values of grade II 0.622 and grade IV 0.604. This dilemma cannot be
overcome by fuzzy comprehensive evaluation alone.

Table 8.18 The clustering process of maximum remainder algorithms

Land parcel	$T^{(1)}$	$K^{(1)}$	Cluster 1	$T^{(2)}$	$K^{(2)}$	Cluster 2	$T^{(3)}$	$K^{(3)}$	Cluster 3	$T^{(4)}$	$K^{(4)}$	Cluster 4
1	16.253	0.954		16.253	0.982		16.253	0.998	1			
2	16.28	0.956		16.28			16.280	1.000	2			
3	17.005	0.998	3									
4	16.28	0.956		16.28	0.984		16.280	1.000	4			
5	16.516	0.969		16.516	0.998	5						
6	16.22	0.952		16.220	0.98		16.220	0.996	6			
7	16.104	0.945		16.104	0.973		16.104	0.989		16.104	1.000	7
8	16.267	0.955		16.267	0.983		16.267	0.999	8			
9	15.928	0.935		15.928	0.963		15.928	0.978		15.928	0.991	9
10	17.005	0.998	10									
11	17.005	0.998	11									
12	16.967	0.996	12									
13	15.858	0.931		15.858	0.958		15.858	0.974		15.858	0.99	13
14	16.182	0.95		16.182	0.978		16.182	0.994	14			
15	17.036	1	15									
16	17.036	1	16									
17	16.939	0.994	17									
18	16.939	0.994	18									
19	17.033	0.999	19									
20	16.545	0.971		16.545	1.000	20						
$T^{(1)}_{\max}$	17.036		$T^{(2)}_{\max}$	16.545		$T^{(3)}_{\max}$	16.280		$T^{(4)}_{\max}$	16.104		

Note (1) $T^{(1)} = \sum\limits_{i=1}^{20} t_{ij}$, $i \neq j$, $i,j = 1,2,L,20$

(2) $T^{(2)} = T^{(1)}_{\text{remainder}}$, $T^{(3)} = T^{(2)}_{\text{remainder}}$, $T^{(4)} = T^{(3)}_{\text{remainder}}$

(3) $K^p_j = \dfrac{T^{(p)}}{T^p_{\max}}$, $j = 1,2,L,20$, $p = 1,2,3,4$,

Table 8.19 The results of maximum characteristics algorithm-based grading

Grade	I	II	III	IV
Cluster	Cluster 3	Cluster 4	Cluster 1	Cluster 2
Land parcel	1, 2, 4, 6, 8, 14	7, 9, 13	3, 10, 11, 12, 15, 16, 17, 18, 19	5, 20
Property	Old urban districts, with good shops, infrastructures, Nanning commercial center, golden land	Near city center with perfect shops infrastructures, but most are obstructed by Yong river, railway from grade I land parcels	Locate at the neighbor of the land parcels with grade I, grade II Mainly resident, culture, and education land	Distribute in the periphery of land grade III. All infrastructures are bad except the ratio of green cover is big

The results of fuzzy comprehensive clustering in Table 8.19 and Fig. 8.12 show that the grade of each land parcel is unique. The results include information from the essential factors and the subordinate factors. Furthermore, the shortcoming of the fuzzy maximum membership is complemented via the quantitative fuzzy clustering without subjective thresholds. The obstructers of river, lake, road, railway, and the spread of land position advantage are all considered when analyzing data. The results obviously show the grades of land parcels 1 ~ 20 given by the fuzzy clustering were in the context of Nanning city land parcels. That is, encircling the center of city, the quality of land in Nanning city decreases as the distance away from the city center spreads like radiation. From the center of the city to the fringe, land grades change from high to low gradually. The high-grade land is centralized in the city center with good infrastructure, while the low-grade land is distributed around the fringes of the city. This reflects the mapping relationships between land quality and land location. Because of the obstructers of the railway and the Yong River, the grade of land parcels close to the city center are better than the grade of land parcels away from the city center, such as the Dragon Palace Hotel and the Provisional Procuratorate.

The transportation system is one of the essential factors that impact land quality. On the two sides of a main road, the land grade shows a decreasing trend as the distance away from the road increases, such as the Nanning Railway Station and the Baisha Paper Mill. The resulting rules matched the real values and characteristics of the Nanning city land parcels.

8.4.3 Mathematical Morphology Clustering

Mathematical morphology clustering (MMC) can handle clusters in arbitrary shapes for point, linear, and planar targets. The spatial objects are processed by a closing operation to connect adjacent regions, and the connected regions are regarded as a cluster. MMC assumes that the distance of any point to the adjacent point within the same cluster is greater than its distance to an arbitrary point within other clusters (i.e., objects belonging to the same cluster are comparatively aggregated in the spatial distribution). Each object first is set as a cluster. The structural element increases from small to large with the number of clusters gradually decreasing from more to less. The optimal number of clusters is heuristically determined. When the closing operation is carried out with large enough structural elements, all the objects are merged into one cluster. During the process of clustering, MMC can also detect outliers and holes simultaneously (Di 2001). As mathematical morphology was originally for raster images, MMC can be directly applied to raster data. Vector data can be converted into raster data with a sample space to preserve neighboring topological relationships.

Algorithm 8.1: MMC algorithm

Input: The interested data X in spatial database (background is 0, target is l).
Output: The clustering result Y (the same value within the same cluster and different values in different clusters).

Steps:

(1) Initialize $i = 1$;
(2) Construct the circular structural element B_i, and its radius as i;
(3) Calculate a closing operation $Y_i = X \cdot B_i$;
(4) Count connected regions in Y_i, i.e., the number of clusters n_i, if $n_i > 1$, then $i = i + 1$, go to step (2); Otherwise, return to step (4);
(5) Calculate the optimal number of clusters n_k according to n_i and obtain the radius k of the corresponding structural element;
(6) Calculate $Y = X \cdot B_k$; and
(7) Show the connected regions in Y with a distinct color for different clustering.

There is a huge disparity within the diameter of the clustering in data a and the distance between them is relatively close; data b and c are concave clustering; and data d is the outcome of data c plus extra noises. For data a, b, c, d, MMC

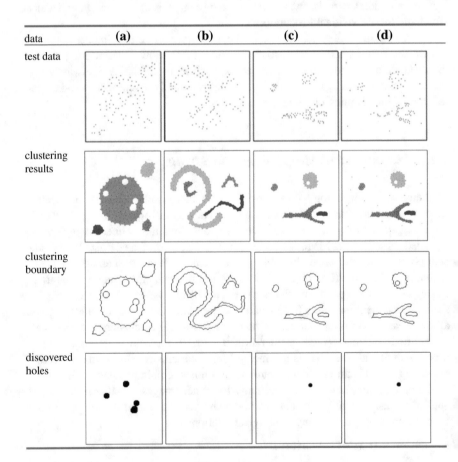

Fig. 8.13 Mathematical morphology clustering (MMC)

automatically determines the optimal number of clusters, which is 4; for data d, the respective values of 3, 4 or 5 are used as the minimum clustering number for the threshold filtering out noises; and the corresponding radius of the structural element for the closing operation is 6, 5, 4, 4 separately. The clustering results are shown in Fig. 8.13. The colored connected regions are the clusters, and the different colors represent different clusters, which are consistent with human observation. Figure 8.13 also shows the clustering boundary and the discovered data holes (no holes in data b). The holes reveal the structural characteristics of data in clustering (i.e., there are no data in the holes), while data outside the holes are more closely distributed in the clustering area. These data holes can be filtered only if an area greater than a certain threshold is regarded as a hole.

8.5 Landslide Monitoring

A landslide can lead to significant hazards if there is excessive movement. To avoid these hazards, landslide displacements are monitored regularly. However, the vast amounts of monitoring data are accumulating continuously and far exceed human ability to completely interpret and use it. Landslide monitoring data mining is a typical and important sample of SDM. The deformation of a landslide is a comprehensive result of many factors, and its outer appearance is the observed data when the landslide is being monitored. Thus, it is necessary to look for a suitable technique to analyze the monitoring dataset from different perspectives, such as SDM view (Wang 2002).

Baota landslide monitoring data mining is utilized as a case study here to apply cloud model and data field under SDM view. The Baota landslide is located in Yunyang, China, and it has been monitored since June 1997. Wang (1999), Zeng (2000), and Jiang and Zhang (2002) studied these monitoring data, the results of which are used for comparison. The deformation values in the database are numerical displacements (i.e., dx, dy, and dh). Respectively, the values of dx, dy, and dh are the measurements of displacements in the direction of X, Y and H of the monitoring points.

$$dx = x_{i+1} - x_0, \quad dy = y_{i+1} - y_0, \quad dh = h_{i+1} - h_0 \qquad (8.1)$$

where 0 is the first observed date, and i is the ith observed date. Respectively, the properties of dx, dy, and dh, are the measurements of displacements in X direction, Y direction, and H direction of the landslide monitoring points.

8.5.1 SDM Views of Landslide Monitoring Data Mining

In landslide monitoring data mining, the background knowledge is the landslide stability monitoring data. There are three basic standpoints when observing the

Fig. 8.14 SDM views of landslide monitoring dataset and their pan-hierarchical relationship

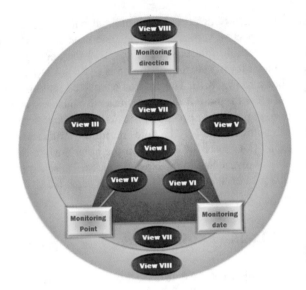

dataset: monitoring point, monitoring date, and moving direction of the landslide deformation. Each basic standpoint has two values, same and different.

- Point-Set: {same monitoring point, different monitoring point}
- Date-Set: {same monitoring date, different monitoring date}
- Direction-Set: {same moving direction, different moving direction}

The different combinations of these three basic standpoints may produce all the SDM views of landslide monitoring data mining. The number of SDM views is

$$C_2^1 \cdot C_2^1 \cdot C_2^1 = 8 \tag{8.2}$$

Among these eight SDM views, each view has a different meaning when mining a dataset from a specific perspective. All the SDM views show a pan-hierarchical relationship (Fig. 8.14) by using a cloud model.

- View I: same monitoring point, same monitoring date, and same moving direction
- View II: same monitoring point, same monitoring date, and different moving direction
- View III: same monitoring point, different monitoring date, and same moving direction
- View IV: same monitoring point, different monitoring date, and different moving direction
- View V: different monitoring point, same monitoring date, and same moving-direction
- View VI: different monitoring point, same monitoring date, and different moving direction

- View VII: different monitoring point, different monitoring date, and same moving direction
- View VIII: different monitoring point, different monitoring date, and different moving direction

In Fig. 8.14, from central SDM view I to outer view VIII, the observed distance expands farther and farther while the mining granularity gets bigger and bigger. In contrast, from the outer SDM view VIII to central view I, the observed distance becomes closer and closer while the mining granularity gets bigger and bigger. This occurs because SDM views I, II, III, and IV focus on the different attributes of the same monitoring point and subsequently discover the individual knowledge of the landslide in a conceptual space; in contrast, SDM views V, VI, VII, and VIII pay attention to the different characteristics of multiple monitoring points and may discover the common knowledge of the landslide in a characteristic's space.

In the eight SDM views, SDM view I is the basic SDM view because the other seven SDM views may be derived from it when one, two, or three basic standpoints of SDM view I change. While the landslide monitoring data are mined in the context of SDM view I, the objective is a single datum of an individual monitoring point on a given date. If three moving-directions dx, dy, dh are taken as a tuple (dx, dy, dh), then SDM views IV, VI, and VIII also may be derived from view II when the monitoring point and/or the monitoring-date change. Thus, SDM view II is called the basic composed SDM view where the focus is on the total displacement of an individual monitoring point on a given date. In the visual field of the basic SDM view or basic composed SDM view, the monitoring data are single isolated data instead of piles of data. In SDM, they are only the fundamental units or composed units but are not the final destination. So the practical SDM views are the remaining six SDM views III, IV, V, VI, VII, and VIII. Tables 8.20 and 8.21 describe the examples of the basic SDM view and basic composed SDM view from cloud model and data field.

Because the Yangtze River flows in the west-east direction, the moving direction of the Baota landslide is geologically north-south; and the displacements of landslide mainly appear in X-direction (north-south), which matches the conditions of SDM view III.

Table 8.20   Basic SDM view

SDM view I	Numerical characters	"landslide stability" cloud	Data field
Monitoring point: BT21 Monitoring-date: 1997.06 Displacement in X direction: dx	$Ex = -3$ $En = 0$ $He = 0$		

Table 8.21 Basic composed SDM view

SDM view II		Numerical character			"landslide stability" cloud	Data field
Monitoring point: BT21 Monitoring-date: 1997.01		*Ex*	*En*	*He*		
Different displacement direction	d*x*	−3	0	0		
	d*y*	4	0	0		
	d*h*	−2	0	0		
Total displacement	d	5.4	0	0		

Algorithms 8.2: Landslide-monitoring data mining under SDM view VIII

Input: Baota landslide monitoring dataset
Output: Baota landslide monitoring knowledge, SDM view VIII
Steps:

(1) Discover the three numerical characters {*Ex, En, He*} from the Baota landslide–monitoring dataset with backward cloud generator;

(2) Interpret {*Ex, En, He*} qualitatively on the basis of some annotation rules;

(3) Create the visual knowledge cloud with forward cloud generator; and

(4) Evaluate the satisfactory degree of the discovered knowledge.

8.5.2 Pan-Concept Hierarchy Tree

It is noted that all of the spatial knowledge that follows was discovered from the databases with the properties of d*x*, d*y*, and d*h*; and the properties of d*x* constitute the major examples. The linguistic terms of different displacements on d*x*, d*y*, and d*h* may be depicted by the pan-concept hierarchy tree (Fig. 8.15) in the conceptual space, which were formed by cloud model (Fig. 8.16). For example, the nodes "very small" and "small" both have the same node "9 mm around."

8.5.3 Numerical Characters and Rules

According to the landslide monitoring characteristics, let the linguistic concepts of "smaller (0–9 mm), small (9–18 mm), big (18–27 mm), bigger (27–36 mm), very big (36–50 mm), extremely big (>50 mm)" with *Ex*; "lower (0–9), low (9–18), high (18–27), higher (27–36), very high (36–50), extremely big (>50)"with *En*; and "more stable (0–9), stable (9–18), unstable (18–27), more unstable (27–36), very unstable (36–50), extremely unstable (>50)" with *He* depict the movements, scattering levels, and stability levels of the displacements, respectively. The certainty of cloud drop (dx_i, μ(dx_i)) is defined as,

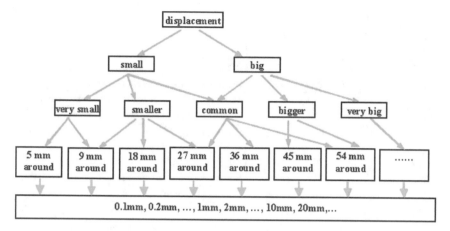

Fig. 8.15 Pan-concept hierarchy tree of landslide displacements

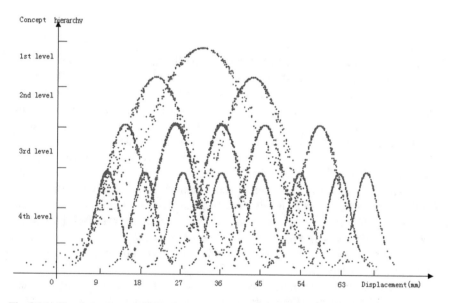

Fig. 8.16 Cloud pan-concept hierarchy trees of different displacements

$$\mu(dx_i) = \frac{dx_i - \min(dx)}{\max(dx) - \min(dx)}$$

where, $\max(dx)$ and $\min(dx)$ are the maximum and minimum of $dx = \{dx_1, dx_2, ..., dx_i, ..., dx_n\}$. Then, the rules for Baota landslide monitoring in X direction can be discovered from the databases in the conceptual space (Table 8.22).

Table 8.22 Quantitative and qualitative knowledge under SDM view VIII

Monitoring Points	Numerical characters			Rules on the displacements
	Ex	En	He	
BT11	−25	18.1	19	Big in south, high scattered and instable
BT12	−22.1	19.4	41.7	Big in south, high scattered and very instable
BT13	−9.3	8.8	8	Small in south, lower scattered and more stable
BT14	−0.3	3.7	6.7	Smaller in south, lower scattered and more stable
BT21	−92.8	66.4	145.8	Extremely big in south, extremely high scattered and extremely instable
BT22	−27	20.8	21.1	Bigger in south, high scattered and instable
BT23	−26.5	21.6	53	Big in south, high scattered and extremely instable
BT24	−20.5	20.2	27.4	Big in south, high scattered and more instable
BT31	−40.3	28.4	92.2	Very big in south, higher scattered and very instable
BT32	−22.9	18.7	38.2	Big in south, low scattered and more instable
BT33	−25	22.2	26.4	Big in south, high scattered and very instable
BT34	−20.9	20.7	32.8	Big in south, high scattered and more instable

8.5.3.1 Rules Visualization

Figure 8.17 visualizes the displacing rules of the Baota landslide displacements on all of its monitoring points. At each point, the rule is visualized with 30,000 pieces of cloud-drops, where the symbol of "+" is the original position of a monitoring point, the different rules are represented via the different pieces of cloud, and the level of color in each piece of cloud denotes the discovered rules of a monitoring point. Figure 8.17 indicates that all the monitoring points move in the direction of the Yangtze River, i.e., south, or the negative axle of X. Moreover, the displacements are different from each other. The BT21 displacements are extremely big in the south, extremely high scattered, and extremely unstable, followed by BT31. On the other hand, the BT14 displacements are smaller south, lower scattered, and more stable. It can be concluded that the displacements of the back part of the Baota landslide are bigger than those of the front part in respect to the Yangtze River, and the biggest exceptions are in BT21 and therefore may be taken as the microcosmic knowledge.

Figure 8.18 shows the more generalized rules of the Baota landslide displacements in three monitoring cross-sections. When compared with every two data-sets of three monitoring cross-sections, the level of the displacement, the degree of scattering deformation and the level of the monitoring of cross-section 2 are the largest, highest, and most unstable. A new sense of the displacement of three cross-sections in one direction (X direction, Y direction, or H direction) is attained by resetting to the displacement of cross-section 1, cross-section 2, and cross-section 3 in three directions (X direction, Y direction, or H direction). Thus, all the monitoring points of three cross-sections have different degrees of displacement. The rule of their range of displacement, scattering degree of displacement and levels of monitoring is X direction $> Y$ direction $> H$ direction, and in H direction

Fig. 8.17 Microcosmic knowledge on all of landslide-monitoring points

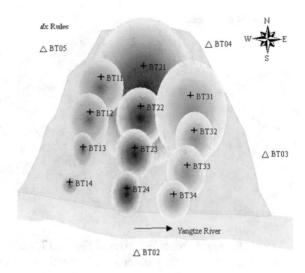

Fig. 8.18 Mid-cosmic knowledge on landslide monitoring cross-sections

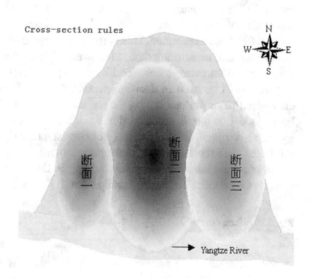

and Y direction: cross-section 2 > cross-section 1 > cross-section 3. Here forward, monitoring point BT21 of the cross-section 2 in the X direction and Y direction changes most greatly about the degree of the displacement, the scattering degree of displacement, and the levels of monitoring.

Figure 8.19 is the most generalized result at a much higher hierarchy than that of Figs. 8.17 and 8.18 in the characteristic space, i.e., the displacement rule of the whole landslide. It is "the whole displacement of Baota landslide is bigger in the south (to Yangtze River), more highly scattered, and extremely instable." Therefore, they can be taken as the macrocosmic knowledge.

Fig. 8.19 Macrocosmic
knowledge

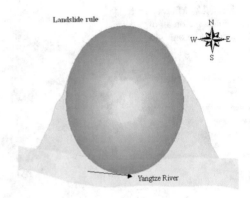

To obtain a comprehensive view of Figs. 8.17, 8.18, 8.19 and 8.20, a conclusion
was drawn to help in the subliming of the spatial knowledge granularity along with
the acquaintanceship level, that is, "the Baota landslide moves towards the south
by west (Yangzi River), along with a small quantity of sinking, and also the displacement of the back part is longer than the displacement of the front part, but
point BT21 is an exception." This is the most generalized conclusion thus far of the
Baota landslide monitoring data. Moreover, it is also a portion of the most concen-
trated spatial knowledge described by conceptual language close to human thinking and could be used in decision-making directly. Baota landslide knowledge is
discovered from the monitoring data at different levels. It explains further that the
Baota landslide moves southward in the horizon and sinks down in perpendicular-
ity. The horizontal displacement is not consistent and fluctuates irregularly, which
illustrates that the horizontal deformation of the most monitoring points of the
Baota landslide are similar, mainly moving in the direction of the Yangzi River. It
is a press landslide. The space around point BT21 is the high incidence area of the

Fig. 8.20 Exception of
landslide

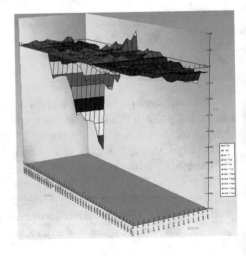

small-scale landslide disaster. Actually, the terrain of the Baota landslide is perpendicular and the obliquity is top-precipitous like chairs. These landslide characteristics are identical to the above knowledge, which explains that the internal forces, which include landslide material character, geological structure, and gradient are the main causes of the formation of landslide disasters. Contemporarily, the information from the patrolling of landslide areas, the natural realism in the area, and the spatial knowledge discovered from the SDM views are very similar.

8.5.4 Rules Plus Exceptions

All the above landslide-monitoring points may further create their data field and isopotential lines spontaneously in a characteristic space. Intuitively, these points are grouped into clusters to represent different kinds of spatial objects naturally. Figure 8.21 visualizes the clustering graph from the landslide monitoring points' potential on dx in the characteristic space.

In Fig. 8.21, all the points' potentials form the potential field and the isopotential lines spontaneously. When the hierarchy increases from Level 1 to Level 5, i.e., from the fine granularity world to the coarse granularity world, these landslide monitoring points are intuitively grouped naturally into different clusters at different hierarchies of various levels as follows.

- No clusters at the hierarchy of Level 1. The displacements of landslide monitoring points are separate at the lowest hierarchy.
- Four clusters at the hierarchy of Level 2: cluster BT14, cluster A (BT13, BT23, BT24, BT32, BT34), cluster B (BT11, BT12, BT22, BT31, BT33), and cluster BT21. At the lower hierarchy, the displacements of landslide -monitoring points

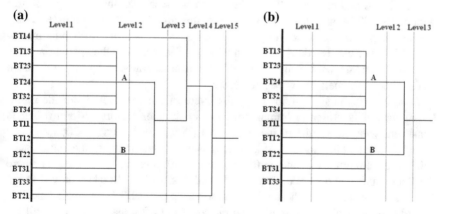

Fig. 8.21 Clustering graph from landslide monitoring points' potential. **a** All points' cluster graph, **b** Points' cluster graph without exceptions

(BT13, BT23, BT24, BT32, BT34) have the same trend of "the displacements are small" and the same is true (BT11, BT12, BT22, BT31, BT33) of "the displacements are big"; while BT14 and BT21 show different trends from both of them and each other i.e., the exceptions, "the displacement of BT14 is smaller" and "the displacement of BT21 is extremely big."

- Three clusters at the hierarchy of Level 3: cluster BT14, cluster (A, B) and cluster BT21. When the hierarchy increases, the displacements of the landslide-monitoring points (BT13, BT23, BT24, BT32, BT34) and (BT11, BT12, BT22, BT31, BT33) have the same trend of "the displacements are small," however, BT14 and BT21 are still unable to be grouped into this trend.
- Two clusters at the hierarchy of Level 4: cluster (BT14, (A, B)) and cluster BT21. When the hierarchy increases, the displacements of landslide -monitoring point BT14 can be grouped into the same trend of (BT13, BT23, BT24, BT32, BT34) and (BT11, BT12, BT22, BT31, BT33), i.e., "the displacements are small," however, BT21 is still an outlier.
- One cluster at the hierarchy of Level 5L cluster ((BT14, (A, B)), BT21). The displacements of landslide monitoring points are unified at the highest hierarchy, i.e., the landslide is moving.

These clusters show different "rules plus exceptions" at different changes from the fine granularity world to the coarse granularity world. The clustering between attributes at different cognitive levels makes many combinations, showing the discovered knowledge with different information granularities. When the exceptions BT14 and BT21 are granted to eliminate, the rules and the clustering process will be more obvious (Fig. 8.21b). Simultaneously, these clusters represent different kinds of landslide monitoring points recorded in the database; and they can naturally form the cluster graphs shown in Fig. 8.21. As can be seen in these two figures, the displacements of landslide- monitoring points (BT13, BT23, BT24, BT32, BT34) and (BT13, BT23, BT24, BT32, BT34) first compose two new clusters, cluster A and cluster B. Then, the two new clusters compose a larger cluster with cluster BT14, and they finally compose the largest cluster with cluster BT21, during the process of which the mechanism of SDM is still "rules plus exceptions." In other words, SDM is particular views for searching the spatial database on Baota landslide displacements by different distances only, and a longer distance leads to pie more meta-knowledge discovery.

In summary, the above knowledge discovered from the Baota landslide monitoring dataset is closer to human thinking in decision-making. Moreover, it is consistent with the results in the references (Wang 1999; Zeng 2000; Jiang and Zhang 2002); and the discovered knowledge very closely matches the actual facts. When the Committee of Yangtze River (Zeng 2000) investigated the region where the Baota landslide occurred, they determined that the landslide had moved to the Yangtze River; and near the landslide monitoring point BT21, a small landslide hazard had taken place. Currently, two pieces of big rift remain, with the wall rift of farmer G. Q. Zhang's house at nearly 15 mm. Therefore, SDM in landslide disaster is not only practical and realistic but essential.

References

Agrawal R, Srikant R (1994) Fast algorithms for mining association rules. In: Proceedings of international conference on very large databases (VLDB), Santiago, Chile, pp 487–499

Clementini E, Felice PD, Koperski K (2000) Mining multiple-level spatial association rules for objects with a broad boundary. Data Knowl Eng 34:251–270

Di KC (2001) Spatial data mining and knowledge discovering. WuHan University Press, WuHan

Dempster A, Laird N, Rubin D (1977) Maximum likelihood from incomplete data via the EM algorithm. J R Stat Soc B 39(1):1–38

Ester M et al (1996) A density-based algorithm for discovering clusters in large spatial databases with noise. In: Proceedings of the 2nd international conference on knowledge discovery and data mining. AAAAI Press, Portland, pp 226–231

Ester M et al (2000) Spatial data mining: databases primitives, algorithms and efficient DBMS support. Data Min Knowl Discovery 4:193–216

Fränti P, Virmajoki O (2006) Iterative shrinking method for clustering problems. Pattern Recogn 39(5):761–765

George K, Han EH, Kumar V (1999) CHAMELEON: a hierarchical clustering algorithm using dynamic modeling. IEEE Comput 27(3):329–341

Grabmeier J, Rudolph A (2002) Techniques of clustering algorithms in data mining. Data Min Knowl Discovery 6:303–360

Guha S, Rastogi R, Shim K (1998) CURE: an efficient clustering algorithm for large databases. In: Proceedings of the ACM SIGMOD international conference on management of data. ACM Press, Seattle, pp 73–84

Han JW, Kamber M, Pei J (2012) Data mining: concepts and techniques, 3rd edn. Morgan Kaufmann Publishers Inc., Burlington

Heller KA, Ghahramani Z (2005) Bayesian hierarchical clustering. In: Proceedings of the 22nd international conference on machine learning, Bonn, Germany

Hinneburg A, Keim D (2003) A general approach to clustering in large databases with noise. Knowl Inf Syst 5:387–415

Horng SJ et al (2011) A novel intrusion detection system based on hierarchical clustering and support vector machines. Expert Syst Appl 38(1):306–313

Jiang Z, Zhang ZL (2002) Model recognition of landslide deformation. Geomatics Inf Sci Wuhan Univ 27(2):127–132

Kerber R (1992) ChiMerge: discretization of numeric attributes. In: Proceedings of AAAI-92, the 9th international conference on artificial intelligence. AAA Press/The MIT Press, San Jose, pp 123–128

Knorr EM, Ng RT (1996) Finding aggregate proximity relationships and commonalities in spatial data mining. IEEE Trans Knowl Data Eng 8(6):884–897

Li DR, Wang SL, Li DY (2006) Theory and application of spatial data mining, 1st edn. Science Press, Beijing

Li J, Wang K, Xu L (2008) Chameleon based on clustering feature tree and its application in customer segmentation. Ann Oper Res 168(1):225–245

Li DR, Wang SL, Li DY (2013) Theory and application of spatial data mining, 2nd edn. Science Press, Beijing

Liu H, Rudy S (1997) Feature selection via discretization. IEEE Trans Knowl Discovery Data Eng 9(4):642–645

Lu H, Setiono R, Liu H (1996) Effective data mining using neural networks. IEEE Trans Knowl Data Eng 8(6):957–961

MacQueen J (1967) Some methods for classification and analysis of multivariate observations. In: Proceedings of the fifth Berkeley symposium on mathematical statistics and probability, vol 1, Berkeley, CA, pp 281–297

Malik HH et al (2010) Hierarchical document clustering using local patterns. Data Min Knowl Discovery 21:153–185

Murray AT, Shyy TK (2000) Integrating attribute and space characteristics in choropleth display and spatial data mining. Int J Geogr Inf Sci 14(7):649–667

Ng R, Han J (2002) CLARANS: a method for clustering objects for spatial data mining. IEEE Trans Knowl Data Eng 14(5):1003–1016

Parsons L, Haque E, Liu H (2004) Subspace clustering for high dimensional data: a review. SIGKDD Explor 6(1):90–105

Pawlak Z (1991) Rough sets: theoretical aspects of reasoning about data. Kluwer Academic Publishers, London

Pitt L, Reinke RE (1988) Criteria for polynomial time (conceptual) clustering. Mach Learn 2(4):371–396

Sembiring RW, Zain JM, Embong A (2010) A comparative agglomerative hierarchical clustering method to cluster implemented course. J Comput 2(12):1–6

Sheikholeslami G, Chatterjee S, Zhang A (1998) Wavecluster: a multi-resolution clustering approach for very large spatial databases. In: Proceedings of the 24th very large databases conference (VLDB 98), New York, NY

Silla CN Jr, Freitas AA (2011) A survey of hierarchical classification across different application domains. Data Min Knowl Discovery 22:31–72

Tung A et al (2001) Spatial clustering in the presence of obstacles. IEEE Trans Knowl Data Eng 359–369

Wang SQ (1999) Landslide monitor and forecast on the three gorges of Yangtze River. Earthquake Press, Beijing

Wang SL (2002) Data field and cloud model based spatial data mining and knowledge discovery. PhD thesis. Wuhan University, Wuhan

Wang SL et al (2004) Rough spatial interpretation. Lecture notes in artificial intelligence, 3066, pp 435–444

Wang SL, Chen YS (2014) HASTA: a hierarchical-grid clustering algorithm with data field. Int J Data Warehouse Min 10(2):39–54

Wang ZY, Klir GJ (1992) Fuzzy measure theory. Plenum Press, New York

Wang XZ, Wang SL (1997) Fuzzy comprehensive method and its application in land grading. Geomatics Inf Sci Wuhan Univ 22(1):42–46

Wang J, Yang J, Muntz R (2000) An approach to active spatial data mining based on statistical information. IEEE Trans Knowl Data Eng 12(5):715–728

Wang SL, Gan WY, Li DY, Li DR (2011) Data field for hierarchical clustering. Int J Data Warehouse Min 7(4):43–63

Wang SL, Fan J, Fang M, Yuan HN (2014) HGCUDF: hierarchical grid clustering using data field. Chin J Electr 23(1):37–42

Xie ZP (2001) Concept lattice-based knowledge discovery. PhD thesis. Hefei University of Technology, Hefei

Xu R, Wunsch D (2005) Survey of clustering algorithms. IEEE Trans Neural Netw 16(3):645–678

Zaki MJ, Ogihara M (2002) Theoretical foundations of association rules. In: Proceedings of the 3rd SIGMOD workshop on research issues in data mining and knowledge discovery, pp 1–8

Zeng XP (2000) Research on GPS application to landslide monitoring and its data processing, dissertation of master. Wuhan University, Wuhan

Zhang T, Ramakrishnan R, Livny M (1996) BIRCH: an efficient data clustering method for very large databases. In: Proceedings of the 1996 ACM SIGMOD international conference on management of data, Montreal, Canada, pp 103–114

Chapter 9
Remote Sensing Image Mining

Remote sensing (RS) images are one of the main spatial data sources. The variability of image data types and their complex relationships are the main difficulties that face SDM. This chapter covers a combination of inductive learning and Bayesian classification to classify RS images, the use of rough sets to describe and classify images and extract thematic information; image retrieval based on spatial statistics; and image segmentation, facial expression analysis, and recognition utilizing cloud model and data field mapping. Potential applications the authors have explored are provided in this chapter as well, such as using the brightness of night-time light imagery data as a proxy for evaluating freight traffic in China, assessing the severity of the Syrian Crisis, and indicating the dynamics of different countries in the global sustainability.

9.1 RS Image Preprocessing

RS approximates the real world by using spectral images. The preprocessing of RS images may filter the uncertainties from the following five aspects (Wu 2004):

(1) *Same object with different spectra*: The same type of objects where individuals have different material compositions and structures under different biological, topological, climatic, and temporal characteristics and other environments may reflect the differences in the spectral characteristics in another way.

(2) *Same spectrum from different objects*: The effects of different types of features in the same spectrum with overlapping phenomena and environmental backgrounds that cause similar spectral features but reflect the different actual object type or geographical phenomena.

© Springer-Verlag Berlin Heidelberg 2015
D. Li et al., *Spatial Data Mining*, DOI 10.1007/978-3-662-48538-5_9

(3) *Mixed pixels*: Due to spatial resolution differences, a single pixel in the image may actually be affected synthetically by multiple features.
(4) *Temporal changes*: With a change of time, RS image data of the same geographical position may incur the complexity and uncertainty induced by the information changes due to differences in climate, conditions, human activities, etc.
(5) The mutual relationships of spatial distribution for unit space in RS images are complicated, rather than a simple mathematical relationship.

9.1.1 Rough Set-Based Image Filter

RS images can be regarded as an information table of knowledge expression system $U = (I, R, V, f)$, where U is a collection of objects; I is a collection of object images, such as pixel and patches; R is the equivalence relationship formed by the attributes of an object, such as the texture isomorphic relation comprised of texture features; V is the range of attribute $r \in R$; f is the information function to specify the attribute value of each object $x \in U$; and I and R constitute the approximation space of an image (Wu 2004).

To filter the noise as much as possible, which is in the upper approximation classification, the noisy pixels are selected via the upper approximation, and the statistical parameter $f(i)$ of pixel p_i is defined to be the condition attribute that constructs noise equivalence relationship R_n:

$$R_n(p_i) = \{p_i | f(j) - f(i)| > Q\}$$

where, j is the neighborhood pixel of pixel i. The statistical variables, equivalence relation, and treatment principle of a noise pixel should be determined considering the reasonable need in practical application.

Based on a rough set, the mean filter to preserve the edge leaves the edge unchanged. As for the non-edge, template pixels are selected with minimal non-statistical variance, using the mean value of the template instead of the pixel gray value. In Fig. 9.1, the original image (ERS-2 SAR) is filtered by using the mean shift filter and the mean shift filter with rough set. The result of the mean shift filter with rough set (c) is much better than that of the mean shift filter (b).

9.1.2 Rough Set-Based Image Enhancement

Image enhancement integrates visual characteristics and rough sets (Wu 2004). To suppress noise simultaneously when enhancing images, the noise of a pixel is

(a)   **(b)**   **(c)**

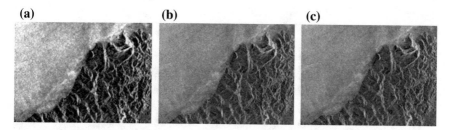

Fig. 9.1 Mean filtering results with rough sets. **a** Original image. **b** Mean shift filter. **c** Mean shift filter with rough set

defined as $C_2 = \{0, 1\}$, of which 0 is a noise pixel and 1 is a non-noise pixel. The equivalence relationship of noise is defined as

$$R_n(S) = Y_i Y_j \{S_{ij} \big| m(S_{ij}) - m(S_{i\pm 1, j\pm 1}) \big| > Q\}$$

where, $R_n(S)$ is the set of all noisy pixels. $S_{i\pm 1, j\pm 1}$ is the adjacent subblocks of subblocks S_{ij}, and S_{ij} and $S_{i\pm 1, j\pm 1}$ are generalized into a macro block. Q is a threshold value—that is, $R_n(S)$ is the absolute value of the difference between S_{ij} and the mean gray value of its adjacent $S_{i\pm 1, j\pm 1}$ if it is rounded to be larger than Q.

All of the above partitioned sub-images are merged to obtain $A_1 = R_t(x) - R_n(s)$ and $A_2 = R_t^-(x) - R_n(s)$. After the noises are eliminated, A_1, A_2 represent the set of all pixels with larger gradients and the set of all pixels with smaller gradients, respectively—both of which need to be enhanced with their characteristics. To supplement the sub-image, the noise pixels are filled with the mean gray value of the macro blocks. In sub-image A_1, the pixels with smaller gradients are filled by using $L/2$ (L is the gray value; for example, $L = 255$). In sub-image A_2, the pixels with larger gradients are filled by using $L/2$.

When enhancing images, an exponential transformation is performed for A_1 to stretch the gray values to increase the overshoot on both sides of the image edge. The noise is augmented, which is not sensed by visual characteristics. A histogram equalization transform is implemented for A_2 to decrease the overshoot on both sides of the edge of the noisy pixels. Noisy pixels with smaller frequencies are merged, thereby weakening the noise. Figure 9.2 is an enhanced image produced by combining the visual characteristics and rough sets. Figure 9.2a–c are, respectively, the original image (TM), the conventional enhancement result of the positive histogram, and the enhancement result of combining the visual characteristics and rough set. Image results (c) are obviously better than that of (b).

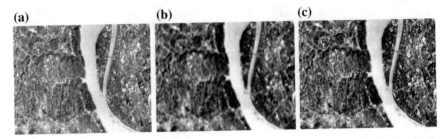

Fig. 9.2 Image enhancement using visual characteristics and rough sets. **a** Original image.
b Histogram equalization. **c** Visual characteristics and rough set

9.2 RS Image Classification

RS image classification is based on the spectral features of spatial objects—that is,
different types of objects have different material compositions and structures, and
they radiate different electromagnetic spectral features. Identical types of objects
have similar electromagnetic spectral features. SDM can support RS image classification for decision-making.

9.2.1 Inductive Learning-Based Image Classification

Bayesian classification (the maximum likelihood method) meets the statistical normal assumption in the spectral data and can derive a minimum classification error
theoretically. It can distinguish water, residential areas, green areas, and other
large area classes from most multi-spectral RS images. However, if further subdivision is needed (e.g., divide water area into rivers, lakes, reservoirs, and ponds or
divide green area into vegetable fields, orchards, forest and grassland), the problems may appear as different objects with the same spectrum and different spectra
from an identical object (Christopher et al. 2000). To solve the problem, inductive
learning is combined with Bayesian classification for using morphological characteristics, distribution rules, and weak spectral differences in the granularity of
the object and the pixel. Figure 9.3 outlines the combination process of inductive
learning and Bayesian classification (Di 2001; Li et al. 2006, 2013a, b, c).

In Fig. 9.3, the inductive learning for spatial data is implemented at two granularities of the spatial object and the pixel, with the class probability value as a
learning attribute from the Bayesian preliminary classification. A GIS database
is essential for providing the training area for Bayesian classification, the polygon and pixel sample data for inductive learning, the control points for RS image
rectification, and the test area for evaluating the accuracy of the image classification results. To maintain representativeness, the training and test areas are selected
interactively. The desired attributes and data format are the same as for learning, but without the class attribute. The results from SDM are a set of rules and a

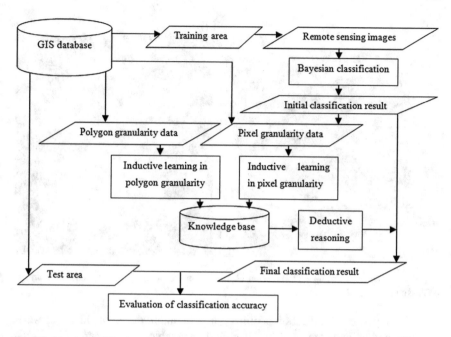

Fig. 9.3 Image classification with inductive learning and Bayesian classification

default class with the confidence between 0 and 1. The strategy of deductive reasoning is as follows:

(1) If only one rule is activated, the rule's output is the resultant class.
(2) If the activation of multiple rules occurs simultaneously, then the class with higher confidence is the resultant class.
(3) If the activation of multiple rules occurs simultaneously and the confidences of the classes are identical, then the output of the rule covered by as many learning samples as possible is the resultant class.
(4) If no rule is activated, the default class is the resultant class.

The experimental image is a portion of a SPOT multispectral image with three bands of a Beijing area acquired in 1996, as shown in Fig. 9.4a. After it is stretched and corrected, the size of $2,412 \times 2,399$ pixels becomes $2,834 \times 2,824$ pixels. The GIS database is the 1:100,000 land use database of the Beijing area before 1996. The GIS software is *ArcView*, the RS image processing software is *ENVI*, and the inductive learning software is See 5 1.10 based on the C5.0 algorithm. Image classification displays the overlay with the GIS layers. The selected GIS layers are the land use layer and the contour layer. Because the contour lines and elevation points are too sparse to be interpolated into a digital elevation model (DEM), the contour layer is processed into an elevated polygon layer—specifically, elevation stripes for less than 50 m (<50 m), 50–100, 100–200, 200–500 m, and greater than 500 m (>500 m).

(a)   **(b)**

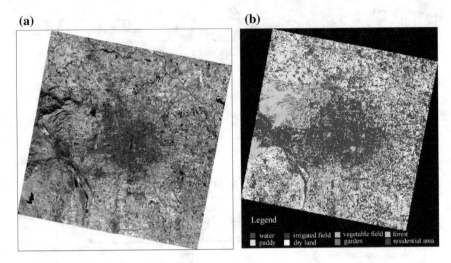

Fig. 9.4 Inductive learning and Bayesian classification (resampled). **a** Original SPOT image. **b** Classification result

Only Bayesian classification is utilized first to obtain the eight classes of water, irrigated paddy, irrigated cropland, dry land, vegetable plot, orchard, forest land, and residential land shown in the confusion matrix of classification (Table 9.1). As can be seen, the classification accuracy is relatively high. The vegetable area is shown in bright green to distinguish it from other green areas. The dry land, orchard, and woodland (shown in dark green) have less spectral difference, which leads to serious reciprocal errors and low classification accuracy. Part of the shadow area in the woodland area is incorrectly classified as water.

Based on the preliminary results of Bayesian classification, inductive learning can be used for two aspects of improving Bayesian classification: (1) subdividing water and distinguishing shadows by using polygon granular learning, and (2) improving the classification accuracy of dry land, orchard, and woodland by using pixel granular learning. For the learning granularity of a polygon to subdivide water, the selected attributes are area, geographical location, compactness, and height. Then, the stripes of 200–500 m and >500 m are merged to >200 m; their class attribute values are 71 for river, 72 for lake, 73 for reservoir, 74 for pond, and 99 for shadow. After 604 polygons are learned with 98.8 % accuracy, 10 production rules are extracted to the spatial distribution and geometric features (Table 9.2).

In Table 9.2, Rule 1 shows that the compactness attribute plays a key role in identifying a river, Rule 2 applies location and compactness for lake recognition, and Rules 9 and 10 manipulate elevation for shadow recognition. When these rules are used to identify shadows and subdivide water areas, the accuracy of shadow recognition is improved to 68 % for woodlands and the error in water recognition is reduced. For the learning granularity of a pixel to distinguish dry land, orchards, and woodlands, the selected attributes include coordinate, elevation stripe, dry land probability,

Table 9.1 The confusion matrix of Bayesian classification

Classified data	Referenced data								
	C1	C2	C3	C4	C5	C6	C7	C8	Sum
C1 water	3.900	0.003	0.020	0.013	0.002	0.021	2.303	0.535	6.797
C2 paddy	0.004	8.496	0.087	0.151	0.141	0.140	0.103	0.712	9.835
C3 irrigated area	0.003	0.016	10.423	0.026	0.012	0.076	0.013	0.623	11.192
C4 dry land	0.063	0.48	0.172	1.709	0.361	2.226	2.292	1.080	8.384
C5 vegetable area	0.001	0.087	0.002	0.114	3.974	0.634	0.435	0.219	5.465
C6 garden	0.010	0.009	0.002	0.325	0.263	4.422	4.571	0.065	9.666
C7 forest	0.214	0.006	0.000	0.271	0.045	1.354	15.671	0.642	18.202
C8 residential area	0.132	0.039	0.127	0.080	0.049	0.168	0.839	29.024	30.459
Sum	4.328	9.135	10.834	2.689	4.846	9.041	26.227	32.901	100
Accuracy (%)	90.113	93.010	96.204	63.580	81.994	48.913	59.754	88.217	

Overall accuracy = 77.6199 %
Kappa coefficient = 0.7474

Table 9.2 Water area segmentation rule resulting from inductive learning

Rule no.	Rule content
Rule 1	(Cover 19) compactness > 7.190882, height = lt50 $\Rightarrow$ class 71 (0.952)
Rule 2	(Cover 5) 453,423.5 < Xcoord $\leq$ 455,898.7442896 < Ycoord $\leq$ 453,423.5, 2.409397 < compactness $\leq$ 7.190882 $\Rightarrow$ class 72 (0.857)
Rule 3	(Cover 33) Xcoord $\leq$ 455,898.74414676 < Ycoord $\leq$ 44,289,582.409397 < compactness $\leq$ 7.190882, height = lt50 $\Rightarrow$ class 72 (0.771)
Rule 4	(Cover 4) area > 500,000, height = 50_100 $\Rightarrow$ class 73 (0.667)
Rule 5	(Cover 144) Ycoord < 4,414,676, compactness $\leq$ 7.190882, height = lt50 $\Rightarrow$ class 74 (0.993)
Rule 6	(Cover 213) Ycoord < 4,428,958, compactness $\leq$ 7.190882, height = lt50, $\Rightarrow$ class 74 (0.986)
Rule 7	(Cover 281) Xcoord > 451,894.7, compactness $\leq$ 7.190882 $\Rightarrow$ class 74 (0.975)
Rule 8	(Cover 38) area > 500,000, height = 50_100 $\Rightarrow$ class 74 (0.950)
Rule 9	(Cover 85) height = gt200 $\Rightarrow$ class 99 (0.989)
Rule 10	(Cover 7) height = gt200 $\Rightarrow$ class 99 (0.778)
Default class: 74	
Evaluation (604 cases): errors 7(1.2 %)	

orchard probability, and woodland probability. The output has three classes: (1) 1 % (2909 pieces) of the massive samples is randomly selected for learning, (2) 63 rules are discovered with the learning accuracy of 97.9 %, and (3) an additional 1 % of the samples are randomly selected for detection, the accuracy of which is 94.4 %.

Figure 9.4b shows the classification results from inductive learning and Bayesian classification (displayed by samplings), and Table 9.3 is the confusion matrix and accuracy indicators for the same test area used for Bayesian classification. Compared with Bayesian classification, the classification accuracy of dry land, orchards, and woodlands are increased to 69.8, 78.5 and 91.8 %; their individual increased rates are 6.2, 29.6 and 32 %, respectively. The overall classification accuracy is improved by 11.2 % (88.8751–77.6199 %) and the kappa coefficient is increased by 0.1245 (0.8719–0.7474). The results also demonstrate that inductive learning can preferably solve the problem of the same spectrum from different objects and the same objects with different spectra.

9.2.2 Rough Set-Based Image Classification

The classification with rough sets is defined to search for an appropriate classification boundary in the boundary region according to the algorithm and parameters, which is the deduction process of the set. Figure 9.5 is an example of RS classification of a volcano with rough sets. The white region is the upper approximate set of the volcano, showing all the possible areas belonging to the volcano; the

Table 9.3   The confusion matrix of inductive learning and Bayesian classification

Classified data	Referenced class								
	C1	C2	C3	C4	C5	C6	C7	C8	Sum
C1 water	3.900	0.003	0.020	0.012	0.002	0.019	0.139	0.535	4.631
C2 paddy	0.004	8.496	0.087	0.151	0.141	0.14	0.103	0.712	9.835
C3 irrigated area	0.003	0.016	10.423	0.026	0.012	0.076	0.013	0.623	11.192
C4 dry land	0.063	0.480	0.172	1.877	0.361	0.205	0.149	1.080	4.386
C5 vegetable area	0.001	0.087	0.002	0.114	3.974	0.634	0.435	0.219	5.465
C6 garden	0.009	0.009	0.002	0.210	0.263	7.102	0.470	0.065	8.131
C7 forest	0.215	0.006	0.000	0.218	0.045	0.696	24.079	0.642	25.899
C8 residential area	0.132	0.039	0.127	0.080	0.049	0.168	0.839	29.024	30.46
Sum	4.328	9.135	10.834	2.689	4.846	9.041	26.227	32.901	100
Accuracy (%)	90.113	93.01	96.204	69.811	81.994	78.561	91.81	88.217	

Overall accuracy = 88.8751 %
Kappa coefficient = 0. 8719

red region is the lower approximation set of the volcano, illustrating all the definite areas as the volcano body. The boundary region between the upper and lower approximate set is the uncertain part of the volcano (Wu 2004).

The classification may improve substantially if a rough set is integrated with other reasonable methods. For example, in a rough neural network integrating rough sets and artificial neural networks (ANNs) (Fig. 9.6), the rough set infers logic reasoning rules from the data on the object's indiscernibility and knowledge simplification, the results of which are used as the input of the knowledge system on ANN.

As discussed in previous chapters, ANN simulates the human visual intuitive thinking mechanism using the idea of nonlinear mapping; it also carries out an implicit function that encodes the input and output correlation knowledge expressed by the neural network structure, which can improve the precision of RS image classification based on a rough set. Figure 9.7 shows the original images (SPOT 5) and classification by the integrating method. In the confusion matrix

Fig. 9.5 The approximate set of volcanic image classification

Fig. 9.6 Rough neural network model for image classification

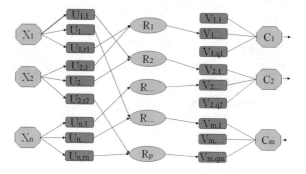

Fig. 9.7 Results of classification with rough neural network

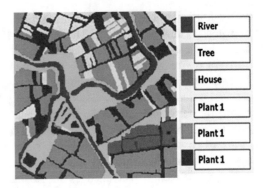

Table 9.4 The confusion matrix of simple rough image classification

Classified data	Referenced data							
	C1	C2	C3	C4	C5	C6	Sum (pixel)	Accuracy (%)
C1 river	62.0	0	19.0	0	0	0	50	11.4
C2 tree	6.0	80.0	5.0	0	5.0	10.0	98	22.3
C3 house	32.0	7.0	63.0	0	0	0	86	19.5
C4 plant1	0	0	0	86.0	1.25	10.0	50	11.4
C5 plant 2	0	11.0	10.0	4.0	82.5	3.3	91	20.7
C6 plant 3	0	2.0	3.0	10.0	11.25	76.7	65	14.8
Sum (pixel)	50	100	100	50	80	60	440	
Accuracy (%)	11.4	22.7	22.7	11.4	18.2	13.6		

Overall accuracy = 74.8 %
Kappa coefficient = 0.821

Table 9.5 The confusion matrix of classification with rough neural networks

Classified data	Referenced data							
	C1	C2	C3	C4	C5	C6	Sum (pixel)	Accuracy (%)
C1 river	88.75	0	0	0	0	0	71	17.75
C2 tree	7.5	92.5	0	0	0	0	43	10.75
C3 house	3.75	7.5	85.0	17.5	0	0	47	11.75
C4 plant1	0	0	15.0	82.5	0	3.6	43	10.75
C5 plant 2	0	0	0	0	92.2	7.3	91	22.75
C6 plant 3	0	0	0	0	7.8	89.1	105	26.50
Sum (pixel)	80	40	40	40	90	110	400	
Accuracy (%)	20.0	10.0	10.0	10.0	22.5	27.5		

Overall accuracy = 91.5 %
Kappa coefficient = 0.895

(Tables 9.4 and 9.5), the overall accuracy of the results of simple rough image classification is 74.8 % and the coefficient kappa is 0.821 (Table 9.4), while the overall accuracy of the results of a multilayer perceptron classification with rough set and the neural network is 91.5 % and the coefficient kappa is 0.895 (Table 9.5). The overall classification accuracy is improved by 16.71 % (91.51–74.8 %) and the kappa coefficient is increased by 0.074 (0.895–0.821). Therefore, the rough neural network takes advantage of the knowledge learning of rough sets and the nonlinear mapping of the neural network for processing takes advantage of the uncertainty of RS images.

9.2.3 Rough Set-Based Thematic Extraction

The thematic extraction of rivers from the RS images is studied for the purpose of monitoring the ecological balance of the environment. Figure 9.8 shows an original image (TM) and its water extraction results. The gray value of the image is used to extract rivers from the images, and a rough topological relationship matrix and a rough membership function are used to deal with the spatial relationships between the class of river and other adjacent classes in the image (Wang et al. 2004).

The rough relationship matrix identifies and propagates the certainties with $Pos(X)$, $Neg(X)$, and uncertainties with $Bnd(X)$. In the matrix, none-empty is 1 and empty is 0.

$$R_{r9}(A, B) = \begin{pmatrix} Pos(A) \cap Pos(B) & Pos(A) \cap Bnd(B) & Pos(A) \cap Neg(B) \\ Bnd(A) \cap Pos(B) & Bnd(A) \cap Bnd(B) & Bnd(A) \cap Neg(B) \\ Neg(A) \cap Pos(B) & Neg(A) \cap Bnd(B) & Neg(A) \cap Neg(B) \end{pmatrix}$$

(a)   **(b)**   **(c)**

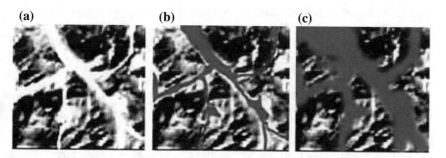

Fig. 9.8 Rough river thematic maps (in continuum) Reprinted from Wang et al. (2004), with kind permission from Springer Science+Business Media. **a** Original image. **b** $L_r(X)$ image. **c** $U_r(X)$ image

The rough membership value is regarded as the probability of $x \in X$ given that x belongs to an equivalence class. $\mu_X(x) + \mu \sim_X(x) = 1$

$$\mu_X(x) = \frac{R_{\text{card}}(X \cap [x]_R)}{R_{\text{card}}([x]_R)} = \begin{cases} 1 & x \in \text{Pos}(x) \\ (0, 1) & x \in \text{Bnd}(x) \\ 0 & x \in \text{Neg}(x) \\ 1 - \mu_{\sim X}(x) & x \in \sim X \end{cases}$$

In Fig. 9.8, the roughness of the river rough classification is $R_d(X) = R_{\text{card}}(B_r(X))$ $/R_{\text{card}}(X) \times 100 \% = 10.37 \%$, which indicates relatively small uncertainty. The lower approximation set $L_r(X)$ is the definitive region of a river, which is the minimum possible watershed; it can be interpreted as the minimum river course during the dry season. This may broaden the biological quantity and activity scope on both sides of the river biosphere, while increasing the difficulty of water consumption. However, the upper approximation set $U_r(x)$ is the indefinite region of a river, which is the maximum possible watershed; this can be interpreted as the maximum inundation area of the river in flood season, producing the largest damage to the ecological environment. The results are close to the actual hydrological characteristics of the river, of which the upper approximation set $U_r(X)$ and the lower approximation set $L_r(X)$ are approximate to the river in winter and summer, respectively. The extraction of the river resources from the RS images enables results with both the maximum possible watershed (the upper approximation set) and the minimum possible watershed (the lower approximation set) at the same time as well as possible errors resulting from multiple descriptive approaches, thereby providing more reliable and feasible results for decision-making.

9.3 RS Image Retrieval

Content-based image retrieval locates the target images by using their visual features, such as color, layout, texture, shape, structure, etc. In a RS image retrieval system, texture, shape, and structure are more favorable features versus levels of gray or false colors (Ma 2002; Zhou 2003).

9.3.1 Features for Image Retrieval

The texture feature is widely applied in image retrieval from the main surface features of the forest, grassland, farmland, and city buildings of RS images. The visual textures—roughness, contrast, orientation, linearity, regularity and coarseness—are used to calculate image similarity. Different scale images lead to changeable textures. At present, co-occurrence matrix, wavelet transform, and Markov random fields are effective in texture analysis.

A shape feature is represented based on the boundary and region, which make use of the shape of the outer boundary and the shape of the region, respectively, in the invariance of movement, translation, rotation, and change. Global shape features have a variety of simple factors, such as roundness, area, eccentricity, principal axis direction, Fourier descriptor, and moment invariants. Local shape features include line segment, arc, corner, and high curvature point. Equipped with professional knowledge and application background in the RS image retrieval system, the shape feature library can be created and the storage space also can be saved by ignoring the shape feature value. The shape matching accuracy and the retrieval rate can be improved if the shape of the ground target combines the metadata of the RS image and the dimension knowledge of typical objects, especially querying ground manmade objects such as vehicles and airplanes for ground reconnaissance and surveillance missions.

The structure feature includes the layout, adjacency, inclusion, etc. for the distribution of objects and regions and their relationships in images. The spatial topological structure among many objects is often implicitly stored in a spatial data structure, the identical index of which supports the relative position and absolute positions simultaneously. The image query can be based on either the position or the feature. Currently, the structure feature is rarely used for image query, and the layout-based query is a simple structure match with the range information between objects.

9.3.2   Semivariogram-Based Parameter to Describe Image Similarity

Among various content-based image retrieval techniques, image comparison is often employed to compare a standard image and candidate images. The measurements for image similarity are usually defined as Euclidean distances, which cannot completely characterize the structural differences between images. Based on a semivariogram, a new parameter is introduced to describe image similarity. It can characterize image structure (texture) better than traditional image similarity measurements. The semivariogram depicts the spatial autocorrelation of measured sample points in spatial statistics. In the spatial domain, the conditions of the second-order stationary random process cannot be satisfied. Therefore, in order to satisfy the random process hypothesis in the spatial domain, an intrinsic random function, which is a random function whose increments are second-order stationary, is proposed in spatial statistics. This new function is characterized by the expectation and variant (Cressie 1991):

$$E(X(t) - X(t + h)) = 0$$
$$\mathrm{var}(X(t) - X(t + h))$$
$$= E(X(t) - X(t + h))^2 - \{E(X(t) - X(t + h))\}^2$$
$$= E(X(t) - X(t + h))^2$$
$$= 2\gamma(h)$$

Both are within the whole definition domain. $\gamma(h)$ is the semivariogram. The semi-variogram is represented in graphics by a variograph—that is, a graph plotted by the semivariogram versus h. Sill, nugget, and range are the three parameters that describe the semivariogram completely (Fig. 9.9a). Each has its own characteristics when the semivariogram is used in the context of image processing. Sill is its limit; nugget reveals noise in an image via an image template window; and range represents the correlation of the RS images in a direction. Inside the range, the image contains spatial correlation that decreases when the range increases; outside the range, its spatial correlation disappears (Franklin et al. 1996). If the size of an image template window is a two-dimensional array of pixels, it can be decided via the range of the horizontal range and the vertical direction by calculating the semi-variogram in the horizontal and vertical directions.

Images with different structural features have different semivariograms, with different variographs (Fig. 9.9b). The semivariogram of an image is

$$\gamma(h) = \frac{1}{2N(h)} \sum_{i=1}^{N} \left[G(x,y) - G(x+h, y+h) \right]^2$$

Here, $G(x,y)$ is the gray value of the image and $N(h)$ is the number of pixel pairs separated by h pixels. Figure 9.9c is an actual variograph of an image.

To define a new parameter to describe image similarity based on a semivariogram,

$$\rho = \frac{\tan \theta}{\tan \beta}$$

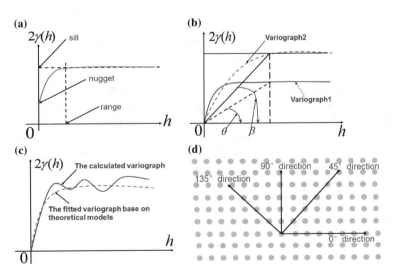

Fig. 9.9 Parameters and variograph of semivariogram. **a** Three parameters of semivariogram. **b** Variograph to represent semivariogram. **c** A real variograph of an image. **d** 4 semivariograms in 4 directions with 4 ρ

where θ corresponding to image1 is the standard image and β corresponding to image2 is the candidate image, which are shown in Fig. 9.9b.

If $|\rho - 1| = 0$, then the two images are completely the same. The larger the parameter $|\rho - 1|$, the larger the difference is between the standard image and the candidate image. For a given set of pixels, there are at least four semivariograms corresponding to four directions: 0°, 45°, 90°, and 135°, along with four parameters ρ. These ρ values reveal structural differences along these four directions.

To further define a ratio of selecting all of the useful images quantifies—that is, pickOratio:

$$\text{pickOratio} = \frac{\text{the number of all useful images}}{\text{the number of all selected images}} \times 100\%$$

Obviously, pickOratio is a quotient between the available useful images and totally selected images. The applications of semivariograms with other spatial statistical models to RS image processing have been extensively explored, while their uses in close-range photographic image processing are more recent applications. Image retrieval training for detecting wheel deformation by using the semivariogram with proposed parameter ρ is presented below as an example.

9.3.3 Image Retrieval for Detecting Train Deformation

When trains stop at a railway station, train workers perform an inspection to determine the health of the trains, focusing on the brakes on each wheel and the connections between two railroad carriages. If a problem is uncovered, it must be reported before the train leaves the station, which is timely troubleshooting; railway stations are eager to automatize this difficult work. The following approach was proposed by the authors. Images first were taken with CCD cameras when the trains entered the station and then the brake of a train wheel was identified and its thickness measured. Nearly 1800 images were calculated because normally it takes 180 s to take images for a train at a station, and about 10 frames of images are taken per second. However, most of the images may be useless because the train wheels may not appear in some of the images and the image features may be blurred by train motion. There are two aspects of the effectiveness of the algorithm: (1) whether or not all the useful images are in the selected images and (2) how the detection ratio is determined.

Traditional methods are difficult. First, feature recognition of a train wheel is time-consuming and was unreliable as a train is in changing motion when it approaches the station. Second, image comparison based on the gray correlation and the histogram intersection proved to be invalid. Third, wavelet transform can recognize the motion of blurred images successfully, but it cannot tell the difference between the standard useful image and images with no train wheels.

To find useful images with train wheels quickly and accurately, semivariogram-based image retrieval was applied by measuring the image similarity parameter ρ.

The useful image includes the features, such as the brake attached to the wheel and the connection of two railroad carriages, while a useless image is blurred by train motion (Fig. 9.10).

The threshold of image similarity plays an important practical role. A larger threshold means useful images may be lost in selection but the computational speed may be faster. A smaller threshold may lead to an increased number of selected images but the computational speed may decrease. This experiment shows that there is no fixed threshold that can separate useful and useless completely in all conditions. In order to search out all the images of train wheels and connections, the similarity threshold can be empirically determined in the context of the actual demands and professional experience:

- If $|\rho - 1| < 0.01$, although all the chosen images are useful, many other useful images cannot be chosen.
- If $|\rho - 1| < 0.1$, all the chosen images are useful, but some useful images are missing.
- If $|\rho - 1| < 0.5$, all the useful images are chosen, along with some useless images.

In the experiment, the train wheel images obtained under different imaging conditions were utilized for detection using the above algorithm to find all the useful images. The results show that all the useful images can be chosen, but with several useless images mixed in. For the image similarity parameters in four directions, the ratio of finding all the useful images was consistently around 90 % for both the detection of train wheels and connections. Figure 9.11 shows the experimental samples of the standard image and candidate images, as well as the image similarities. In Fig. 9.11a, the two images are clearly quite different. In Fig. 9.11b, two images are similar. In Fig. 9.11c, two images are similar but the lighting conditions are different. The selection procedure can be automatically finished within one minute, which satisfies the practical requirement of less than 6 min.

This technique was successfully applied in the actual system (Figs. 9.12 and 9.13). Supported by the system, a professional operator can easily make an accurate final judgment of whether or not the train wheels are normal.

(a)   **(b)**   **(c)**

Fig. 9.10 Useful images versus useless images for train health. **a** Useful image with the brake attached to the wheel. **b** Useful image with the connection of two railroad carriages. **c** Useless image blurred by train motion

standard image	candidate image	
		(a) |ρ0°-1|=0.023 |ρ45°-1|=0.047 |ρ90°-1|=0.035 |ρ135°-1|=0.102
		(b) |ρ0°-1|=0.083 |ρ45°-1|=0.027 |ρ90°-1|=0.073 |ρ135°-1|=0.094
		(c) |ρ0°-1|=0.023 |ρ45°-1|=0.047 |ρ90°-1|=0.035 |ρ135°-1|=0.102

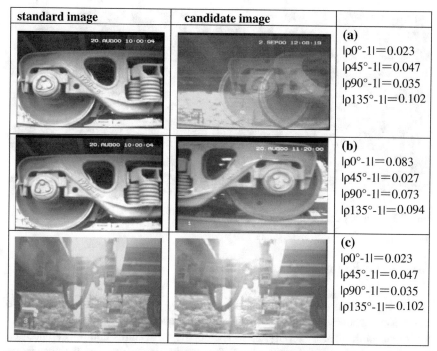

Fig. 9.11   Sampled image retrieval on standard image and candidate image

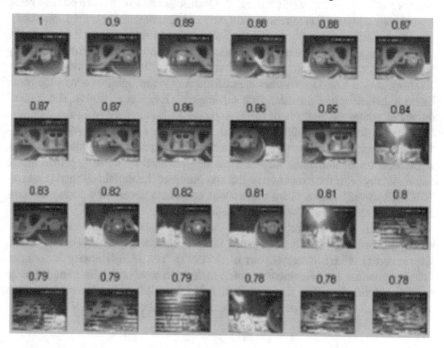

Fig. 9.12   Selected images from the preceding 200 images of a train

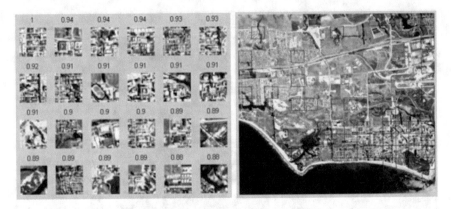

Fig. 9.13   Arbitrary content discovered from RS images

9.4 Facial Expression Image Mining

Facial expressions are identifiable by extracting features, such as eyes, nose, mouth, cheekbones, and jaw, while neglecting everything else, such as bodies, trees, and buildings. Generally, there are geometric features and appearance features for analyzing facial expressions. The geometric features are the shapes and positions of facial components and are the locations of facial feature points (Wang et al. 2009). However, face recognition may concentrate more on the motion of a number of feature points but ignore the information in skin texture changes. At the same, the appearance features include skin motion and texture changes that may be more susceptible to changes in illumination and differences between individuals. Appearance-based approaches are utilized for face recognition across the pose, Eigen light-fields, and Bayesian face sub-regions (Ralph et al. 2004). The state of the art of automatic analysis of facial expressions is summarized in the literature (Pantic and Rothkrantz 2000; Ralph et al. 2004; Zeng et al. 2009). Age is synthesized by rendering the face image aesthetically with natural aging and rejuvenating effects on an individual face and can be estimated by labeling a face image automatically with the exact age or the age group of the individual face (Fu et al. 2010). However, various facial expressions can cause uncertainty during the process of face recognition (Pantic and Rothkrantz 2000).

In the following section, the cloud model and data fields are used in facial expression mining. The test data are taken from the Japanese Female Facial Expression (JAFFE) library (Lyons et al. 1998). The JAFFE library is an open facial expression image database, which includes a total of 10 different Japanese women identified as KA, KL, KM, KR, MK, NA, NM, TM, UY, and YM; each woman has a total of seven different expressions identified as AN (angry), DI (disgusted), FE (frightened), HA (happy), NE (non-emotional), SA (sad), and SU (surprised). There are three to four sample images of each expression, for a total of 213 facial images. The original size is 256 × 256 pixels.

9.4.1 Cloud Model-Based Facial Expression Identification

In the {*Ex*, *En*, *He*} of the cloud model for facial expression images, *Ex* reveals the common features and expressions of the women. Constructing the standard facial expression can reflect the average facial expression state as follows: *En* reveals the degree of deviation of different expressions from the standard facial expression, which can reflect the degree to which face recognition is impacted by individual characteristics and different internal and external factors; *He* reveals the degree of dispersion of the different expressions from the standard facial expression (i.e., the degree to which individual characteristics and environmental factors affect a facial expression; Table 9.6).

9.4.1.1 Identical Facial Expressions from Different Individuals

First, the AN expression was selected in the image library as well as the set of 10 different Japanese women (KA, KL, KM, KR, MK, NA, NM, TM, UY, and YM), whose AN face images become the original images—that is, the input of cloud drops, with the non-certainty reverse cloud generator algorithm. The 10 different individual facial images of {*Ex*, *En*, *He*} when they are at AN status (i.e., the output of the cloud numerical characters) are shown in the first column of Table 9.4 along with the results. The cloud numerical characters for the six remaining expressions (DI, FE, HA, NE, SA, and SU) were obtained; their results are shown in the 2nd, 3rd, 4th, 5th, 6th, and 7th rows, respectively, in Table 9.6.

From Table 9.6, it can be seen that the cloud drops reveal that the input (i.e., the original selected images) reflects the different personalities of the 10 different individuals for one facial expression; the digital features of an image of {*Ex*, *En*, *He*} as output reflect the common features of 10 individuals who show one expression. Although there are 10 different individuals, 10 different human expressions can become part of the basis for the common features for one expression by adding them to the database. *Ex* reveals the basic common features of one expression as the standard facial expression, which can reflect the average expression state; *En* shows the degree of deviation from the standard expression for 10 different individuals for one type of expression; and *He* reveals the dispersion of the degree of deviation for one type of expression from the standard facial expression for 10 different individuals with different characteristics and influences by internal and external factors.

Table 9.6   Human facial expression mining with cloud model

9.4.1.2 Different Facial Expressions from Identical Individuals

The KA woman in the image library and the set of her seven face images (AN, DI, FE, HA, NE, SA, and SU) were selected as the original images (i.e., the input of cloud drops), with the non-certainty reverse cloud generator algorithm, the image of {Ex, En, He} of seven facial images obtained for KA (i.e., the output cloud numerical characters); the results are shown in the first row of Table 9.4. The cloud numerical characters for the remaining nine women (KL, KM, KR, MK, NA, NM, TM, UY, and YM) were obtained by adopting the method for KA, the results of which are shown in the 2nd, 3rd, 4th, 5th, 6th, 7th, 8th, 9th, 10th rows, respectively, of Table 9.6.

From Table 9.6, it can be seen that the cloud drop of the original image as the image input reflects the different personality characteristics of seven types of expressions of one individual; the numerical character image of {Ex, En, He} as the output reflects the common features of the identical person. Although there are seven different expressions, the different expressions are based on the common features of one expression, which can be increased by the features of different individuals. Ex reveals the basic common features of one person, which is set as the standard facial expression, and reflects the undisturbed calm state of one person; En reveals the degree of deviation for different expressions from the standard facial expression for one person and reflects the individual's degree of mood fluctuation influenced by internal and external factors; He reveals the degree of difference that expressions deviate from the standard facial expression for one person and reflects the degree of difference between the individual's mood fluctuations influenced by internal and external factors (i.e., the degree of psychological stability).

9.4.1.3 Different Facial Expressions from Different Individuals

From Table 9.4, it can be seen that the original input image can reflect the different features of different expressions, the output digital feature image of {Ex, En, He}, and reflects the common features of different expressions. Although the input images are different facial expression images, these input images are based on one common feature, which are expanded by the addition of different personality features. Ex reveals the basic common features of a facial expression, which is set as its standard facial expression and reflects the average state of facial expression; En reveals the degree of difference that expressions deviate from the standard facial expression (i.e., the degree of influence by individual characteristics and environmental factors); and He reveals that the dispersion of the degree of deviation of the different expressions from the standard facial expression and reflects the degree of difference that expressions deviate from the standard facial expression (i.e., the effect degree of personal characteristics and environment factors on facial expression).

9.4.2 Data Field-Based Human Facial Expression Recognition

A new facial feature abstraction and recognition method is introduced here to represent the influence at different levels for different faces. During the process of automatic face recognition, face detection and orientation detect a human face from a complex background and segments it before processing. Feature abstraction and recognition is from a unified human face image with geometry normalization and gray normalization.

9.4.2.1 Data Fields in Facial Images

In the data space $\Omega \subseteq R^J$ ($J = P \times Q$) of the universe of discourse for facial images, dataset $D = \{X_1, X_2, ..., X_I\}$ is a set of facial expression images. If a portion of a facial expression image is treated as a grid with P rows and Q columns, X_i ($i = 1, 2, ..., I$) is extended:

$$X_{i_{pq}} = \begin{pmatrix} x_{i_{11}} & x_{i_{12}} & \cdots & x_{i_{1Q}} \\ x_{i_{21}} & x_{i_{22}} & \cdots & x_{i_{2Q}} \\ \cdots & \cdots & \cdots & \cdots \\ x_{i_{P1}} & x_{i_{P2}} & \cdots & x_{i_{PQ}} \end{pmatrix}$$

Put $x_{ipq} = (i = 1, 2, ..., I; p = 1, 2, ..., P; q = 1, 2, ..., Q)$ into $\Omega \subseteq R^J$. $x_{ipq} = (x_{i11}, x_{i11}, ..., x_{i1Q}; x_{i21}, x_{i22}, ..., x_{i2Q}; ...; x_{iP1}, x_{iP2}, ..., x_{iPQ})$. Let g_{ipq} be the value of pixel X_{ipq} with the pth row and the qth column of the ith facial expression image in D. X_{ipq} creates a single data field covering the whole universe of facial expression images X_i. In facial expression image X_i, the potential of an arbitrary point x_i on the pixel X_{ipq} can be computed by

$$\varphi(x_i) = g_{ipq} \times e^{-\frac{\|x_i - x_{ipq}\|^2}{2\sigma^2}}$$

where $\|x_i, x_{ipq}\|$ is the distance between point x_i and pixel object x_{ipq}, g_{ipq} denotes the energy of x_{ipq}, and σ indicates the impact factor. In face recognition, it is suggested that $\|x_i, x_{ipq}\|$ is the Euclidean norm and g_{ipq} is the gray of pixel X_{ipq} when the image is black and white, varying from black at the weakest intensity to white at the strongest. The potential of point x_i is the sum of the potentials from each object X_{ipq}.

$$\Psi(x_i) = \sum_{p=1}^{P} \sum_{q=1}^{Q} \varphi(x_i) = \sum_{p=1}^{P} \sum_{q=1}^{Q} g_{ipq} \times e^{-\frac{\|x_i - x_{ipq}\|^2}{2\sigma^2}}$$

If the points with the same potential value are lined up together, equipotential lines come into being. Furthermore, all of the equipotential lines depict the interesting topological relationships among the objects, which visually indicate the interacted characteristics of the objects. Figure 9.14 shows a facial image that is portioned by a grid with p row ($p = 1, 2, ..., P$) and q ($q = 1, 2, ..., Q$) column (a) and its equipotential lines with the data field in two-dimensional space (b) and three-dimensional space (c).

In $\Omega \subseteq R^J$ on face recognition, the data field of face X_i to recognize on dataset $D = \{X_1, X_2, ..., X_I\}$ receives a more summarized potential.

$$\Psi(X_i) = \sum_{i=1}^{I} \varphi(x_i) = \sum_{i=1}^{I} \sum_{p=1}^{P} \sum_{q=1}^{Q} g_{i_{pq}} \times e^{-\frac{\left\|x_i \cdot x_{i_{pq}}\right\|^2}{2\sigma^2}}$$

9.4.2.2   Facial Expression Recognition in Eigenface Space

Human facial expression recognition with a data field first normalizes the original face image to obtain a standardized 32×32 pixel facial image. Specifically, the image is rotated, cut, and scaled based on the center of the left and right eyes of the original facial image as reference data, along with the elliptical mask to eliminate the effects of hair and background. Then, the gray transformation for the standard face image is carried out to extract the important feature points of each face image using the feature extraction method with the data field, which forms the simplified human face image. Third, the deviation matrix is simplified, and the public eigenface space is constructed, the simplified human face image is projected to the public eigenface space, with the corresponding projection coefficient as the logical features of the human face image. Finally, according to the logical features, all of the face images in the new feature space create a second-order data field, and the clustering recognition of human face images is realized under the interaction between the data and the self-organizing aggregation.

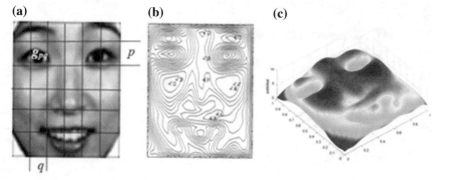

Fig. 9.14 Facial data field of KA's happy face: **a** partitioned facial image **b** 2D field equipotential **c** 3D field equipotential

A total of 213 face images in the JAFFE database were processed uniformly, and the clustering algorithm with the data field then were applied to cluster projection data in the "Eigen face" space (Fig. 9.15).

Figure 9.15a shows the standardized facial images. Figure 9.15b shows the equipotential of the data field on the facial images with the impact factor of 0.05. The high potential area in the equipotential line distribution clearly concentrates on the cheek, forehead, and nose, whose gray is relatively larger. In the distribution of the data field, each local point with the maximum potential value possesses all of the contribution radiated from adjacent pixels, and the maximum potential value and position of these local points can be regarded as the logical feature of the human face image. The simplified human face image was extracted by the algorithm based on the features of facial data field (Fig. 9.15c). The eigenface image corresponding to the six preceding principal eigenvectors is processed by K-L transform from the set of simplified human face images. The simplified face images are projected into the public eigenface space, and the resulting first two principal eigenvectors that constitute a two-dimensional eigenface space whose projected data distribution are shown in Fig. 9.15d.

By all appearances, there is preferable separability in the two-dimensional eigenface space for simplified face images representing different human facial expressions. The final recognition results are shown in Table 9.7, which show that this method provides a favorable recognition rate.

Specifically, 10 different expressions of frontal face gray images chosen from the JAFFE database, consisting of seven images from the same person and the other three images from three strangers, are obtained facial topological structures by natural clustering. It is apparent that the clustering speed of the three strangers I, H, and J was the slowest. The process is shown in Fig. 9.16.

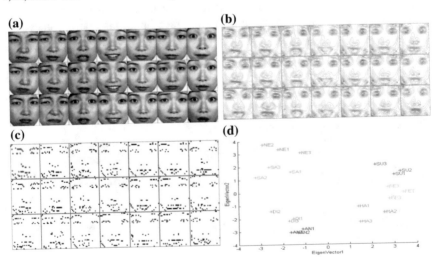

Fig. 9.15 Facial expression recognition in the eigen-face space. **a** Standardized facial images. **b** Equipotential of data fields. **c** Simplified facial images. **d** Recognized faces on the projection

Table 9.7 Clustered recognition results of JAFFE facial expression

Facial expression	Anger	Disgusted	Fearful	Happy	Neutral	Sad	Surprise	Sum
Sample amount	30	29	32	31	30	31	30	213
Recognition rate (%)	63.3	51.7	71.8	80.6	70.0	61.2	93.3	70.6

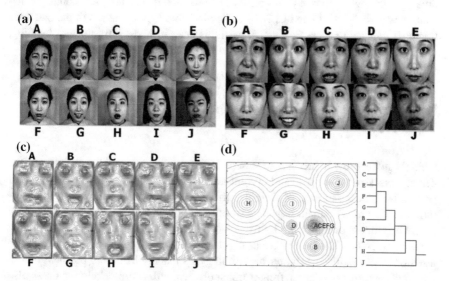

Fig. 9.16 Face recognition with facial data fields. **a** Different facial images and their normalized images. **b** Equipotential of facial images and their hierarchical recognition result

9.5 Brightness of Nighttime Light Images as a Proxy

Satellite-observed RS images have been used as a supplement to ground observations because RS creates an objective data source that is widely used in evaluation. Compared to traditional census-based data, RS data are more objective and timely and further enhance spatial information. However, high-resolution images are expensive when applied across broad regions (Oda and Maksyutov 2011; Sulik and Edwards 2010). In contrast, coarse-resolution nighttime light images are newly emerging as less expensive data sources for regional global evaluation (Li et al. 2013a, b, c).

The brightness of nighttime light (sum light) is not only used as a measure of gross domestic product (GDP) and some socioeconomic factors, such as fossil fuel carbon dioxide emission and electric power consumption, but also serves as a factor to disaggregate socioeconomic data from the national or provincial level to the pixel level (Li et al. 2013a, b, c; Zhao et al. 2012; Bharti et al. 2011; Letu et al. 2010; Sutton et al. 2007; Bernhard et al. 2001). The United States' Defense Meteorological Satellite Program (DMSP)'s Operational Linescan System (OLS) nighttime light image is a distinctive and powerful RS resource to monitor and evaluate anthropogenic activities

(Tian et al. 2014). DMSP-OLS images have been widely applied in socioeconomic studies due to their unique capacity to reflect human activities (Li et al. 2013a, b, c). If luminosity recorded by nighttime light imagery could be used as a proxy (Chen and Nordhaus 2011), not only can the relative amounts of freight traffic (Tian et al. 2014) efficiently increase but areal demands for freight traffic as well. The responses of nighttime light during the Syrian Crisis were monitored and its potential for monitoring the conflict was analyzed (Li and Li 2014). The nighttime light dynamics of the countries (regions) indicate that the Belt and Road Initiative will boost infrastructure building, financial cooperation, and cultural exchanges in those regions.

9.5.1 Brightness of Nighttime Lights As a Proxy for Freight Traffic

China's huge growth has led to a rapid increase in demand for freight traffic. Timely assessments of past and current amounts of freight traffic are the basis for predicting future demands of freight traffic and appropriately allocating transportation resources. The main objective of this study was to investigate the feasibility of the brightness of nighttime lights as a proxy for freight traffic demand in China (Tian et al. 2014). The brightness of nighttime light as a proxy for freight traffic in a region with relatively brighter nighttime lights usually has more business activities and consequently larger freight transport demand. More developed areas also usually have brighter nights and larger freight traffic demand.

This section provides a description of the used method for extracting the sum light from nighttime light images. The sum light of each province/municipality was extracted from the re-valued annual image composite, which is equal to the total digital number (DN) values of all the pixels in the province/municipality. Then, the sum light was regressed on total freight traffic (TFT) consisting of railway freight traffic (RFT), highway freight traffic (HFT), and waterway freight traffic (WFT) at the province level; three groups of regression functions between the sum light and TFT, RFT, and HFT were developed. The standard error of the estimates (SEEs) was calculated to measure the accuracy of predictions using the regression functions. Third, each province/municipality's HFT to each pixel was disaggregated based on the pixel's DN value for downscale HFT from the province level to the pixel level and a 1 km × 1 km resolution map for 2008 was produced. For the DN of the pixels of the nighttime lights, a minimum threshold was set to eliminate the effects of blooming and to mask large rural lit areas where dim nighttime lights can be detected but there are no small business activities and freight traffic. Because electric power consumption and freight traffic are both typical socioeconomic indicators that have strong correlations with the GDP, we selected 10 as the threshold value and re-valued the annual image composites (Zhao et al. 2012).

This study was mainly derived from the following data sources. Three version 4 DMSP-OLS stable lights annual image composites for the years 2000, 2004, and

2008 were obtained from the National Oceanic and Atmospheric Administration's (NOAA) National Geophysical Data Center (NGDC) (Earth Observation Group, 2011). The annual image composites were produced by all of the available cloud-free images from satellites F14, F15, and F16 for their corresponding calendar years in the NGDC's digital archives. The DN values of the annual image composites represent the brightness of the nighttime stable lights varying from 0 to 63. Each stable light's annual image composite has a spatial resolution of 1 km × 1 km. Acquired from the National Bureau of Statistics of China (NBSC) (2001, 2005, 2009, 2011), freight traffic data (China Statistical Yearbook 2009, 2005, 2001) were used as baseline data, and the GDP was supporting data (Zhou et al. 2011).

The brightness of the nighttime light is a good proxy for TFT and HFT at the province level. Sum light can be used to estimate the GDP, and freight transport demand correlates strongly to the GDP. The experiments in Tian et al. (2014) show strong relationships between the natural logarithm of sum light and the natural logarithm of the GDP and the natural logarithm of the GDP and the natural logarithm of TFT; sum light more strongly correlates to HFT than to RFT even though the relationships between sum light and RFT are significant at the 0.01 level. Rail transport is the traditional primary form of transportation in China. Normally, goods that are large in mass and volume (e.g., coal, steel, timber) are shipped via railways, but the price of such goods per unit weight may be very low comparatively (e.g., a software CD vs. a ton of coal). In a further anomaly, Heilongjiang and Shanxi are two moderately developed provinces and provide most of the timber and coal in China. The timber and coal are transported out of Heilongjiang and Shanxi mainly via railways. Guangdong and Jiangsu are two of the most developed provinces in China; high-tech and light industries are their pillar economic activities. Materials and finished goods are shipped into and out of Guangdong and Jiangsu mainly through highways. Although the GDPs and sum light of Guangdong and Jiangsu are far larger than those of Heilongjiang and Shanxi, the RFT amounts of Guangdong and Jiangsu are much smaller than those of Heilongjiang and Shanxi. Thus, when RFT does not have a strong correlation to the GDP, sum light is not a sound measure of RFT. By contrast, road transport is a burgeoning transportation method in China and has replaced rail transport as the primary mode for freight. The more developed Chinese provinces or municipalities normally have more developed highway networks and consequently more goods and materials are shipped on the highways. Therefore, sum light can be used as a proxy of HFT. Sum light does not strongly correlate to RFT, but the ratios of RFT to TFT are very small; consequently, sum light still has strong relationships with TFT in these provinces.

TFT and HFT are also socioeconomic factors that significantly correlate with the GDP. Because the brightness of nighttime light can reflect a region's population and economic level, the brightness can be used as a proxy of HFT at the pixel level. To show a specific application of the brightness as a proxy of HFT at the pixel level, this study produced a Chinese HFT map for 2008 by disaggregating each province/municipality's HFT to each pixel in proportion to the DN value of the pixels of the 2008 nighttime light image composite (Fig. 9.17).

Fig. 9.17 Chinese HFT map

The spatial variation of HFT demand is clearly shown by the map in Fig. 9.17. The amount of HFT gradually decreases from the urban core regions to suburban regions. In Sichuan and Chongqing, where the population is very high but the developed area is limited, the HFTs of urban core regions are extremely large. In Beijing, Shanghai, and Guangzhou, where economic development and urbanization levels are both very high, the HFTs of urban core regions are moderate because their large total HFTs have been distributed to relatively broader developed areas. Therefore, compared to the traditional census-based HFT data, the nighttime-light-derived HFT data can provide more detailed geographic information for freight traffic demand. Compared to an undeveloped region, a developed region normally has more and brighter nighttime lights and consequently a larger sum light. Additionally, the developed region usually has a larger population and produces and consumes more goods, which lead to a larger freight transport demand.

9.5.2 Evaluating the Syrian Crisis with Nighttime Light Images

This study aimed to analyze the responses of nighttime light to the Syrian Crisis and evaluated their potential to monitor the conflict (Li and Li 2014). For the country and all of its provinces, a direct method of evaluation was determined as the amount of nighttime light for the country and each provincial region. First, the sum of nighttime lights (SNL) of Syria was calculated for each month based on the digital values of all the pixels in a region so that the SNL had no physical unit. Second, the SNL of each province was calculated. Third, the change in the proportions of SNL was calculated for different regions. Fourth, nighttime light variation was correlated with statistics by summarizing the humanitarian scale of this crisis. To show the nighttime light trend in a continuous spatial dimension, a data clustering method was further developed for time series nighttime light images; while the trend is irrelevant to the overall magnitude of the nighttime lights, similar trends of nighttime light were grouped into one class. The method has three steps: extracting the lit pixels, normalizing the time series nighttime light, and clustering the time series nighttime light images. Because the lit areas at night were the only focus of this study, the dark areas were excluded at first. For each pixel, if its value was smaller than the threshold of every month, then it was labelled as a dark pixel or a lit pixel. The threshold was set to three in this analysis. Each pixel in the lit area was normalized. The K-means algorithm was used to cluster the normalized nighttime light data into classes in the lit areas, and the dark areas were labelled as the "dark region" class.

The ongoing Syrian Crisis has caused severe humanitarian disasters, including more than 190,000 deaths (Cumming-Bruce 2014). This study mainly was derived from three data sources: administrative boundaries, satellite-observed nighttime light images, and statistical data from human rights groups. The international and provincial boundaries of Syria were derived from the Global Administrative Areas (http://www.gadm.org/). Syria adjoins Turkey to the north, Iraq to the east, Jordan to the south, and Israel and Lebanon to the west. Syria is divided into 14 provincial regions, with Damascus, Aleppo, and Homs as its major economic centers. The DMSP/OLS monthly composites between January 2008 and February 2014 were selected as the nighttime light images for the analysis (Fig. 9.18). A total of 38 monthly composites were used for this analysis. The March 2011 image was selected as the base image for the intercalibration (Wu et al. 2013).

Here, it is analyzed by administrative regions. Figure 9.18 shows that the nighttime light in Syria has very sharply declined since the crisis, and many small lit patches have gone dark. The city of Aleppo, the site of fierce battles (AAAS 2013), has lost most of its lit areas. Although the cities of Homs and Damascus also lost a lot of nighttime lights, the intensity of their losses appear to be less than that of Aleppo. The SNLs between January 2008 and February 2014 in Syria show some fluctuations before March 2011, but the nighttime light in Syria has continuously declined since March 2011. The SNL of each province was calculated between January 2008 and February 2014. The Golan Heights, as a part of Quneitra, has been controlled by Israel since 1967, the nighttime lights of which were excluded when calculating the SNL in Quneitra. The SNLs between January 2008 and February 2014 in different provinces of Syria show a sharply declining trend of nighttime lights for all the provinces since March 2011.

The change in the proportions of SNL was calculated in different regions between March 2011 and February 2014. In addition, the size of the lit area, defined as the area where the light value is greater than 3, was also retrieved for each region, and its change in proportion during the period was also calculated. The two indices are illustrated in Fig. 9.19.

In Fig. 9.19, it can be seen that Syria has lost about 74 and 73 % of its nighttime lights and lit areas, respectively. In addition, most of the provinces lost >60 %

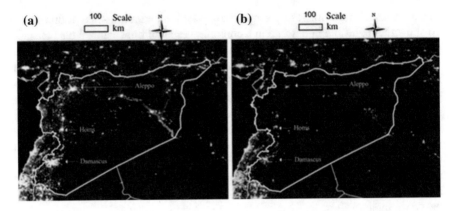

Fig. 9.18 The nighttime light monthly composites Copyright © 2013, Taylor & Francis Group.
a March 2011. **b** February 2014

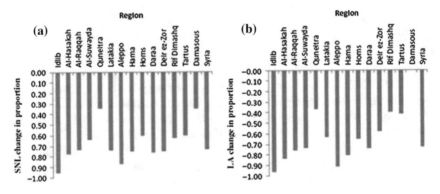

Fig. 9.19 The nighttime light change in proportions for Syria and all the provinces between
March 2011 and February 2014 Copyright © 2013, Taylor & Francis Group. **a** SNL change in
proportion. **b** Lit area (LA) change in proportion

of their nighttime lights and lit area. Damascus, the capital of Syria, is an exception; it only lost about 35 % of the nighttime lights and no lit areas during the
period, and the nighttime light has fluctuated notably, which is different from most
of the provinces that experienced continuous decline. However, Rif Dimashq, in
the countryside of Damascus, lost about 63 % of its lights, showing that its security situation is much more severe than in Damascus. This finding is consistent
with the fact that the Assad regime has strongly controlled the capital, although
the battles around the capital were intense (Barnard 2013). Quneitra, another
exception, also lost only 35 % of its lights during the period. Idlib and Aleppo

Table 9.8 The number of IDPs and SNL loss of all the Syrian regions between March 2011 and December 2013

Province	Number of IDPs	SNL loss
Idlib	569,000	33,198
Al-Hasakah	230,000	63,787
Al-Raqqah	251,000	40,176
Al-Suwayda	52,000	13,075
Quneitra	78,000	19,912
Latakia	222,000	29,991
Aleppo	1,735,000	92,041
Hama	423,000	39,722
Homs	588,000	56,131
Daraa	372,000	32,075
Deirez-zor	420,000	79,306
Rif Dimashq and Damascus	1,080,000	101,067
Tartus	500,000	126,199

are the provinces with the most severe decline in nighttime light, losing 95 and 88 % of the nighttime lights and 96 and 91 % of the lit areas, respectively. In fact, the battles in these two provinces were particularly fierce (Al-Hazzaa 2014; AAAS 2013). We also found that Deirez-Zor and Rif Dimashq lost 63 and 60 % of their nighttime lights, respectively, but only 40 and 42 % of their lit area. We can infer that basic power supplies in these two provinces still were working in most of the areas, although the battles there also were intense.

Can the nighttime light variation be correlated with statistics summarizing the humanitarian scale of this crisis? The number of internally displaced persons (IDPs) and the SNL loss of all the Syrian regions between March 2011 and December 2013 are presented in Table 9.8. These two data groups and linear regression analysis showed that the nighttime light decline was correlated to the number of displaced persons during the Syrian Crisis. This finding supports the assumption that multi-temporal nighttime light brightness is a proxy for population dynamics (Bharti et al. 2011). In brief, nighttime light variation can reflect the humanitarian disasters in the Syrian Crisis.

Therefore, the nighttime light experienced a sharp decline as the crisis broke out. Most of the provinces lost >60 % of the nighttime light and lit areas because of the war, and the amount of the loss of nighttime light was correlated with the number of IDPs. Also, the international border of Syria is a boundary to the nighttime light variation patterns, confirming the previous study's conclusion that the administrative border has the effect of socioeconomic discontinuity (Pinkovskiy 2013).

9.5.3 Nighttime Light Dynamics in the Belt and Road

The "Belt and Road" refers to the Silk Road Economic Belt and the twenty-first-century Maritime Silk Road. In the twenty-first century—a new era marked by the theme of peace, development, cooperation and mutual benefit—it is all the more important for us to carry on the Silk Road Spirit in face of the weak recovery of the global economy and complex international and regional situations. When Chinese President Xi Jinping visited Central Asia and Southeast Asia in September and October 2013, he raised the initiative of jointly building the Silk Road Economic Belt and the twenty-first-century Maritime Silk Road, which have attracted close attention from all over the world. It is a systematic project, which should be jointly built through consultation to meet the interests of all, and efforts should be made to integrate the development strategies of the countries along the Belt and Road. The Chinese government has drafted and published the *Vision and Actions on Jointly Building Silk Road Economic Belt and twenty-first-Century Maritime Silk Road* to promote the implementation of the initiative; instill vigor and vitality into the ancient Silk Road; connect Asian, European and African countries more closely; and promote mutually beneficial cooperation to a new high and in new forms (The National Development and Reform Commission, Ministry of Foreign Affairs, and Ministry of Commerce of the People's Republic of China, with State Council authorization 2015).

The Belt and Road Initiative not only covers the countries along the ancient roads, but it also links the Asia-Pacific, European, African and Eurasian economic circles. There are three directions of the Silk Road Economic Belt: China—Central Asia—Russia—Europe, China—Central Asia—West Asia—Persian Gulf—Mediterranean, and China—Southeast Asia—South Asia—Indian Ocean. Also, there are two directions of the twenty-first-century Maritime Silk Road: China—South China Sea—Indian Ocean—Europe, China—South China Sea—South Pacific. It passes through more than 60 countries and regions with a total population of 4.4 billion, with the purpose of boosting infrastructure building, financial cooperation, and cultural exchanges in those regions.

Nighttime light imagery acquired from DMSP can illustrate the socioeconomic dynamics of countries along the Belt and Road. Comparing the DMSP nighttime light imagery from 1995 to 2013, the urbanizations and economic growth patterns can be uncovered. Most of the countries along the Belt and Road have experienced significant growth (Fig. 9.20).

By further analyzing the countries within a distance of 500 km to the Belt and Road (Fig. 9.21), we found that the nighttime light in China and Southeast Asia (e.g. Cambodia, Laos, Thailand, Vietnam and Burma) has quickly grown, showing rapid urbanization and economic growth in these countries. Particularly, China surpassed Russia to be the country with the largest total nighttime light. The rapid growth of night-time light in some countries (e.g. Bosnia And Herzegovina, Somalia and Afghan) is from the peace process and post-war reconstruction. Nighttime light decline in some developing countries (e.g. Syria and Ukraine) may

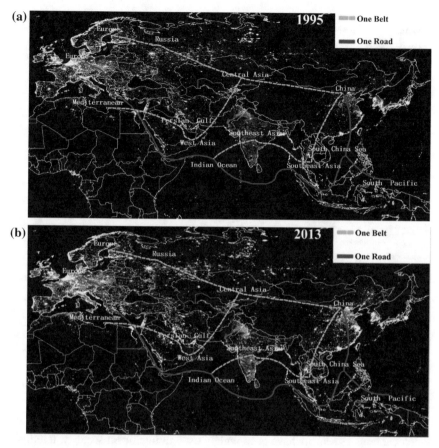

Fig. 9.20 DMSP nighttime light imagery in the belt and road region. **a** DMSP nighttime light imagery in 1995. **b** DMSP nighttime light imagery in 2013

be due to economic decline and war. Nighttime light decline in some developed countries (e.g. Sweden and Denmark) is because of their control of excessive outdoor lighting.

Therefore, in order to resolve regional and global structural contradictions, all nations need to cooperate and the solution to domestic structural contradictions in many countries also depends on cooperating with other nations. The Belt and Road Initiative will play an important role to help establish stability and recovery and lead to the next phase of prosperity in the global economy. The Belt and Road cooperation features mutual respect and trust, mutual benefit and win-win cooperation, and mutual learning between civilizations. As long as all countries along the Belt and Road make concerted efforts to pursue our common goal, there will be bright prospects for the Silk Road Economic Belt and the twenty-first-century Maritime Silk Road, and the people of countries along the Belt and Road can all benefit from this initiative.

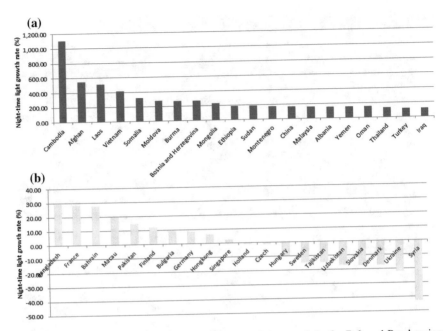

Fig. 9.21 Twenty countries (regions) with nighttime light growth in the Belt and Road region during 1995–2013. **a** Countries (regions) with the highest nighttime light growth. **b** Countries (regions) with the lowest nighttime light growth

9.6 Spatiotemporal Video Data Mining

Video cameras as sensors have been widely used in intelligent security, intelligent transportation, and intelligent urban management along with the development of smart cities. At present, there is a national multilevel online monitoring project, which is a five hierarchical monitoring network. There are more than 20 million ubiquitous cameras for city management, public security, and health care. Video monitoring systems are running in more than 600 cities in China (Li et al. 2006, 2013a, b, c).

Digital video data are acquired in an integrated manner under a large-scale spatiotemporal environment, which includes information about objects, places, behaviors, and time. With the popularization of high definition (HD) video data, the storage of HD video occupies more than several times that of low-definition video. It is difficult to store massive video data for a long period of time and there also is the lack of a fast and effective video retrieval mechanisms to find meaningful video information. Furthermore, the large amount of information on HD video requires much more rapid computing speed. Nowadays, we are unable to make full use of the surveillance video data in a large-scale spatiotemporal environment because its storage cost is prohibitive, the retrieval results are inaccurate, and the data files are too large to analyze. It is critical to find new automatic and real-time techniques for video data intelligent compression, analysis, and processing.

Spatiotemporal video data mining not only extracts information and processes data intelligently, but it also distinguishes abnormal behavior from normal behavior of persons, cars, etc. automatically. In this way, we can delete large amounts of video recordings of the normal activities and private activities of individuals that need to be protected, while keeping the data about suspicious cars and people, as well as data about people who need care and supervision (e.g., elderly with dementia and children with disabilities). There are no popular techniques and products of spatiotemporal video data mining due to their technical difficulties.

9.6.1 Technical Difficulties in Spatiotemporal Video Data Mining

Spatiotemporal video data mining has some technical difficulties: The data are too expensive to store. The amount of data to be processed and stored on a city-level video surveillance network is massive. The amount of geo-spatial data in digital cities reaches the TB level, while the amount of city-scale video data in smart cities may be in the PB level. According to the current technical and industry standards, the data collected by a HD camera in one hour, compressed using the H.264 standard, needs 3.6 GB space to be stored. Videos are always kept for three months. If a 4 TB storage server costs 50,000 yuan, the cost to the Tianjin city security system may be as high as 58.32 billion yuan only for storing the videos, which equals the total GDP of Tibet in 2012. Because the amount of surveillance video data is huge, the vast majority of the cities in China relieve the pressure from construction funds by reducing video time and quality, which also reduces the value of criminal investigation and identification through video data. The rapid growth of storage size and investment has become a problem that is restricting city surveillance systems.

The retrieval of video data is inaccurate. It is difficult for people to locate useful knowledge from massive video data. A large number of experiments have shown that if an operator looks at two monitors at the same time, he will miss 95 % of the surveillance targets after 22 min. Currently, cross-regional crimes have become the main modus operandi for international crimes. The Procuratorial Daily indicated that 70–90 % of the total crimes are committed by moving people. As the areas of crime-related regions expand, increasingly more video data need to be retrieved. However, traditional monitoring systems simply can only sample and record; they cannot analyze or extract critical information and high-level semantic content efficiently. Once crimes have occurred, investigators must manually browse the suspected targets one by one. The results are therefore low, and it is easy to miss the best time to solve the case. To improve the detection rate, it is vital to utilize video retrieval technology to locate useful information from massive video data quickly.

The video data are not being utilized to their full advantage in public security. The lifecycle of social safety incidents generally include cause, planning, implementation, occurrence, and escape. Most of the professional crimes have the characteristics of large behavior spatiotemporal span and high-level concealment. Although the behavior in the crime preparation period is traceable, the existing domestic technology can only analyze existing local data and make simple judgments, and it is difficult to detect the spatiotemporal behavior abnormal events early from multi-scale and multi-type media data. Some crime prediction software can detect criminals' abnormal behavior to a certain extent, which has a great effect on the prevention of crime. In Los Angeles, for example, the use of a set of crime prediction software named "PredPol" reduced the crime rate by 13 %; however, at the same time, the city's crime rate increased by 0.4 %. At present, video monitoring systems are operating throughout China, which obtains PB-level video monitoring data every day, but it cannot produce alerts through analysis to prevent security incidents.

9.6.2 Intelligent Video Data Compression and Cloud Storage

To solve the problem of storage size and investment, video data compression removes redundant information from video data and can reduce the storage size of video data. At the same time, cloud storage can store massive video data as it breaks through the bottleneck of the capacity of traditional storage methods and realizes the linear expansion in performance and capacity. It provides data storage and business access to the external world based on the technology of application clusters, networks, and distributed file systems. Application software enables a large variety of storage devices to work together in the network. In its virtual storage, all the equipment is completely transparent to cloud users. Any cloud or any authorized user can connect to the cloud storage through a cable. The user has a storage capacity equivalent to the entire cloud. The monitored district is distributed widely, and each monitoring point creates a great deal of data. Using cloud storage systems can realize distributed management easily and can extend capacity at any time.

9.6.3 Content-Based Video Retrieval

As a medium to express information, video has its own independent structure. Generally, a video is composed of scenes that describe some independent story units. A scene is composed of shots that have related composition. A shot is composed of consecutive frames, which can be represented by one or more key frames.

Content-based video retrieval is a commonly used retrieval technology in video surveillance. In the context of video content, it can automatically extract and describe the feature and content of videos without human intervention. First, the video is divided into various lenses to achieve feature extraction of each lens; this makes it possible to obtain an image as fully as possible to reflect the content of the feature space, which will serve as the basis of the video retrieval. The extraction of features includes the motion features of key frames and the visual features of lenses. The key frames are static images, which can be extracted from the video data and can summarize the content of the lens. Through a certain algorithm, we can achieve the extraction of these visual features of static images by their color, texture, shape, etc. The extraction of camera motion features can be achieved by the motion analysis of the lens (e.g., changes in lens movement, changes in the size of moving objects and the trajectories of video objects). The methods are based mainly on the optical flow equation, block, pixel recursion, and Bayesian. Then, the index is determined according to certain static characteristics on the dynamic characteristics of the lens and key frames. Eventually, the user can browse and retrieve video in a simple and convenient way.

Although content-based video retrieval can solve the problem of inaccurate retrieval to a certain extent, increasingly more video data are rapidly growing, which need to be retrieved because the areas of crime-related regions are expanding quickly as well. The false alarm data from small range of search results can be found and eliminated manually, but the rapidly growing scale of involved video data makes the efficiency of the retrieval system too low to be accepted, and the scale of outputted false alarm data is now beyond the limitations of manual processing. To solve the problem, spatiotemporal video data mining can be used to reduce the scale of false alarms by taking advantage of spatial and temporal characteristics for filtering invalid video data.

9.6.4 Video Data Mining Under Spatiotemporal Distribution

To successfully detect the abnormal behavior of a criminal, it is necessary to detect abnormal events that consist of a series of abnormal behaviors. Video data can play an important role in early warning through video monitoring. Behavior analysis and event detection is the basic technology of video monitoring for abnormal behavior detection and prediction.

9.6.4.1 Behavior Analysis

The behavior analysis of video objects is the basis of event detection by the pre-image processing of video and conducting research about the temporal and spatial data of the target of interest in the video scenes (sometimes in relation to background objects) and finally understanding or interpreting the video target's actions during a particular period of time, which can aid decision system response. Abnormal analysis of spatial correlations in video data is a comprehensive and interactive application of video data and geo-spatial information on the spatial distribution of geographic information.

From the view of data processing, because the sequence of video tracks information in the order of time, it can be simply understood as a multidimensional time signal. Therefore, the behavior analysis of video objects is time-series piecewise. Sometimes, it is necessary to calculate the spatial relationships between each time segment and the background target, and then matching it with the classic behavior model in the model library, thus completing the classification identification. The current behavior analysis methods for video objects include analysis of the cumulative template technology behavior, Bayesian network behavior analysis method, finite state machine method, the declarative model (logic-based method), and a Petri net-based method.

By using video data mining software, key information about human behavior is automatically stored in the video, such as aggregation, running, climbing over a wall, and wandering (Fig. 9.22), which is conducive to investigators focused on observing and analyzing a suspect. The Boston Marathon bombing was solved by finding the criminals' abnormal behavior on video images by going inversely to the crowds after the explosion.

9.6.4.2 Event Detection

Event detection is used to uncover interesting or apparently unusual event from the video, then make a proper analysis of the behavior with sequential videos distributed in spatiotemporal conditions (Fig. 9.23). A video semantic event refers to one behavior or a series of behaviors with certain semantic information in the spatiotemporal video sequence; behavior refers to a series of actions with certain semantic and temporal continuity of action in the video sequence.

(a)   **(b)**   **(c)**   **(d)**

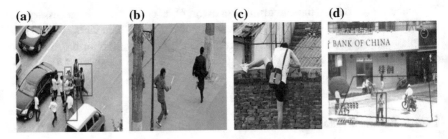

Fig. 9.22 Abnormal behavior automatic mining. **a** Gathering. **b** Running. **c** Over the wall. **d** Wandering

Fig. 9.23 Spatiotemporal video sequence based event detection

Semantic event detection is mainly implemented on a predefined event model, learning event model, and clustering analysis. The predefined event model is used to construct the event model by using predefined rules or constraints; due to the need for relevant knowledge about the environment, it can only be used in specific areas. Therefore, its use is very limited. The learning event model learns from training data after the feature extraction by using hidden Markov models or dynamic Bayesian networks to analyze the relationship between the characteristic value of each key frame and then mining the semantic relationship between each lens and detecting some typical events. Clustering analysis includes the temporal derivative and collaborative embedded prototype. The event fragments of a weight matrix are detected by partitioning the spectral graph, and the weighting matrix is determined by computing the similarity between video clips.

9.6.4.3 Spatiotemporal Correlation Analysis

The integration of GIS data and media data (video, audio) can enable the monitoring of a city from four spatiotemporal dimensions. By combining video with GIS, continuous information is located through continuous automatic video data mining and then obtaining spatial information by GIS. The combination of the two methods can conduct meaningful spatiotemporal correlation analysis and abnormal behavior analysis (Fig. 9.24).

Traditional geographic information analysis is time-consuming, labor-intensive, and unstable. Comparatively, spatiotemporal analysis from both geo-spatial data and video data offers universality, humanity, and intellectualization. Real-time analysis and data mining can obtain the background data of static space and continuous data about humans, cars, etc. Through spatiotemporal analysis of geographic information and video showing fixed monitory points on a map, surveillance on a moving car's position, and warning prompts, users can observe a monitor's operating state clearly and obtain the graphic information of monitored sites quickly and intuitively. If necessary, users can switch to the monitor site instantly and then provide a basis for the remote command. It also can be applied to various fields such as automatic protection system, emergency response, road maintenance, river improvement, city management, mobile surveillance, tourism, etc.

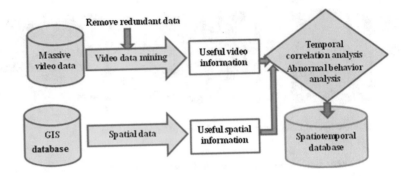

Fig. 9.24 Spatiotemporal analysis on correlation and abnormal behavior

References

AAAS (American Association for the Advancement of Science) (2013) Conflict in Aleppo, Syria: a retrospective analysis. http://www.aaas.org/aleppo_retrospective

Al-Hazzaa HA (2014) Opposition and regime forces split Idlib province after ISIS withdrawal. http://www.damascusbureau.org/?p=6622

Barnard A (2013) Syrian forces recapture damascus suburb from rebels. http://www.nytimes.com/2013/11/14/world/middleeast/syrian-forces-recapture-damascus-suburb-from-rebels.html?_r=0

Bernhard S, John CP, John AT, Alex JS (2001) Estimating the support of a high-dimensional distribution. Neural Comput 13(7):1443–1471

Bharti N, Tatem AJ, Ferrari MJ, Grais RF, Djibo A, Grenfell BT (2011) Explaining seasonal fluctuations of measles in niger using nighttime lights imagery. Science 334:1424–1427

Chen X, Nordhaus WD (2011) Using luminosity data as a proxy for economic statistics. Proc Natl Acad Sci USA 108:8589–8594

Christopher BJ, Ware JM, Miller DR (2000) Bayesian probabilistic methods for change detection with area-class maps. Proc Accuracy 2000:329–336

Cressie N (1991) Statistics for spatial data. Wiley, New York

Cumming-Bruce N (2014) Death toll in Syria estimated at 191,000. http://www.nytimes.com/2014/08/23/world/middleeast/un-raises-estimate-of-dead-in-syrian-conflict-to-191000.html?_r=0

Di KC (2001) Spatial data mining and knowledge discovering. WuHan University Press, WuHan

Franklin SE, Wulder MA, Lavigne MB (1996) Automated derivation of geographic window sizes for use in remote sensing digital image texture analysis. Comput Geosci 22(6):665–673

Fu Y, Guo G, Huang TS (2010) Age synthesis and estimation via faces: a survey. IEEE Trans Pattern Anal Mach Intell 32(11):1965–1976

Letu H, Hara M, Yagi H, Naoki K, Tana G, Nishio F, Shuher O (2010) Estimating energy consumption from night-time DMSP/OLS imagery after correcting for saturation effects. Int J Remote Sens 31:4443–4458

Li X, Li DR (2014) Can night-time light images play a role in evaluating the Syrian Crisis? Int J Remote Sens 35(18):6648–6661

Li DR, Wang SL, Li DY (2006) Theory and application of spatial data mining, 1st edn. Science Press, Beijing

Li X, Chen F, Chen X (2013a) Satellite-observed nighttime light variation as evidence for global armed conflicts. IEEE J Sel Topics Appl Earth Obs Remote Sens 6:2302–2315

Li X, Ge L, Chen X (2013b) Detecting Zimbabwe's decadal economic decline using nighttime light imagery. Remote Sens 5:4551–4570

Li DR, Wang SL, Li DY (2013c) Theory and application of spatial data mining, 2nd edn. Science Press, Beijing

Lyons et al. (1998) Coding facial expression with gabor wavelets. In: Proceedings of 3rd IEEE international conference on automatic face and gesture recognition, Nara, Japan

Ma HC (2002) An intrinsic random process-based texture analysis and its application, [postdoctoral report]. Wuhan University, Wuhan

National Bureau of Statistics of China (2001) China statistical yearbook 2001. http://www.stats.gov.cn/english/statisticaldata/yearlydata/YB2001e/ml/indexE.htm

National Bureau of Statistics of China (2005) China statistical yearbook 2005. http://www.stats.gov.cn/tjsj/ndsj/2011/indexeh.htm

National Bureau of Statistics of China (2009) China statistical yearbook 2009. http://www.stats.gov.cn/tjsj/ndsj/2009/indexeh.htm

National Bureau of Statistics of China (2011) China statistical yearbook 2011. http://www.stats.gov.cn/tjsj/ndsj/2011/indexeh.htm

Oda T, Maksyutov S (2011) A very high-resolution (1 km × 1 km) global fossil fuel CO_2 emission inventory derived using a point source database and satellite observations of nighttime lights. Atmos Chem Phys 11:543–556

Pantic M, Rothkrantz L (2000) Automatic analysis of facial expressions—the state of the art. IEEE Trans Pattern Anal Mach Intell 22(12):1424–1445

Pinkovskiy M (2013) Economic discontinuities at borders: evidence from satellite data on lights at night. http://economics.mit.edu/files/7271

Ralph G et al (2004) Face recognition across pose and illumination. In: Li SZ, Jain AK (eds) Handbook of face recognition. Springer, Berlin

Sulik JJ, Edwards S (2010) Feature extraction for darfur: geospatial applications in the documentation of human rights abuses. Int J Remote Sens 31:2521–2533

Sutton PC, Elvidge CD, Ghosh T (2007) Estimation of gross domestic product at sub-national scales using nighttime satellite imagery. Int J Ecol Econ Stat 8:5–21

The National Development and Reform Commission (2015) Ministry of Foreign Affairs, and Ministry of Commerce of the People's Republic of China, with state council authorization, vision and actions on jointly building silk road economic belt and 21st-Century Maritime Silk Road

Tian J, Zhao N, Samson EL, Wang SL (2014) Brightness of nighttime lights as a proxy for freight traffic: a case study of China. IEEE J Sel Topics Appl Earth Obs Remote Sens 7(1):206–212

Wang SL et al (2004) Rough spatial interpretation. Lecture notes in artificial intelligence, 3066, pp 435–444

Wang SL et al (2009) Localization and extraction on the eyes, nose and mouth of human face. In: Proceedings of IEEE international conference on granular computing (IEEE GrC 2009), Nanchang, China, 16–18 May, pp 561–564

Wu ZC (2004) Rough set and its applications in remote sensing image classification. PhD thesis. Wuhan University, Wuhan

Wu J, He S, Peng J, Li W, Zhong X (2013) Intercalibration of DMSP-OLS night-time light data by the invariant region method. Int J Remote Sens 34:7356–7368

Zeng Z et al (2009) A survey of affect recognition methods: audio, visual, and spontaneous expressions. IEEE Trans Pattern Anal Mach Intell 31(1):4–37

Zhao N, Ghosh T, Samson EL (2012) Mapping spatiotemporal changes of Chinese electric power consumption using nighttime imagery. Int J Remote Sens 33:6304–6320

Zhou Y (2003) Content-based remote sensing image retrieval (Postdoctoral report). Wuhan University, Wuhan

Zhou J, Chen Y, Wang J, Zhan W (2011) Maximum nighttime urban heat island (UHI) intensity simulation by integrating remotely sensed data and meteorological observations. IEEE J Sel Topics Appl Earth Obs Remote Sens 4:138–146

Chapter 10
SDM Systems

SDM, which was developed on the principles of software engineering as well as professional geo-spatial knowledge, can be used as an isolated system, as a plug-in software attachment to a spatial information system, or as an integral part of intelligent systems. Because spatial data are mainly GIS or imagery data, the systems we developed—*GISDBMiner* for GIS data and *RSImageMiner* for imagery data (Li et al. 2006, 2013)—are introduced first in this chapter, followed by spatiotemporal video data mining and the *EveryData* system.

10.1 GISDBMiner for GIS Data

Spatial data are stored in the form of file documents, while the attribute data are stored in databases. *GISDBMiner* is a GIS data-based SDM system (Fig. 10.1) that supports the common standard data formats. The discovery algorithms are executed automatically, and there is geographic human–computer interaction. The user can define the subset of interest in a dataset, provide the context knowledge, set the threshold, and select the knowledge representation format. If the needed parameters are not provided, the system will automatically use default parameters.

In *GISDBMiner*, the user first must trigger the knowledge discovery command, which signals the spatial database management to allow access to the data subset of interest from the spatial database (i.e., the task-related data). Then, the mining algorithms are executed to discover knowledge from the task-related data according to the user's requirements and professional knowledge. Finally, the discovered knowledge is output for the user's application or is added to the knowledge base, which is used for future knowledge discovery tasks. Generally, knowledge discovery requires repeated interactions to obtain satisfactory final results. To start the knowledge discovery process, the following steps are necessary: (1) the user

© Springer-Verlag Berlin Heidelberg 2015

D. Li et al., *Spatial Data Mining*, DOI 10.1007/978-3-662-48538-5_10

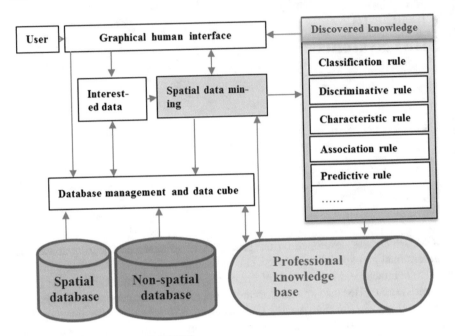

Fig. 10.1   The framework of *GISDBMiner*

interactively selects the data of interest from the spatial database, (2) a visual query and the search results are displayed to the user, and (3) the data subset of interest is gradually refined. The process of knowledge discovery begins thereafter.

GISDBMiner has several key functions, which include data preprocessing, association rule discovery, clustering analysis, classification analysis, sequence analysis, variance analysis, visualization, and spatial knowledgebase. The association rule discovery process employs the Apriori algorithm and concept lattice, the clustering analysis adopts the K-means clustering algorithm and the concept clustering algorithm, and the classification analysis applies decision trees and neural networks.

10.2 RSImageMiner for Image Data

RSImageMiner is a system for remote sensing imagery data (Fig. 10.2). This imagery feature is frequently used in remote sensing image classification and object recognition and is an important aspect of image mining for discovering feature knowledge. For example, a typical difference between remote sensing images and natural images is the multi-spectral (hyper-spectral) features (e.g., TM image with 7 bands and SPOT image with 4 bands). Data mining can determine the association rules between different spectral values for image classification and different spectral reflection laws, the maximum degree of correlation between certain

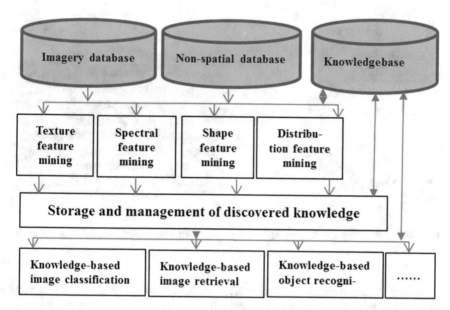

Fig. 10.2   The architecture of *RSImageMiner*

bands, and the classification rules by which some object classes are established from certain band intervals. Moreover, there are a large number of complex spatial objects in remote sensing images that possess complex relationships and distribution rules between the objects, for which the spatial distribution rule mining feature of *RSImageMiner* was designed.

RSImageMiner includes image management and data mining on the image features of spectrum (color), texture, shape, and spatial distribution pattern, as well as image knowledge storage and management and knowledge-based image classification, retrieval, and object recognition. During the *RSImageMiner* process, if the image is too large, the data are divided into a number of relatively small sample areas. The extent to which the subsample image is cut is set in accordance with the sample image extent setting. Image features reflect the relationships between the local regions of different imagery gray values, which should be normalized. After the normalization process, the feature data of the image are stored in a database in the form of a table. The various functionalities of image processing in the system are based on integral management within the same project. The functions do not interfere with each other, and different views can be designed for different purposes by switching views. Saving the file is executed by entering the file name in the pop-up dialog box. The image is stored by using the binary long object mode, the knowledge is stored as a file, the rules in the knowledge base are stored in the form of relational tables for data records, and the field text of knowledge is stored as a whole in the relational table.

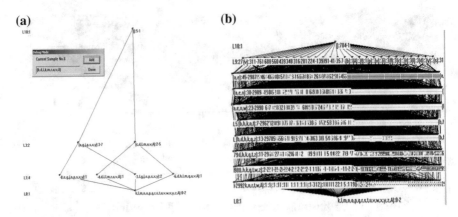

Fig. 10.3 Generating the *Hasse* diagram of concept lattice **a** manual generation, **b** automatic generation

Image management may open, store, delete, display, preprocess, and browse images. Image and feature data are integrally stored and managed by using a relational database management system as a binary long object mode.

Texture feature mining is implemented in the imagery texture feature, which includes cutting samples, normalizing, storing feature data, generating *Hasse* diagrams, and generating rules. The *Hasse* diagrams are generated by setting the parameters of the support threshold and the confidence threshold, the table file name and text file name of redundant rules, and the table file name and text file name of the non-redundant rules. The *Hasse* diagram in Fig. 10.3, which was created based on the mining algorithm of concept lattice, reflects the hierarchical relationship within the data manually or automatically. The manual mode inputs each sample point by hand, which reflects the illustration process of constructing the incremental concept lattice and the *Hasse* diagram. The automatic mode calls all data records in the database to generate the complete concept lattice by itself. Based on the generated *Hasse* diagram, the rules are generated as the list for display (Fig. 10.4). Some of the frequent nodes have a degree of support value greater than the support threshold value. Redundant rules are generated according to the frequent nodes directly, and the non-redundant rules are generated according to the algorithm for non-redundant rules.

Shape feature mining is implemented by the imagery shape feature, which includes detecting the edge, extracting the shape feature, generating *Hasse* diagrams, and generating rules. Boundary extraction delineates the typical sample area and extracts the shape feature based on those boundaries. Shape feature extraction calculates the shape feature of the boundary polygon in different sample areas.

Spectral feature mining is applied to the imagery spectral feature and includes opening the image; acquiring its spectral values by outlining the sample area and its schema; storing the spectral values in the sample area; normalizing the spectral values, such as the degree of brightness (1, 2, 3, 4, or 5 corresponding to very dark, dark, general, relatively bright, and very bright); generating the *Hasse*

Fig. 10.4 Texture feature rules

```
{{0,0,2},{1,0,2}}==>{{-1,0,2}},0.364583,0.958904
{{-1,0,2},{0,0,2}}==>{{1,0,2}},0.364583,0.955631
{{-1,0,2},{1,0,2}}==>{{0,0,2}},0.364583,0.851064
{{1,-1,2},{1,0,2}}==>{{-1,0,2}},0.256510,0.807377
{{1,-1,2},{1,0,2}}==>{{-1,-1,2}},0.256510,0.807377
{{1,-1,2},{-1,0,2}}==>{{1,0,2}},0.256510,0.827731
{{1,-1,2},{-1,0,2}}==>{{-1,-1,2}},0.270833,0.873950
{{-1,-1,2},{1,0,2}}==>{{-1,0,2}},0.259115,0.872807
{{-1,-1,2},{1,0,2}}==>{{1,-1,2}},0.256510,0.864035
{{-1,-1,2},{1,-1,2}}==>{{0,-1,2}},0.364583,0.851064
{{1,0,2},{1,1,2}}==>{{-1,0,2}},0.256510,0.810700
{{0,-1,2},{1,-1,2}}==>{{-1,-1,2}},0.364583,0.958904
{{-1,-1,2},{0,-1,2}}==>{{1,-1,2}},0.364583,0.955631
{{-1,0,2},{1,1,2}}==>{{1,0,2}},0.256510,0.867841
{{-1,0,2},{1,1,2}}==>{{-1,1,2}},0.257813,0.872247
{{-1,1,2},{1,1,2}}==>{{0,1,2}},0.356771,0.867089
{{1,0,2},{-1,1,2}}==>{{-1,0,2}},0.270833,0.885106
{{1,0,2},{-1,1,2}}==>{{1,1,2}},0.252604,0.825532
{{0,1,2},{1,1,2}}==>{{-1,1,2}},0.356771,0.951389
{{-1,1,2},{0,1,2}}==>{{1,1,2}},0.356771,0.951389
{{1,0,2},{0,1,2}}==>{{-1,0,2}},0.255208,0.875000
{{-1,0,2},{0,1,2}}==>{{1,0,2}},0.255208,0.875000
```

diagrams; and generating the non-redundant association rules of the spectral feature according to the generated *Hasse* diagram. According to the mining algorithm for association rules, the spectrum of each band is divided into several sections, and each spectral band is a data item. The spectrum of each band value corresponds to the interval for each value and establishes a multiple-value context table.

Spatial distribution rule mining is implemented by the imagery distribution feature and includes cutting the sample, normalizing, storing feature data, generating *Hasse* diagrams, and generating rules. To determine the spatial distribution rule, all the data should be mined. However, the processing speed for large amounts of data is very slow, so the system provides the option of choosing a sample training area to uncover the spatial distribution rule of the sample area. First, a classified remote sensing image is opened; for example, the image is categorized into four types: green field, water, residential area, and bare land. Second, the extent of a sub-sample image is selected. Third, the sample image is cut, and the resulting image is saved. The gray value of the classified image is then normalized, such as 0, 1, 2, 3, respectively, representing four types of ground features; and the spatial relationship data of the sample image is stored as a record into the database. Finally, in accordance with the generated *Hasse* diagram, the non-redundant association rules are generated, reflecting the spatial distribution.

The knowledge management function integrally stores and manages the discovered knowledge. When the knowledge is generated, it is stored in the knowledge base; each item of the knowledge from the image is stored as a relational data record and the knowledge documents are stored in the form of text as a whole. This function includes edit, query, comparison, and classification. Knowledge addition refers to a specific rule from the data that is added as a record into the knowledge base. Knowledge deletion refers to a specific rule that is deleted from the repository. Knowledge query searches one or more rules and compares multiple knowledge bases to extract the common rules and stores them in a file for common knowledge. Multiple knowledge files can be used to generate a knowledge classifier.

The discovered knowledge—that is, texture association rules, spectral (color) knowledge, shape knowledge, and spatial distribution rules—can be used to assist image classification, retrieval, and object recognition. The discovered knowledge also can be applied along with professional knowledge for comprehensive results. Additionally, various texture association rules are used to determine the image class by calculating the match degree of a texture image from the sample area.

10.3 Spatiotemporal Video Data Mining

Spatiotemporal video data mining can analyze abnormal motion trajectory, suspicious objects, moving object classes, video in spatiotemporal multi-camera, and video from different periods.

(1) *Analysis of abnormal motion trajectory.* The linear trajectory and characteristics are extracted, identified. and classified. Then, the problem of trajectory cross or separation also addressed as well as the problem of analyzing multi-objective overlapping anomaly.

(2) *Target analysis based on the hybrid model.* Target analysis includes color model analysis, shape model analysis, and feature blocks or feature point analysis, such as a pedestrian classification based on a hog operator (see Fig. 10.5).

Fig. 10.5 Pedestrian classification based on a hog operator

(3) *Classification combined with pattern recognition.* SVM, ANN, and Boost rapid classification technology are used to classify and identify moving targets.

(4) *Spatiotemporal analysis of multi-camera video.* This feature captures a target from multiple views at the same time. Where multiple cameras are installed at a fixed location, the times for all cameras at which a target appears are interconnected. With this association information, the image target matching technology can be combined to conduct temporal and spatial correlation analysis.

(5) *Camera video analysis in different periods.* Extraction is not dependent on the time-related information of image characteristics, such as environmental light, the contrast ratio, and other information. Pedestrians or vehicles that appear in different periods are automatically retrieved and recognized.

10.4 EveryData

The *EveryData* system is a web-based platform for users to collect, pre-process, analyze, and visualize the results for spatial data and non-spatial data (Fig. 10.6). The website uses J2EE-related frameworks like Struts2, Hibernate3, and Spring 2 to deal with the database and to handle http requests. The front page uses d3, BootStrap framework, JS, JSP, and CSS. The data mining algorithms are implemented using Java, R, and Weka. Some of its available features to date are shown in Fig. 10.7.

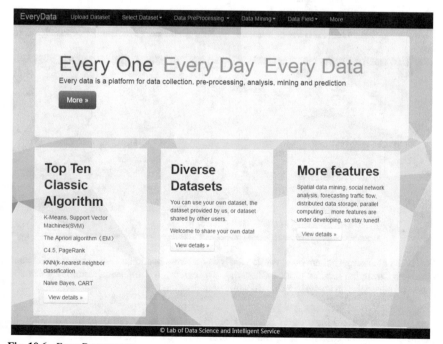

Fig. 10.6 *EveryData* system

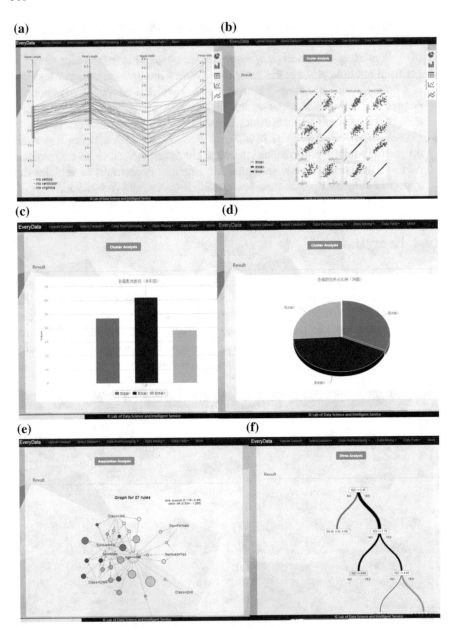

Fig. 10.7 *EveryData* system on clustering analysis and classification **a** Clustering (parallel coordinate). **b** Clustering (scatted plot). **c** Clustering (Histogram). **d** Clustering (partition). **e** Classification with association analysis. **f** Classification with decision tree

Users can upload their dataset files. Before uploading, users are required to set the parameters for the dataset (e.g., the delimiter of the dataset) and indicate whether the file has missing values and whether the first row is data or the parameter description.

After uploading the data files, users can choose the dataset of interest and preview it. Users then may proceed with the data mining tasks using the selected dataset.

Using the Iris dataset (https://archive.ics.uci.edu/ml/datasets/Iris) as an example, users can use the K-means clustering algorithm to process the data. At the backend, the K-means algorithm is invoked using R. After clustering, users can see the cluster centers shown in the figures, the percentages of each cluster are shown using a pie chart, and the total number of elements in each cluster is shown as a bar chart. Because the Iris dataset has four dimensions (length of sepal, length of petal, width of sepal, and width of petal), the website also uses 16 two-dimensional images to show the clustering results. Users also can use parallel coordinates to obtain a clear view of every cluster. Figure 10.7a–d are interoperable.

KNN is the classification algorithm available in the *EveryData* system. Users can upload a training set and a test set separately, the results of which will be printed in the webpage and will include the parameters of the classifier and the accuracy and classification results of each record in the test set. The KNN algorithm is implemented by invoking functions provided by Weka. A simplified decision tree algorithm and an Apriori algorithm to analyze the association rules then are implemented. Clicking on the nodes can expand or shrink the decision tree. The results of the Apriori algorithm are shown in graph format (see Fig. 10.7e–f).

The clustering algorithm HASTA (Wang and Chen 2014; Wang et al. 2015) and several image segmentation algorithms by the developers, including the GDF-Ncut

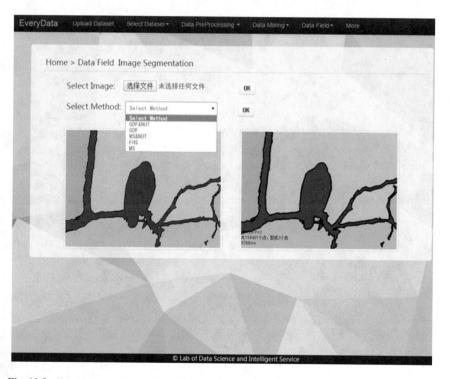

Fig. 10.8 *EveryData* system on data field

algorithm and the mean shift algorithm (Wang and Yuan 2014; Yuan et al. 2014), also are available on the website. After uploading and choosing the proper dataset, the user is required to set three parameters for the algorithms in order to produce clustering results (Fig. 10.8).

References

Li DR, Wang SL, Li DY (2006) Theory and application of spatial data mining, 1st edn. Science Press, Beijing

Li DR, Wang SL, Li DY (2013) Theory and application of spatial data mining, 2nd edn. Science Press, Beijing

Wang SL, Chen YS (2014) HASTA: a hierarchical-grid clustering algorithm with data field. Int J Data Warehousing Min 10(2):39–54

Wang SL, Yuan HN (2014) Spatial data mining: a perspective of big data. Int J Data Warehousing Min 10(4):50–70

Wang SL, Chen YS, Yuan HN (2015) A novel method to extract rocks from Mars images. Chin J Electr 24(3):455–461

Yuan HN, Wang SL, Li Y, Fan JH (2014) Feature selection with data field. Chin J Electr 23(4):661–665